Graduate Texts in Mathematics 172

Springer Science+Business Media, LLC

Graduate Texts in Mathematics

continued after index

Reinhold Remmert

Classical Topics in Complex Function Theory

Translated by Leslie Kay

With 19 Illustrations

Springer

Reinhold Remmert
Mathematisches Institut
Westfälische Wilhelms–Universität Münster
Einsteinstrasse 62
Münster D-48149
Germany

Leslie Kay (Translator)
Department of Mathematics
Virginia Polytechnic Institute
and State University
Blacksburg, VA 24061-0123
USA

Mathematics Subject Classification (1991): 30-01, 32-01

Library of Congress Cataloging-in-Publication Data
Remmert, Reinhold.
[Funktionentheorie. 2. English]
Classical topics in complex function theory / Reinhold Remmert : translated by Leslie Kay.
p. cm. — (Graduate texts in mathematics ; 172)
Translation of: Funktionentheorie II.
Includes bibliographical references and indexes.
ISBN 978-1-4419-3114-6 ISBN 978-1-4757-2956-6 (eBook)
DOI 10.1007/978-1-4757-2956-6
1. Functions of complex variables. I. Title. II. Series.
QA331.7.R4613 1997
515′.9—dc21 97-10091

Printed on acid-free paper.

Originally published by Springer-Verlag New York, Inc. in 1998
Softcover reprint of the hardcover 1st edition 1998

Production managed by Lesley Poliner; manufacturing supervised by Jeffrey Taub.
Photocomposed pages prepared from the author's TEX files.

9 8 7 6 5 4 3 2 1

ISBN 978-1-4419-3114-6 SPIN 10536728

Max Koecher
in memory

Preface

Preface to the Second German Edition

In addition to the correction of typographical errors, the text has been materially changed in three places. The derivation of Stirling's formula in Chapter 2, §4, now follows the method of Stieltjes in a more systematic way. The proof of Picard's little theorem in Chapter 10, §2, is carried out following an idea of H. König. Finally, in Chapter 11, §4, an inaccuracy has been corrected in the proof of Szegö's theorem.

Oberwolfach, 3 October 1994 *Reinhold Remmert*

Preface to the First German Edition

> Wer sich mit einer Wissenschaft bekannt machen will, darf nicht nur nach den reifen Früchten greifen — er muß sich darum bekümmern, wie und wo sie gewachsen sind. (Whoever wants to get to know a science shouldn't just grab the ripe fruit — he must also pay attention to how and where it grew.)
>
> — J. C. Poggendorf

Presentation of function theory with vigorous connections to historical development and related disciplines: This is also the leitmotif of this second volume. It is intended that the reader experience function theory personally and participate in the work of the creative mathematician. Of course, the scaffolding used to build cathedrals cannot always be erected afterwards; but a textbook need not follow Gauss, who said that once a good building is completed its scaffolding should no longer be seen.[1] Sometimes even the framework of a smoothly plastered house should be exposed.

The edifice of function theory was built by Abel, Cauchy, Jacobi, Riemann, and Weierstrass. Many others made important and beautiful contributions; not only the work of the kings should be portrayed, but also the life of the nobles and the citizenry in the kingdoms. For this reason, the bibliographies became quite extensive. But this seems a small price to pay. "Man kann der studierenden Jugend keinen größeren Dienst erweisen als wenn man sie zweckmäßig anleitet, sich durch das Studium der Quellen mit den Fortschritten der Wissenschaft bekannt zu machen." (One can render young students no greater service than by suitably directing them to familiarize themselves with the advances of science through study of the sources.) (letter from Weierstrass to Casorati, 21 December 1868)

Unlike the first volume, this one contains numerous glimpses of the function theory of several complex variables. It should be emphasized how independent this discipline has become of the classical function theory from which it sprang.

In citing references, I endeavored — as in the first volume — to give primarily original works. Once again I ask indulgence if this was not always successful. The search for the first appearance of a new idea that quickly becomes mathematical folklore is often difficult. The *Xenion* is well known:

> Allegire der Erste nur falsch, da schreiben ihm zwanzig
> Immer den Irrthum nach, ohne den Text zu besehn. [2]

The selection of material is conservative. The Weierstrass product theorem, Mittag-Leffler's theorem, the Riemann mapping theorem, and Runge's approximation theory are central. In addition to these required topics, the reader will find

- Eisenstein's proof of Euler's product formula for the sine;
- Wielandt's uniqueness theorem for the gamma function;
- an intensive discussion of Stirling's formula;
- Iss'sa's theorem;

---

[1]Cf. W. Sartorius von Waltershausen: *Gauß zum Gedächtnis*, Hirzel, Leipzig 1856; reprinted by Martin Sändig oHG, Wiesbaden 1965, p. 82.

[2]Just let the first one come up with a wrong reference, twenty others will copy his error without ever consulting the text. [The translator is grateful to Mr. Ingo Seidler for his help in translating this couplet.]

- Besse's proof that all domains in $\mathbb{C}$ are domains of holomorphy;
- Wedderburn's lemma and the ideal theory of rings of holomorphic functions;
- Estermann's proofs of the overconvergence theorem and Bloch's theorem;
- a holomorphic imbedding of the unit disc in $\mathbb{C}^3$;
- Gauss's expert opinion of November 1851 on Riemann's dissertation.

An effort was made to keep the presentation concise. One worries, however:

> Weiß uns der Leser auch für unsre Kürze Dank?
> Wohl kaum? Denn Kürze ward durch Vielheit leider! lang. [3]

Oberwolfach, 3 October 1994 *Reinhold Remmert*

[3]Is the reader even grateful for our brevity? Hardly? For brevity, through abundance, alas! turned long.

Gratias ago

It is impossible here to thank by name all those who gave me valuable advice. I would like to mention Messrs. R. B. Burckel, J. Elstrodt, D. Gaier, W. Kaup, M. Koecher, K. Lamotke, K.-J. Ramspott, and P. Ullrich, who gave their critical opinions. I must also mention the Volkswagen Foundation, which supported the first work on this book through an academic stipend in the winter semester 1982–83.

Thanks are also due to Mrs. S. Terveer and Mr. K. Schlöter. They gave valuable help in the preparatory work and eliminated many flaws in the text. They both went through the last version critically and meticulously, proofread it, and compiled the indices.

Advice to the reader. Parts A, B, and C are to a large extent mutually independent. A reference 3.4.2 means Subsection 2 in Section 4 of Chapter 3. The chapter number is omitted within a chapter, and the section number within a section. Cross-references to the volume *Funktionentheorie I* refer to the *third* edition 1992; the Roman numeral I begins the reference, e.g. I.3.4.2.[4] No later use will be made of material in small print; chapters, sections and subsections marked by * can be skipped on a first reading. Historical comments are usually given after the actual mathematics. Bibliographies are arranged at the end of each chapter (occasionally at the end of each section); page numbers, when given, refer to the editions listed.

Readers in search of the older literature may consult A. Gutzmer's German-language revision of G. Vivanti's *Theorie der eindeutigen Funktionen*, Teubner 1906, in which 672 titles (through 1904) are collected.

[4][In this translation, references, still indicated by the Roman numeral I, are to *Theory of Complex Functions* (Springer, 1991), the English translation by R. B. Burckel of the second German edition of *Funktionentheorie I*. Trans.]

The correction to this publication is available online at
https://doi.org/10.1007/978-1-4757-2956-6_15

Contents

B Mapping Theory 145

C Selecta 223

Part A

Infinite Products and Partial Fraction Series

1
Infinite Products of Holomorphic Functions

> Allgemeine Sätze über die Convergenz der unendlichen Producte sind zum grossen Theile bekannt. (General theorems on the convergence of infinite products are for the most part well known.)
>
> — Weierstrass, 1854

Infinite products first appeared in 1579 in the work of F. Vieta (*Opera*, p. 400, Leyden, 1646); he gave the formula

$$\frac{2}{\pi} = \sqrt{\frac{1}{2}} \cdot \sqrt{\frac{1}{2} + \frac{1}{2}\sqrt{\frac{1}{2}}} \cdot \sqrt{\frac{1}{2} + \frac{1}{2}\sqrt{\frac{1}{2} + \frac{1}{2}\sqrt{\frac{1}{2}}}} \cdots$$

for π (cf. [Z], p. 104 and p. 118). In 1655 J. Wallis discovered the famous product

$$\frac{\pi}{2} = \frac{2 \cdot 2}{1 \cdot 3} \cdot \frac{4 \cdot 4}{3 \cdot 5} \cdot \frac{6 \cdot 6}{5 \cdot 7} \cdot \ldots \cdot \frac{2n \cdot 2n}{(2n-1) \cdot (2n-1)} \cdot \ldots,$$

which appears in "Arithmetica infinitorum," *Opera* I, p. 468 (cf. [Z], p. 104 and p. 119). But L. Euler was the first to work systematically with infinite products and to formulate important product expansions; cf. Chapter 9 of his *Introductio.* The first convergence criterion is due to Cauchy, *Cours d'analyse*, p. 562 ff. Infinite products had found their permanent place in analysis by 1854 at the latest, through Weierstrass ([Wei], p. 172 ff.).[1]

---

[1] In 1847 Eisenstein, in his long-forgotten work [Ei], had already systematically used infinite products. He also uses *conditionally convergent* products (and

One goal of this chapter is the derivation and discussion of Euler's product

$$\sin \pi z = \pi z \prod_{\nu=1}^{\infty} \left(1 - \frac{z^2}{\nu^2}\right)$$

for the sine function; we give two proofs in Section 3.

Since infinite products are only rarely treated in lectures and textbooks on infinitesimal calculus, we begin by collecting, in Section 1, some basic facts about infinite products of numbers and of holomorphic functions. *Normally convergent* infinite products $\prod f_\nu$ of functions are investigated in Section 2; in particular, the important theorem on *logarithmic differentiation of products* is proved.

§1. Infinite Products

We first consider infinite products of sequences of complex numbers. In the second section, the essentials of the theory of compactly convergent products of functions are stated. A detailed discussion of infinite products can be found in [Kn].

1. Infinite products of numbers. If $(a_\nu)_{\nu\geq k}$ is a sequence of complex numbers, the sequence $\left(\prod_{\nu=k}^{n} a_\nu\right)_{n\geq k}$ of partial products is called a(n) *(infinite) product* with the *factors* a_ν. We write $\prod_{\nu=k}^{\infty} a_\nu$ or $\prod_{\nu\geq k} a_\nu$ or simply $\prod a_\nu$; in general, $k = 0$ or $k = 1$.

If we now — by analogy with series — were to call a product $\prod a_\nu$ convergent whenever the sequence of partial products had a limit a, undesirable pathologies would result: for one thing, a product would be convergent with value 0 if just one factor a_ν were zero; for another, $\prod a_\nu$ could be zero even if not a single factor were zero (e.g. if $|a_\nu| \leq q < 1$ for all ν). We will therefore take precautions against zero factors and convergence to zero. We introduce the partial products

$$p_{m,n} := a_m a_{m+1} \dots a_n = \prod_{\nu=m}^{n} a_\nu, \quad k \leq m \leq n,$$

and call the *product* $\prod a_\nu$ *convergent* if there exists an index m such that the sequence $(p_{m,n})_{n\geq m}$ has a limit $\widehat{a}_m \neq 0$.

---

series) and carefully discusses the problems, then barely recognized, of conditional and absolute convergence; but he does not deal with questions of compact convergence. Thus logarithms of infinite products are taken without hesitation, and infinite series are casually differentiated term by term; this carelessness may perhaps explain why Weierstrass nowhere cites Eisenstein's work.

We then call $a := a_k a_{k+1} \ldots a_{m-1}\widehat{a}_m$ the *value* of the product and introduce the suggestive notation

$$\prod a_\nu := a_k a_{k+1} \ldots a_{m-1}\widehat{a}_m = a.$$

The number a is independent of the index m: since $\widehat{a}_m \neq 0$, we have $a_n \neq 0$ for all $n \geq m$; hence for each fixed $l > m$ the sequence $(p_{l,n})_{n\geq l}$ also has a limit $\widehat{a}_l \neq 0$, and $a = a_k a_{k+1} \ldots a_{l-1}\widehat{a}_l$. Nonconvergent products are called *divergent.* The following result is immediate:

A product $\prod a_\nu$ is convergent if and only if at most finitely many factors are zero and the sequence of partial products consisting of the nonzero elements has a limit $\neq 0$.

The restrictions we have found take into account as well as possible the special role of zero. Just as for finite products, the following holds (by definition):

A convergent product $\prod a_\nu$ is zero if and only if at least one factor is zero.

We note further:

If $\prod_{\nu=0}^{\infty} a_\nu$ converges, then $\widehat{a}_n := \prod_{\nu=n}^{\infty} a_\nu$ exists for all $n \in \mathbb{N}$. Moreover, $\lim \widehat{a}_n = 1$ and $\lim a_n = 1$.

Proof. We may assume that $a := \prod a_\nu \neq 0$. Then $\widehat{a}_n = a/p_{0,n-1}$. Since $\lim p_{0,n-1} = a$, it follows that $\lim \widehat{a}_n = 1$. The equality $\lim a_n = 1$ holds because, for all n, $\widehat{a}_n \neq 0$ and $a_n = \widehat{a}_n/\widehat{a}_{n+1}$. □

Examples. a) Let $a_0 := 0$, $a_\nu := 1$ for $\nu \geq 1$. Then $\prod a_\nu = 0$.

b) Let $a_\nu := 1 - \frac{1}{\nu^2}$, $\nu \geq 2$. Then $p_{2,n} = \frac{1}{2}(1 + \frac{1}{n})$; hence $\prod_{\nu\geq 2} a_\nu = \frac{1}{2}$.

c) Let $a_\nu := 1 - \frac{1}{\nu}$, $\nu \geq 2$. Then $p_{2,n} = \frac{1}{n}$; hence $\lim p_{2,n} = 0$. The product $\prod_{\nu\geq 2} a_\nu$ is divergent (since no factor vanishes) although $\lim a_n = 1$.

In 4.3.2 we will need the following generalization of c):

d) Let $a_0, a_1, a_2, \ldots$ be a sequence of real numbers with $a_n \geq 0$ and $\sum(1 - a_\nu) = +\infty$. Then $\lim \prod_{\nu=0}^{n} a_\nu = 0$.

Proof. $0 \leq p_{0,n} = \prod_0^n a_\nu \leq \exp[-\sum_0^n (1 - a_\nu)]$, $n \in \mathbb{N}$, since $t \leq e^{t-1}$ for all $t \in \mathbb{R}$. Since $\sum(1 - a_\nu) = +\infty$, it follows that $\lim p_{0,n} = 0$. □

It is *not* appropriate to introduce, by analogy with series, the concept of absolute convergence. If we were to call a product $\prod a_\nu$ absolutely convergent whenever $\prod |a_\nu|$ converged, then convergence would always imply absolute convergence — but $\prod(-1)^\nu$ would be absolutely convergent without being convergent! The first comprehensive treatment of the convergence theory of infinite products was given in 1889 by A. Pringsheim [P].

Exercises. Show:

a) $\prod_{\nu=2}^{\infty} \frac{\nu^3 - 1}{\nu^3 + 1} = \frac{2}{3}, \quad \prod_{\nu=1}^{\infty} \frac{\nu + (-1)^{\nu+1}}{\nu} = 1,$

b) $\prod_{\nu=2}^{\infty} \cos \frac{\pi}{2^\nu} = \frac{2}{\pi}$ (Vieta's product).

2. Infinite products of functions. Let X denote a *locally compact metric* space. It is well known that the concepts of *compact convergence* and *locally uniform convergence* coincide for such spaces; cf. I.3.1.3. For a sequence $f_\nu \in \mathcal{C}(X)$ of continuous functions on X with values in $\mathbb{C}$, the (infinite) product $\prod f_\nu$ is called *compactly convergent* in X if, for every compact set K in X, there is an index $m = m(K)$ such that the sequence $p_{m,n} := f_m f_{m+1} \cdots f_n$, $n \geq m$, converges *uniformly* on K to a *nonvanishing* function $\widehat{f}_m$. Then, for each point $x \in X$,

$$f(x) := \prod f_\nu(x) \in \mathbb{C}$$

exists (in the sense of Subsection 1); we call the function $f : X \to \mathbb{C}$ the *limit of the product* and write

$$f = \prod f_\nu; \quad \text{then, on } K, \quad f|K = (f_0|K) \cdot \ldots \cdot (f_{m-1}|K) \cdot \widehat{f}_m.$$

The next two statements follow immediately from the continuity theorem I.3.1.2.

a) If $\prod f_\nu$ converges compactly to f in X, then f is continuous in X and the sequence f_ν converges compactly in X to 1.

b) If $\prod f_\nu$ and $\prod g_\nu$ converge compactly in X, then so does $\prod f_\nu g_\nu$:

$$\prod f_\nu g_\nu = \left(\prod f_\nu\right)\left(\prod g_\nu\right).$$

We are primarily interested in the case where X is a domain[2] in $\mathbb{C}$ and all the functions f_ν are holomorphic. The following is clear by the Weierstrass convergence theorem (cf. I.8.4.1).

c) *Let G be a domain in $\mathbb{C}$. Every product $\prod f_\nu$ of functions f_ν holomorphic in G that converges compactly in G has a limit f that is holomorphic in G.*

Examples. a) The functions $f_\nu := (1+\frac{2z}{2\nu-1})(1+\frac{2z}{2\nu+1})^{-1}$, $\nu \geq 1$, are holomorphic in the unit disc $\mathbb{E}$. We have

$$p_{2,n} = \left(1 + \frac{2}{3}z\right)\left(1 + \frac{2z}{2n+1}\right)^{-1} \in \mathcal{O}(\mathbb{E}); \quad \text{hence} \quad \lim p_{2,n} = 1 + \frac{2}{3}z,$$

---

[2][As defined in *Funktionentheorie I*, a region ("Bereich" in German) is a nonempty open subset of $\mathbb{C}$; a domain ("Gebiet" in German) is a connected region. In consulting *Theory of Complex Functions*, the reader should be aware that there "Bereich" was translated as "domain" and "Gebiet" as "region." Trans.]

and the product $\prod_{\nu=1}^{\infty} f_\nu$ therefore converges compactly in $\mathbb{E}$ to $1+2z$.

b) Let $f_\nu(z) \equiv z$ for all $\nu \geq 0$. The product $\prod_{\nu=0}^{\infty} f_\nu$ does not converge (even pointwise) in the unit disc $\mathbb{E}$, since the sequence $p_{m,n} = z^{n-m+1}$ converges to zero for every m.

We note an important sufficient

Convergence criterion. *Let $f_\nu \in \mathcal{C}(X)$, $\nu \geq 0$. Suppose there exists an $m \in \mathbb{N}$ such that every function f_ν, $\nu \geq m$, has a logarithm $\log f_\nu \in \mathcal{C}(X)$. If $\sum_{\nu \geq m} \log f_\nu$ converges compactly in X to $s \in \mathcal{C}(X)$, then $\prod f_\nu$ converges compactly in X to $f_0 f_1 \ldots f_{m-1} \exp s$.*

Proof. Since the sequence $s_n := \sum_{\nu=m}^{n} \log f_\nu$ converges compactly to s, the sequence $p_{m,n} = \prod_{\nu=m}^{n} f_\nu = \exp s_n$ converges compactly in X to $\exp s$. As $\exp s$ does not vanish, the assertion follows.[3] □

§2. Normal Convergence

The convergence criterion 1.2 is hardly suitable for applications, since series consisting of logarithms are generally hard to handle. Moreover, we need a criterion — by analogy with infinite series — that ensures the compact convergence of *all partial products* and *all rearrangements*. Here again, as for series, "normal convergence" proves superior to "compact convergence." We recall this concept of convergence for series, again assuming the space X to be locally compact: then $\sum f_\nu$, $f_\nu \in \mathcal{C}(X)$, is normally convergent in X if and only if $\sum |f_\nu|_K < \infty$ for every compact set $K \subset X$ (cf. I.3.3.2). Normally convergent series are compactly convergent; normal convergence is preserved under passage to partial sums and arbitrary rearrangements (cf. I.3.3.1).

The factors of a product $\prod f_\nu$ are often written in the form $f_\nu = 1 + g_\nu$; by 1.2 a), the sequence g_ν converges compactly to zero if $\prod f_\nu$ converges compactly.

1. Normal convergence. A product $\prod f_\nu$ with $f_\nu = 1 + g_\nu \in \mathcal{C}(X)$ is called *normally convergent in* X if the series $\sum g_\nu$ converges normally in X. It is easy to see that

if $\prod_{\nu \geq 0} f_\nu$ converges normally in X, then

- *for every bijection $\tau : \mathbb{N} \to \mathbb{N}$, the product $\prod_{\nu \geq 0} f_{\tau(\nu)}$ converges normally in X;*

[3] The simple proof that the compact convergence of s_n to s implies the compact convergence of $\exp s_n$ to $\exp s$ can be found in I.5.4.3 (composition lemma).

– *every subproduct* $\prod_{j\geq 0} f_{\nu_j}$ *converges normally in* X;

– *the product converges compactly in* X.

We will see that the concept of normal convergence is a good one. At the moment, however, it is not clear that a normally convergent product even has a limit. We immediately prove this and more:

Rearrangement theorem. *Let* $\prod_{\nu\geq 0} f_\nu$ *be normally convergent in* X. *Then there is a function* $f: X \to \mathbb{C}$ *such that for every bijection* $\tau: \mathbb{N} \to \mathbb{N}$ *the rearrangement* $\prod_{\nu\geq 0} f_{\tau(\nu)}$ *of the product converges compactly to* f *in* X.

Proof. For $w \in \mathbb{E}$ we have $\log(1+w) = \sum_{\nu\geq 1} \frac{(-1)^{\nu-1}}{\nu} w^\nu$. It follows that

$$|\log(1+w)| \leq |w|(1+|w|+|w|^2+\cdots);\ \text{hence}\ |\log(1+w)| \leq 2|w|\ \text{if}\ |w| \leq 1/2.$$

Now let $K \subset X$ be an arbitrary compact set and let $g_n = f_n - 1$. There is an $m \in \mathbb{N}$ such that $|g_n|_K \leq \frac{1}{2}$ for $n \geq m$. For all such n,

$$\log f_n = \sum \frac{(-1)^{\nu-1}}{\nu} g_n^\nu \in \mathcal{C}(K), \quad \text{where} \quad |\log f_n|_K \leq 2|g_n|_K.$$

We see that $\sum_{\nu\geq m} |\log f_\nu|_K \leq \sum_{\nu\geq m} |g_\nu|_K < \infty$. Hence, by the rearrangement theorem for series (cf. I.0.4.3), for every bijection σ of $\mathbb{N}_m := \{n \in \mathbb{N} : n \geq m\}$ the series $\sum_{\nu\geq m} \log f_{\sigma(\nu)}$ converges uniformly in K to $\sum_{\nu\geq m} \log f_\nu$. By 1.2, it follows that for such σ the products $\prod_{\nu\geq m} f_{\sigma(\nu)}$ and $\prod_{\nu\geq m} f_\nu$ converge uniformly in K to the same limit function. But an arbitrary bijection τ of $\mathbb{N}$ (= permutation of $\mathbb{N}$) differs only by finitely many transpositions (which have no effect on convergence) from a permutation $\sigma': \mathbb{N} \to \mathbb{N}$ with $\sigma'(\mathbb{N}_m) = \mathbb{N}_m$. Hence there exists a function $f: X \to \mathbb{C}$ such that every product $\prod_{\nu\geq 0} f_{\tau(\nu)}$ converges compactly in X to f. □

Corollary. *Let* $f = \prod_{\nu\geq 0} f_\nu$ *converge normally in* X. *Then the following statements hold.*

1) *Every product* $\widehat{f}_n := \prod_{\nu\geq n} f_\nu$ *converges normally in* X, *and*

$$f = f_0 f_1 \cdots f_{n-1} \widehat{f}_n.$$

2) *If* $\mathbb{N} = \bigcup_1^\infty N_\kappa$ *is a (finite or infinite) partition of* $\mathbb{N}$ *into pairwise disjoint subsets* $N_1, \ldots, N_\kappa, \ldots$, *then every product* $\prod_{\nu\in N_\kappa} f_\nu$ *converges normally in* X *and*

$$f = \prod_{\kappa=1}^{\infty} \left(\prod_{\nu\in N_\kappa} f_\nu \right).$$

Products can converge compactly without being normally convergent, as is shown, for example, by $\prod_{\nu\geq 1}(1+g_\nu)$, $g_\nu := (-1)^{\nu-1}/\nu$. It is always true that $(1+g_{2\nu-1})(1+g_{2\nu}) = 1$; hence $p_{1,n} = 1$ for even n and $p_{1,n} = 1+\frac{1}{n}$ for odd n. The product $\prod_{\nu\geq 1}(1+g_\nu)$ thus converges compactly in $\mathbb{C}$ to 1. In this example the subproduct $\prod_{\nu\geq 1}(1+g_{2\nu-1})$ is not convergent!

All later applications (sine product, Jacobi's triple product, Weierstrass's factorial, general Weierstrass products) will involve normally convergent products.

Exercises. 1) Prove that if the products $\prod f_\nu$ and $\prod \widetilde{f}_\nu$ converge normally in X, then the product $\prod(f_\nu \widetilde{f}_\nu)$ also converges normally in X.

2) Show that the following products converge normally in the unit disc $\mathbb{E}$, and prove the identities

$$\prod_{\nu\geq 0}(1+z^{2^\nu}) = \frac{1}{1-z}, \qquad \prod_{\nu\geq 1}[(1+z^\nu)(1-z^{2\nu-1})] = 1.$$

2. Normally convergent products of holomorphic functions. The zero set $Z(f)$ of any function $f \neq 0$ holomorphic in G is locally finite in G;[4] hence $Z(f)$ is at most countably infinite (see I.8.1.3).

For *finitely* many functions $f_0, f_1, \ldots, f_n \in \mathcal{O}(G)$, $f_\nu \neq 0$,

$$Z(f_0 f_1 \ldots f_n) = \bigcup_0^n Z(f_\nu) \quad \text{and} \quad o_c(f_0 f_1 \ldots f_n) = \sum_0^n o_c(f_\nu), \quad c \in G,$$

where $o_c(f)$ denotes the order of the zero of f at c (I.8.1.4). For infinite products, we have the following result.

Proposition. *Let $f = \prod f_\nu$, $f_\nu \neq 0$, be a normally convergent product in G of functions holomorphic in G. Then*

$$f \neq 0, \quad Z(f) = \bigcup Z(f_\nu), \quad o_c(f) = \sum o_c(f_\nu) \quad \textit{for all } c \in G.$$

Proof. Let $c \in G$ be fixed. Since $f(c) = \prod f_\nu(c)$ converges, there exists an index n such that $f_\nu(c) \neq 0$ for all $\nu \geq n$. By Corollary 1,1), $f = f_0 f_1 \cdots f_{n-1}\widehat{f_n}$, where $\widehat{f_n} := \prod_{\nu\geq n} f_\nu \in \mathcal{O}(G)$ by the Weierstrass convergence theorem. It follows that

$$o_c(f) = \sum_0^{n-1} o_c(f_\nu) + o_c(\widehat{f_n}), \quad \text{with} \quad o_c(\widehat{f_n}) = 0 \quad (\text{since } \widehat{f_n}(c) \neq 0).$$

[4]Let G be an open subset of $\mathbb{C}$. A subset of G is locally finite in G if it intersects every compact set in G in only a finite number of points. Equivalently, a subset of G is locally finite in G if it is discrete and closed in G.

This proves the addition rule for infinite products. In particular, $Z(f) = \cup Z(f_\nu)$. Since each $f_\nu \neq 0$, all the sets $Z(f_\nu)$ and hence also their countable union are countable; it follows that $f \neq 0$. □

Remark. The proposition is true even if the convergence of the product in G is only compact. The proof remains valid word for word, since it is easy to see that for every n the tail end $\widehat{f}_n = \prod_{\nu \geq n} f_\nu$ converges compactly in G.

We will need the following result in the next section.

If $f = \prod f_\nu$, $f_\nu \in \mathcal{O}(G)$, is normally convergent in G, then the sequence $\widehat{f}_n = \prod_{\nu \geq n} f_\nu \in \mathcal{O}(G)$ converges compactly in G to 1.

Proof. Let $\widehat{f}_m \neq 0$. Then $A := Z(\widehat{f}_m)$ is locally finite in G. All the partial products $p_{m,n-1} \in \mathcal{O}(G)$, $n > m$, are nonvanishing in $G \setminus A$ and

$$\widehat{f}_n(z) = \widehat{f}_m(z) \cdot \left(\frac{1}{p_{m,n-1}(z)} \right) \quad \text{for all} \quad z \in G \setminus A.$$

Now the sequence $1/p_{m,n-1}$ converges compactly in $G \setminus A$ to $1/\widehat{f}_m$. Hence, by the sharpened version of the Weierstrass convergence theorem (see I.8.5.4), this sequence also converges compactly in G to 1. □

Exercise. Show that $f = \prod_{\nu=1}^{\infty} \cos(z/2\nu)$ converges normally in $\mathbb{C}$. Determine $Z(f)$. Show that for each $k \in \mathbb{N} \setminus \{0\}$ there exists a zero of order k of f and that

$$\prod_{\nu=1}^{\infty} \cos \frac{z}{2\nu} = \prod_{\nu=1}^{\infty} \left(\frac{2\nu - 1}{z} \sin \frac{z}{2\nu - 1} \right).$$

3. Logarithmic differentiation. The *logarithmic derivative* of a meromorphic function $h \in \mathcal{M}(G)$, $h \neq 0$, is by definition the function $h'/h \in \mathcal{M}(G)$ (see also I.9.3.1, where the case of nonvanishing holomorphic functions is discussed). For finite products $h = h_1 h_2 \ldots h_m$, $h_\mu \in \mathcal{M}(G)$, we have the

Addition formula: $\displaystyle \frac{h'}{h} = \frac{h_1'}{h_1} + \frac{h_2'}{h_2} + \cdots + \frac{h_m'}{h_m}.$

This formula carries over to infinite products of holomorphic functions.

Differentiation theorem. *Let $f = \prod f_\nu$ be a product of holomorphic functions that converges normally in G. Then $\sum f_\nu'/f_\nu$ is a series of meromorphic functions that converges normally in G, and*

$$\frac{f'}{f} = \sum \frac{f_\nu'}{f_\nu} \in \mathcal{M}(G).$$

Proof. 1) For all $n \in \mathbb{N}$ (by Corollary 1,1)),

$$f = f_0 f_1 \cdots f_{n-1} \widehat{f}_n, \text{ with } \widehat{f}_n := \prod_{\nu \geq n} f_\nu; \text{ hence } \frac{f'}{f} = \sum_{\nu=1}^{n-1} \frac{f'_\nu}{f_\nu} + \frac{\widehat{f}'_n}{\widehat{f}_n}.$$

Since the sequence $\widehat{f}_n$ converges compactly in G to 1 (cf. 2), the derivatives $\widehat{f}'_n$ converge compactly in G to 0 by Weierstrass. For every disc B with $\overline{B} \subset G$ there is thus an $m \in \mathbb{N}$ such that all f_n, $n \geq m$, are nonvanishing in B and the sequence $\widehat{f}'_n / \widehat{f}_n \in \mathcal{O}(B)$, $n \geq m$, converges compactly in B to zero. This shows that $\sum f'_\nu / f_\nu$ converges compactly in G to f'/f.

2) We now show that $\sum f'_\nu / f_\nu$ converges normally in G. Let $g_\nu := f_\nu - 1$. We must assign an index m to every compact set K in G so that every pole set $P(f'_\nu / f_\nu)$, $\nu \geq m$, is disjoint from K and

$$(*) \qquad \sum_{\nu \geq m} \left| \frac{f'_\nu}{f_\nu} \right|_K = \sum_{\nu \geq m} \left| \frac{g'_\nu}{f_\nu} \right|_K < \infty \quad \text{(cf. I.11.1.1).}$$

We choose m so large that all the sets $Z(f_\nu) \cap K$, $\nu \geq m$, are empty and $\min_{z \in K} |f_\nu(z)| \geq \frac{1}{2}$ for all $\nu \geq m$ (this is possible, since the sequence f_ν converges compactly to 1). Now, by the Cauchy estimates for derivatives, there exist a compact set $L \supset K$ in G and a constant $M > 0$ such that $|g'_\nu|_K \leq M |g_\nu|_L$ for all ν (cf. I.8.3.1). Thus $|g'_\nu / f_\nu|_K \leq |g'_\nu|_K \cdot (\min_{z \in K} |f_\nu(z)|)^{-1} \leq 2M |g_\nu|_L$ for $\nu \geq m$. Since $\sum |g_\nu|_L < \infty$ by hypothesis, $(*)$ follows. □

The differentiation theorem is an important tool for concrete computations; for example, we use it in the next subsection to derive Euler's product for the sine, and we give another application in 2.2.3. The theorem holds verbatim if the word "normal" is replaced by "compact." (Prove this.)

The differentiation theorem can be used to prove:

If f is holomorphic at the origin, then f can be represented uniquely in a disc B about 0 as a product

$$f(z) = bz^k \prod_{\nu=1}^{\infty} (1 + b_\nu z^\nu), \quad b, b_\nu \in \mathbb{C}, \quad k \in \mathbb{N},$$

which converges normally in B to f.

This theorem was proved in 1929 by J. F. Ritt [R]. It is not claimed that the product converges in the *largest* disc about 0 in which f is holomorphic. There seem to be no compelling applications of this product expansion, which is a multiplicative analogue of the Taylor series.

§3. The Sine Product $\sin \pi z = \pi z \prod_{\nu=1}^{\infty}(1 - z^2/\nu^2)$

The product $\prod_{\nu=1}^{\infty}(1-z^2/\nu^2)$ is normally convergent in $\mathbb{C}$, since $\sum_{\nu=1}^{\infty} z^2/\nu^2$ converges normally in $\mathbb{C}$. In 1734 Euler discovered that

$$(1) \qquad \boxed{\sin \pi z = \pi z \prod_{\nu=1}^{\infty}\left(1 - \frac{z^2}{\nu^2}\right), \quad z \in \mathbb{C}.}$$

We give two proofs of this formula.

1. Standard proof *(using logarithmic differentiation and the partial fraction decomposition for the cotangent)*. Setting $f_\nu := 1 - z^2/\nu^2$ and $f(z) := \pi z \prod_{\nu=1}^{\infty} f_\nu$ gives

$$f_\nu'/f_\nu = \frac{2z}{z^2 - \nu^2}, \quad \text{and thus} \quad f'(z)/f(z) = \frac{1}{z} + \sum_{\nu=1}^{\infty} \frac{2z}{z^2 - \nu^2}.$$

Here the right-hand side is the function $\pi \cot \pi z$ (cf. I.11.2.1). As this is also the logarithmic derivative of $\sin \pi z$, we have[5] $f(z) = c \sin \pi z$ with $c \in \mathbb{C}^\times$. Since $\lim_{z\to 0} \frac{f(z)}{\pi z} = 1 = \lim_{z \to 0} \frac{\sin \pi z}{\pi z}$, it follows that $c = 1$.

Substituting special values for z in (1) yields interesting (and uninteresting) formulas. Setting $z := \frac{1}{2}$ gives the product formula

$$\frac{\pi}{2} = \frac{2}{1} \cdot \frac{2}{3} \cdot \frac{4}{3} \cdot \frac{4}{5} \cdot \frac{6}{5} \cdot \frac{6}{7} \cdot \ldots = \prod_{\nu=1}^{\infty} \frac{2\nu}{2\nu - 1} \cdot \frac{2\nu}{2\nu + 1} \quad \text{(Wallis, 1655).}$$

For $z := 1$, one obtains the trivial equality $\frac{1}{2} = \prod_{\nu=2}^{\infty}(1 - \frac{1}{\nu^2})$ (cf. Example 1.1, b); on the other hand, setting $z := i$ and using the identity $\sin \pi i = \frac{i}{2}(e^\pi - e^{-\pi})$ give the bizarre formula

$$\prod_{\nu=1}^{\infty}\left(1 + \frac{1}{\nu^2}\right) = \frac{e^\pi - e^{-\pi}}{2\pi}.$$

Using the identity $\sin z \cos z = \frac{1}{2} \sin 2z$ and Corollary 2.1, one obtains

$$\begin{aligned} \cos \pi z \sin \pi z &= \pi z \prod_{\nu=1}^{\infty}\left(1 - \left(\frac{2z}{\nu}\right)^2\right) \\ &= \pi z \prod_{\nu=1}^{\infty}\left(1 - \left(\frac{2z}{2\nu}\right)^2\right) \prod_{\nu=1}^{\infty}\left(1 - \left(\frac{2z}{2\nu - 1}\right)^2\right), \end{aligned}$$

[5] Let $f \not\equiv 0$, $g \not\equiv 0$ be two meromorphic functions on a domain G which have the same logarithmic derivative. Then $f = cg$, with $c \in \mathbb{C}^\times$. To prove this, note that $f/g \in \mathcal{M}(G)$ and $(f/g)' \equiv 0$.

and hence Euler's product representation for the cosine:

$$\boxed{\cos\pi z = \prod_{\nu=1}^{\infty}\left(1 - \frac{4z^2}{(2\nu-1)^2}\right), \quad z \in \mathbb{C}.}$$

In 1734–35, with his sine product, Euler could in principle compute all the numbers $\zeta(2n) := \sum_{\nu=1}^{\infty}\nu^{-2n}$, $n = 1, 2, \ldots$ (cf. also I.11.3.2). Thus it follows immediately, for example, that $\zeta(2) = \frac{\pi^2}{6}$: Since $f_n(z) := \prod_{\nu=1}^{n}(1 - z^2/\nu^2) = 1 - (\sum_{\nu=1}^{n}\nu^{-2})z^2 + \cdots$ tends compactly to $f(z) := (\sin\pi z)/(\pi z) = 1 - \frac{\pi^2 z^2}{6} + \cdots$, it follows that $\frac{1}{2}f_n''(0) = -\sum_{\nu=1}^{n}\nu^{-2}$ converges to $\frac{1}{2}f''(0) = -\frac{1}{6}\pi^2$. □

Wallis's formula permits an elementary calculation of the Gaussian error integral $\int_0^{\infty} e^{-x^2}dx$. For $I_n := \int_0^{\infty} x^n e^{-x^2}dx$, we have

$$2I_n = (n-1)I_{n-2}, \quad n \geq 2 \quad \text{(integration by parts!)}.$$

Since $I_1 = \frac{1}{2}$, an induction argument gives

$$(\circ) \qquad 2^k I_{2k} = 1\cdot 3\cdot 5\cdot\ldots\cdot(2k-1)I_0, \quad 2I_{2k+1} = k!, \quad k \in \mathbb{N}.$$

Since $I_{n+1} + 2tI_n + t^2 I_{n-1} = \int_0^{\infty} x^{n-1}(x+t)^2 e^{-x^2}dx$ for all $t \in \mathbb{R}$, it follows that

$$I_n^2 < I_{n-1}I_{n+1}; \quad \text{hence} \quad 2I_n^2 < nI_{n-1}^2.$$

With ($\circ$) we now obtain

$$\frac{(k!)^2}{4k+2} = \frac{2}{2k+1}I_{2k+1}^2 < I_{2k}^2 < I_{2k-1}I_{2k+1} = \frac{(k!)^2}{4k}.$$

This can also be written

$$I_{2k}^2 = \frac{(k!)^2}{4k+2}(1+\varepsilon_k), \quad \text{with} \quad 0 < \varepsilon_k < \frac{1}{2k}.$$

Using ($\circ$) to substitute I_0 into this yields

$$2I_0^2 = \frac{[2\cdot 4\cdot 6\cdot\ldots\cdot(2k)]^2}{[1\cdot 3\cdot 5\cdot\ldots\cdot(2k-1)]^2(2k+1)}(1+\varepsilon_k).$$

From $\lim\varepsilon_k = 0$ and Wallis's formula, it follows that $2I_0^2 = \frac{1}{2}\pi$ and hence that $\int_0^{\infty} e^{-x^2}dx = \frac{1}{2}\sqrt{\pi}$. □

This derivation was given by T.-J. Stieltjes: Note sur l'intégrale $\int_0^{\infty} e^{-u^2}du$, *Nouv. Ann. Math.* 9, 3rd ser., 479–480 (1890); *Œuvres complètes* 2, 2nd ed., Springer, 1993, 263–264.

Exercises. Prove:

1) $\lim \frac{2\cdot 4\cdot 6\cdot\ldots\cdot 2n}{3\cdot 5\cdot 7\cdot\ldots\cdot(2n+1)}\sqrt{n} = \frac{1}{2}\sqrt{\pi}$;

2) $\frac{1}{4}\pi = \prod_{\nu=1}^{\infty}\left(1 - \frac{1}{(2\nu+1)^2}\right)$;

3) $e^{az} - e^{bz} = (a-b)ze^{\frac{1}{2}(a+b)z}\prod_{\nu=1}^{\infty}\left(1 + \frac{(a-b)^2 z^2}{4\nu^2\pi^2}\right)$;

4) $\cos(\frac{1}{4}\pi z) - \sin(\frac{1}{4}\pi z) = \prod_{n=1}^{\infty}\left(1+\frac{(-1)^n z}{2n-1}\right)$.

2. Characterization of the sine by the duplication formula. We characterize the sine function by properties that are easy to verify for the product $z\prod(1-z^2/\nu^2)$. The equality $\sin 2z = 2\sin z \cos z$ is a

Duplication formula: $\sin 2\pi z = 2\sin \pi z \sin \pi(z+\frac{1}{2}), \quad z \in \mathbb{C}.$

In order to use it in characterizing the sine, we first prove a lemma.

Lemma (*Herglotz, multiplicative form*).[6] *Let $G \subset \mathbb{C}$ be a domain that contains an interval $[0,r)$, $r > 1$. Suppose that $g \in \mathcal{O}(G)$ has no zeros in $[0,r)$ and satisfies a multiplicative duplication formula*

$$(*)\ g(2z) = cg(z)g(z+\tfrac{1}{2}) \quad \textit{when} \quad z,\ z+\tfrac{1}{2},\ 2z \in [0,r) \quad (\textit{with } c \in \mathbb{C}^\times).$$

Then $g(z) = ae^{bz}$ with $1 = ace^{\frac{1}{2}b}$.

Proof. The function $h := g'/g \in \mathcal{M}(G)$ is holomorphic throughout $[0,r)$, and $2h(2z) = 2g'(2z)/g(2z) = h(z)+h(z+\frac{1}{2})$ whenever $z,\ z+\frac{1}{2},\ 2z \in [0,r)$. By Herglotz's lemma (additive form), h is constant.[6] It follows that $g' = bg$ with $b \in \mathbb{C}$; hence $g(z) = ae^{bz}$. By $(*)$, $ace^{\frac{1}{2}b} = 1$. □

The next theorem now follows quickly.

Theorem. *Let f be an odd entire function that vanishes in $[0,1]$ only at 0 and 1, and vanishes to first order there. Suppose that it satisfies the*

Duplication formula: $f(2z) = cf(z)f(z+\frac{1}{2}), \quad z \in \mathbb{C}, \quad$ *where* $c \in \mathbb{C}^\times$.

Then $f(z) = 2c^{-1}\sin \pi z$.

Proof. The function $g(z) := f(z)/\sin \pi z$ is holomorphic and nowhere zero in a domain $G \supset [0,r)$, $r > 1$; we have $g(2z) = \frac{1}{2}cg(z)g(z+\frac{1}{2})$. By Herglotz, $f(z) = ae^{bz}\sin \pi z$ with $ace^{\frac{1}{2}b} = 2$. Since $f(-z) = f(z)$, it also follows that $b = 0$. □

---

[6] We recall the following lemma, discussed in I.11.2.2:

Herglotz's lemma (*additive form*). *Let $[0,r) \subset G$ with $r > 1$. Let $h \in \mathcal{O}(G)$ and assume that the additive duplication formula $2h(2z) = h(z) + h(z+\frac{1}{2})$ holds when $z,\ z+\frac{1}{2},\ 2z \in [0,r)$. Then h is constant.*

Proof. Let $t \in (1,r)$ and $M := \max\{|h'(z)| : z \in [0,t)\}$. Since $4h'(2z) = h'(z) + h'(z+\frac{1}{2})$ and $\frac{1}{2}z$ and $\frac{1}{2}(z+1)$ always lie in $[0,t]$ whenever z does, it follows that $4M \le 2M$, and hence that $M = 0$. By the identity theorem, $h' = 0$; thus $h = \text{const}$. □

We also use the duplication formula for the sine to derive an integral that will be needed in the appendix to 4.3 for the proof of Jensen's formula:

$$(1) \qquad \int_0^1 \log \sin \pi t dt = -\log 2.$$

Proof. Assuming for the moment that the integral exists, we have

$$(\circ) \qquad \int_0^{\frac{1}{2}} \log \sin 2\pi t dt = \tfrac{1}{2}\log 2 + \int_0^{\frac{1}{2}} \log \sin \pi t dt + \int_0^{\frac{1}{2}} \log \sin \pi(t + \tfrac{1}{2}) dt.$$

Setting $\tau := 2t$ on the left-hand side and $\tau := t + \frac{1}{2}$ in the integral on the extreme right immediately yields (1). The second integral on the right in ($\circ$) exists whenever the first one does (set $t + \frac{1}{2} = 1 - \tau$). The first integral exists since $g(t) := t^{-1} \sin \pi t$ is continuous and nonvanishing in $[0, \frac{1}{2}]$.[7]

3. Proof of Euler's formula using Lemma 2. The function

$$s(z) := z \cdot \prod_{\nu=1}^{\infty}(1 - z^2/\nu^2)$$

is *entire* and *odd* and has zeros precisely at the points of $\mathbb{Z}$, and these are first-order zeros. Since $s'(0) = \lim_{z\to 0} s(z)/z = 1$, Theorem 2 implies that $\sin \pi z = \pi s(z)$ whenever s satisfies a duplication formula. This can be verified immediately. Since s converges normally, it follows from Corollary 2.1 that

$$(+) \qquad \begin{aligned} s(2z) &= 2z \cdot \prod_{\nu=1}^{\infty}\left(1 - \frac{(2z)^2}{(2\nu)^2}\right) \cdot \prod_{\nu=1}^{\infty}\left(1 - \frac{4z^2}{(2\nu - 1)^2}\right) \\ &= 2s(z)\prod_{\nu=1}^{\infty}\left(1 - \frac{4z^2}{(2\nu - 1)^2}\right). \end{aligned}$$

A computation (!) gives

$$\left(1 - \frac{1}{4\nu^2}\right)\left(1 - \frac{4z^2}{(2\nu - 1)^2}\right) = \frac{1 + 2z/(2\nu - 1)}{1 + 2z/(2\nu + 1)}\left(1 - \frac{(2z+1)^2}{4\nu^2}\right), \quad \nu \geq 1.$$

If we take Example a) of 1.2 into account, this yields

$$\begin{aligned} \prod_{\nu=1}^{\infty}\left(1 - \frac{1}{4\nu^2}\right)\prod_{\nu=1}^{\infty}\left(1 - \frac{4z^2}{(2\nu - 1)^2}\right) &= (1 + 2z)\prod_{\nu=1}^{\infty}\left(1 - \frac{(2z+1)^2}{4\nu^2}\right) \\ &= 2s(z + \tfrac{1}{2}). \end{aligned}$$

---

[7] *Let* $f(t) = t^{-n}g(t)$, $t \in \mathbb{N}$, *where* g *is continuous and nonvanishing in* $[0, r]$, $r > 0$. Then $\int_0^r \log f(t)\, dt$ exists. This is clear since $\int_0^r \log t\, dt$ exists ($x \log x - x$ is an antiderivative, and $\lim_{\delta \searrow 0} \delta \log \delta = 0$).

Thus (+) is a duplication formula: $s(2z) = 4a^{-1}s(z)s(z+\frac{1}{2})$, where $a := \prod(1-1/4\nu^2) \neq 0$. □

This multiplicative proof dates back to the American mathematician E. H. Moore; a number of computations are carried out in his 1894 paper [M]. The reader should note the close relationship with Schottky's proof of the equation

$$\pi \cot \pi z = \frac{1}{z} + \sum_{\nu=-\infty}^{\infty}{}' \left(\frac{1}{z+\nu} - \frac{1}{\nu}\right)$$

in I.11.2.1; Moore probably did not know Schottky's 1892 paper.

4*. Proof of the duplication formula for Euler's product, following Eisenstein. Long before Moore, Eisenstein had proved the duplication formula for $s(z)$ in passing. In 1847 ([Ei], p. 461 ff.), he considered the apparently complicated product

$$E(w,z) := \prod_{\nu=-\infty}^{\infty}{}_e \left(1+\frac{z}{\nu+w}\right) = \left(1+\frac{z}{w}\right) \lim_{n\to\infty} \prod_{\nu=-n}^{n}{}' \left(1+\frac{z}{\nu+w}\right)$$

of two variables $(w,z) \in (\mathbb{C}\backslash\mathbb{Z}) \times \mathbb{C}$; here $\prod_e = \lim_{n\to\infty} \prod_{\nu=-n}^{n}$ denotes the Eisenstein multiplication (by analogy with the Eisenstein summation $\sum_e$, which we introduced in I.11.2). Moreover, $\prod'$ indicates that the factor with index 0 is omitted. The Eisenstein product $E(w,z)$ is normally convergent in the (w,z)-space $(\mathbb{C}\backslash\mathbb{Z}) \times \mathbb{C}$, since

$$\prod_{\nu=-n}^{n}{}' \left(1+\frac{z}{\nu+w}\right) = \prod_{\nu=1}^{n} \left(1-\frac{z^2+2wz}{\nu^2-w^2}\right)$$

and $\sum_{\nu=1}^{\infty} 1/(w^2-\nu^2)$ converges normally in $\mathbb{C}\backslash\mathbb{Z}$ (cf. I.11.1.3). The function $E(z,w)$ is therefore continuous in $(\mathbb{C}\backslash\mathbb{Z}) \times \mathbb{C}$ and, for fixed w, holomorphic in each $z \in \mathbb{C}$. *Computations can be carried out elegantly with* $E(z,w)$, and the following is immediate.

Duplication formula. $E(2w,2z) = E(w,z)E(w+\frac{1}{2},z)$.

Proof.

$$\begin{aligned} E(2w,2z) &= \prod_{\nu=-\infty}^{\infty}{}_e \left(1+\frac{2z}{2\nu+2w}\right) \cdot \prod_{\nu=-\infty}^{\infty}{}_e \left(1+\frac{2z}{2\nu+1+2w}\right) \\ &= E(w,z)E(w+\tfrac{1}{2},z). \end{aligned}$$

□

Eisenstein used the (trivial, but astonishing!) formula

$$(*) \qquad 1+\frac{z}{\nu+w} = \left(1+\frac{w+z}{\nu}\right) \Big/ \left(1+\frac{w}{\nu}\right) \qquad \text{(Eisenstein's trick)}$$

to reduce his "double product" to Euler's product:

$$E(w,z) = \frac{s(w+z)}{s(w)}, \quad \text{where} \quad s(z) = z\prod_{\nu\geq 1}\left(1-\frac{z^2}{\nu^2}\right).$$

Proof.

$$\begin{aligned} E(w,z) &= \frac{w+z}{w} \lim_{n\to\infty} \prod_{\nu=-n}^{n}{}' \left(1+\frac{w+z}{\nu}\right) \Big/ \lim_{n\to\infty} \prod_{\nu=-n}^{n}{}' \left(1+\frac{w}{\nu}\right) \\ &= (w+z)\prod_{\nu=1}^{\infty}\left(1-\frac{(w+z)^2}{\nu^2}\right) \Big/ \left(w\prod_{\nu=1}^{\infty}\left(1-\frac{w^2}{\nu^2}\right)\right) \\ &= \frac{s(w+z)}{s(w)}. \end{aligned}$$ □

The duplication formula for $s(z)$ is now contained in the equation

$$\frac{s(2w+2z)}{s(2w)} = E(2w,2z) = E(w,z)E(w+\tfrac{1}{2},z) = \frac{s(w+z)}{s(w)} \cdot \frac{s(w+\frac{1}{2}+z)}{s(w+\frac{1}{2})}.$$

Since s is continuous and $\lim_{w\to 0} \frac{s(2w)}{s(w)} = 2$, it follows that

$$s(2z) = \lim_{w\to 0} \frac{s(2w)}{s(w)} s(w+z) \frac{s(w+\frac{1}{2}+z)}{s(w+\frac{1}{2})} = 2s(\tfrac{1}{2})^{-1}s(z)s(z+\tfrac{1}{2}).$$ □

The elegance of Eisenstein's reasoning is made possible by the second variable w. Eisenstein also notes (loc. cit.) that E is periodic in w: $E(w+1,z) = E(w,z)$ (proved by substituting $\nu+1$ for ν); he uses E and s to prove the quadratic reciprocity law; the duplication formula appears there at the bottom of p. 462. Eisenstein calls the identity $E(w,z) = s(w+z)/s(w)$ the *fundamental formula* and writes it as follows (p. 402; the interpretation is left to the reader):

$$\prod_{m\in\mathbb{Z}} \left(1 - \frac{z}{\alpha m + \beta}\right) = \frac{\sin \pi(\beta - z)/\alpha}{\sin \pi\beta/\alpha}, \quad \alpha,\ \beta \in \mathbb{C}, \quad \beta/\alpha \notin \mathbb{Z}.$$

5. On the history of the sine product. Euler discovered the *cosine* and *sine products* in 1734–35 and published them in the famous paper "De Summis Serierum Reciprocarum" ([Eu], I-14, pp. 73–86); the formula

$$\begin{aligned} 1 - \frac{s^2}{1\cdot 2\cdot 3} + \frac{s^4}{1\cdot 2\cdot 3\cdot 4\cdot 5} - \frac{s^6}{1\cdot 2\cdot 3\cdot 4\cdot 5\cdot 6\cdot 7} + \cdots \\ = \left(1-\frac{s^2}{p^2}\right)\left(1-\frac{s^2}{4p^2}\right)\left(1-\frac{s^2}{9p^2}\right)\left(1-\frac{s^2}{16p^2}\right)\cdots \end{aligned}$$

(with $p := \pi$) appears on p. 84. As justification Euler asserts that the zeros of the series are p, $-p$, $2p$, $-2p$, $3p$, $-3p$, etc., and that the series is therefore (by analogy with polynomials) divisible by $1-\frac{s}{p}$, $1+\frac{s}{p}$, $1-\frac{s}{2p}$, $1+\frac{s}{2p}$ etc.!

In a letter to Euler dated 2 April 1737, Joh. Bernoulli emphasizes that this reasoning would be legitimate only if one knew that the function $\sin z$

had no zeros in $\mathbb{C}$ other than $n\pi$, $n \in \mathbb{Z}$: "demonstrandum esset nullam contineri radicem impossibilem" ([C], vol. 2, p.16); D. and N. Bernoulli made further criticisms; cf. [Weil], pp. 264–265. These objections, acknowledged to some extent by Euler, were among the factors giving incentive to his discovery of the formula $e^{iz} = \cos z + i \sin z$; from this Euler, in 1743, derived his product formula, which then gives him *all* the zeros of $\cos z$ and $\sin z$ as a byproduct.

Euler argues as follows: since $\lim(1 + z/n)^n = e^z$ and $\sin z = (e^{iz} - e^{-iz})/2i$,

$$\sin z = \frac{1}{2i} \lim p_n\left(\frac{iz}{n}\right), \quad \text{where} \quad p_n(w) := (1+w)^n - (1-w)^n.$$

For every even index $n = 2m$, it follows that

$$(*) \qquad p_n(w) = 2nw(1 + w + \cdots + w^{n-2}).$$

The roots ω of p_n are given by $(1+\omega) = \zeta(1-\omega)$, where $\zeta = \exp(2\nu\pi i/n)$ is any nth root of unity; hence p_{2m}, as an odd polynomial of degree $n-1$, has the $n-1$ distinct zeros $0, \pm\omega_1, \ldots, \pm\omega_{m-1}$, where

$$\omega_\nu = \frac{\exp(2\nu\pi i/n) - 1}{\exp(2\nu\pi i/n) + 1} = i \tan\frac{\nu\pi}{n}, \quad \nu = 1, \ldots, m-1.$$

The factorization

$$p_{2m}(w) = 2nw \prod_{\nu=1}^{m-1}\left(1 - \frac{w}{\omega_\nu}\right)\left(1 + \frac{w}{\omega_\nu}\right) = 2nw \prod_{\nu=1}^{m-1}\left(1 + w^2 \cot^2\frac{\nu\pi}{n}\right)$$

then follows from $(*)$. Thus

$$\sin z = z \lim_{n\to\infty} \prod_{\nu=1}^{\frac{1}{2}n-1}\left(1 - z^2\left(\frac{1}{n}\cot\frac{\nu\pi}{n}\right)^2\right).$$

Since $\lim_{n\to\infty}\left(\frac{1}{n}\cot\frac{\nu\pi}{n}\right) = \frac{1}{\nu\pi}$, interchanging the limits yields the product formula. This last step can, of course, be rigorously justified (cf., for example, [V], p. 42 and p. 56). An even simpler derivation of the sine product, based on the same fundamental idea, is given in [Nu], 5.4.3.

§4*. Euler Partition Products

Euler intensively studied the product

$$Q(z,q) := \prod_{\nu\geq 1}(1 + q^\nu z) = (1+qz)(1+q^2z)(1+q^3z)\cdot\ldots$$

as well as the sine product. $Q(z,q)$ converges normally in $\mathbb{C}$ for every $q \in \mathbb{E}$ since $\sum |q|^\nu < \infty$; the product is therefore an entire function in z, which for $q \neq 0$ has zeros precisely at the points $-q^{-1}$, $-q^{-2}, \ldots$, and these are first-order zeros. Setting $z = 1$ and $z = -1$ in $Q(z, q)$ gives, respectively, the products

$$(1+q)(1+q^2)(1+q^3) \cdot \ldots \quad \text{and} \quad (1-q)(1-q^2)(1-q^3) \cdot \ldots, \quad q \in \mathbb{E},$$

which are holomorphic in the unit disc. As we will see in Subsection 1, their power series about 0 play an important role in the theory of partitions of natural numbers. The expansion of $\prod(1-q^\nu)$ contains only those monomials q^ν for which n is a *pentagonal number* $\frac{1}{2}(3\nu^2 \pm \nu)$: this is contained in the famous pentagonal number theorem, which we discuss in Subsection 2. In Subsection 3 we expand $Q(z, q)$ in powers of z.

1. Partitions of natural numbers and Euler products. Every representation of a natural number $n \geq 1$ as a sum of numbers in $\mathbb{N}\backslash\{0\}$ is called a *partition* of n. The number of partitions of n is denoted by $p(n)$ (where two partitions are considered the same if they *differ* only in the order of their summands); for example, $p(4) = 5$, since 4 has the representations $4 = 4$, $4 = 3+1$, $4 = 2+2$, $4 = 2+1+1$, $4 = 1+1+1+1$. We set $p(0) := 1$. The values of $p(n)$ grow astronomically:

n	7	10	30	50	100	200
$p(n)$	15	42	5604	204,226	190,569,292	3,972,999,029,388

In order to study the partition function p, Euler formed the power series $\sum p(\nu)q^\nu$; he discovered the following surprising result.

Theorem ([I], p. 267). *For every* $q \in \mathbb{E}$,

$$(*) \qquad \prod_{\nu=1}^{\infty}(1-q^\nu)^{-1} = \sum_{\nu=1}^{\infty} p(\nu)q^\nu.$$

Sketch of proof. One considers the geometric series $(1-q^\nu)^{-1} = \sum_{k=0}^{\infty} q^{\nu k}$, $q \in \mathbb{E}$, and observes that $\prod_{\nu=1}^{n}(1-q^\nu)^{-1} = \sum_{\nu=1}^{\infty} p_n(k)q^k$, $q \in \mathbb{E}$, $n \geq 1$, where $p_n(0) := 1$ and, for $n \geq 1$, $p_n(k)$ denotes the number of partitions of k whose summands are all $\leq n$. Since $p_n(k) = p(k)$ for $n \geq k$, the assertion follows by passing to the limit. A detailed proof can be found in [HW] (p. 275). □

There are many formulas analogous to (*). The following appears in Euler ([I], pp. 268–269):

Let $u(n)$ (resp., $v(n)$), denote the number of partitions of $n \geq 1$ into odd (resp., distinct) summands. Then, for every $q \in \mathbb{E}$,

$$\prod_{\nu\geq 1}(1-q^{2\nu-1})^{-1} = 1 + \sum_{\nu\geq 1} u(\nu)q^\nu, \quad \prod_{\nu\geq 1}(1+q^\nu) = 1 + \sum_{\nu\geq 1} v(\nu)q^\nu.$$

From this, since

$$\begin{aligned}(1+q)(1+q^2)(1+q^3)\dots &= \frac{1-q^2}{1-q}\cdot\frac{1-q^4}{1-q^2}\cdot\frac{1-q^6}{1-q^3}\cdots\\ &= \frac{1}{1-q}\cdot\frac{1}{1-q^3}\cdot\frac{1}{1-q^5}\cdots,\end{aligned}$$

one obtains the surprising and by no means obvious conclusion

$$u(n)=v(n), \quad n\ge 1. \qquad \square$$

Since Euler's time, every function $f:\mathbb{N}\to\mathbb{C}$ is assigned the formal power series $F(z)=\sum f(\nu)z^\nu$; this series converges whenever $f(\nu)$ does not grow too fast. We call F the *generating function of* f; the products $\prod(1-q^\nu)^{-1}$, $\prod(1-q^{2\nu-1})^{-1}$, and $\prod(1+q^\nu)$ are thus the generating functions of the partition functions $p(n)$, $u(n)$, and $v(n)$, respectively. Generating functions play a major role in number theory; cf., for instance, [HW] (p. 274 ff.).

2. Pentagonal number theorem. Recursion formulas for $p(n)$ and $\sigma(n)$. The search for the Taylor series of $\prod(1-q^\nu)$ about 0 occupied Euler for years. The answer is given by his famous

Pentagonal number theorem. *For all $q\in\mathbb{E}$,*

$$(*)\qquad\begin{aligned}\prod_{\nu\ge1}(1-q^\nu) &= 1+\sum_{\nu\ge1}(-1)^\nu[q^{\frac12(3\nu^2-\nu)}+q^{\frac12(3\nu^2+\nu)}]\\ &= \sum_{\nu=-\infty}^{\infty}(-1)^\nu q^{\frac12(3\nu^2-\nu)}\\ &= 1-q-q^2+q^5+q^7-q^{12}-q^{15}+q^{22}+q^{26}\\ &\quad -q^{35}-q^{40}+q^{51}+\cdots.\end{aligned}$$

We will derive this theorem in 5.2 from Jacobi's triple product identity.

The sequence $\omega(\nu):=\frac12(3\nu^2-\nu)$, which begins with 1, 5, 12, 22, 35, 51, was already known to the Greeks (cf. [D], p. 1). Pythagoras is said to have determined $\omega(n)$ by nesting regular pentagons whose edge length increases by 1 at each stage and counting the number of vertices (see Figure 1.1).

Because of this construction principle, the numbers $\omega(\nu)$, $\nu\in\mathbb{Z}$, are called *pentagonal numbers*; this characterization gave the identity $(*)$ its name.

Statements about the partition function p can be obtained by comparing coefficients in the identity

$$1=\left(\sum_{\nu\ge0}p(\nu)q^\nu\right)\left(1+\sum_{\nu\ge1}(-1)^\nu[q^{\omega(\nu)}+q^{\omega(-\nu)}]\right),$$

which is clear by 1$(*)$ and $(*)$. In fact, Euler obtained the following formula in this way. (Cf. also [HW], pp. 285–286.)

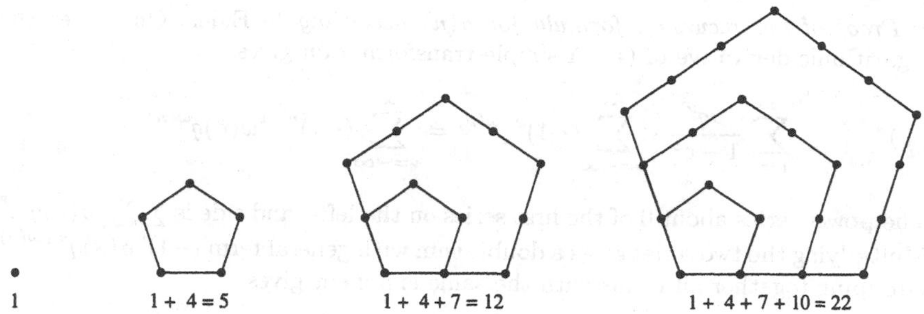

FIGURE 1.1.

Recursion formula for $p(n)$. If we set $p(n) := 0$ for $n < 0$, then

$$\begin{aligned} p(n) &= p(n-1) + p(n-2) - p(n-5) - p(n-7) + \cdots \\ &= \sum_{k \geq 1} (-1)^{k-1} [p(n-\omega(k)) + p(n-\omega(-k))]. \end{aligned}$$

It was a great surprise for Euler when he recognized — and proved, using the pentagonal number theorem — that almost the same formula holds for sums of divisors. Let $\sigma(n) := \sum_{d|n} d$ denote the sum of all positive divisors of the natural number $n \geq 1$. Then we have the

Recursion formula for $\sigma(n)$. If we set $\sigma(\nu) := 0$ for $\nu \leq 0$, then

$$\begin{aligned} \sigma(n) &= \sigma(n-1) + \sigma(n-2) - \sigma(n-5) - \sigma(n-7) + \cdots \\ &= \sum_{k \geq 1} (-1)^{k-1} [\sigma(n-\omega(k)) + \sigma(n-\omega(-k))] \end{aligned}$$

for every natural number $n \geq 1$ that is not a pentagonal number. On the other hand, for every number $n = \frac{1}{2}(3\nu^2 \pm \nu)$, $\nu \geq 1$,

$$\begin{aligned} \sigma(n) &= (-1)^{\nu-1} n + \sigma(n-1) + \sigma(n-2) - \sigma(n-5) - \sigma(n-7) + \cdots \\ &= (-1)^{\nu-1} n + \sum_{k \geq 1} (-1)^{k-1} [\sigma(n-\omega(k)) + \sigma(n-\omega(-k))]. \end{aligned}$$

Often, in the literature, only the first formula is given for *all* $n \geq 1$, with the provision that the summand $\sigma(n-n)$, if it occurs, is given the value n. Euler also stated the formula this way. For $12 = \frac{1}{2}(3 \cdot 3^2 - 3)$, we have

$$\sigma(12) = (-1)^2 12 + \sigma(11) + \sigma(10) - \sigma(7) - \sigma(5) + \sigma(0) = 12 + 12 + 18 - 8 - 6 = 28.$$

Proof of the recursion formula for $\sigma(n)$ according to Euler. One takes the logarithmic derivative of (*). A simple transformation gives

$$(+) \qquad \sum_{\nu=1}^{\infty} \frac{\nu q^\nu}{1-q^\nu} \cdot \sum_{\nu=-\infty}^{\infty} (-1)^\nu q^{\omega(\nu)} = \sum_{\nu=-\infty}^{\infty} (-1)^{n-1} \omega(n) q^{\omega(n)}.$$

The power series about 0 of the first series on the left-hand side is $\sum_{\kappa=1}^{\infty} \sigma(\kappa) q^\kappa$.[8] Multiplying the two series gives a double sum with general term $(-1)^\nu \sigma(\kappa) q^{\kappa+\omega(\nu)}$. Grouping together all terms with the same exponent gives

$$\sum_{n=1}^{\infty} \left[\sum_{k \in \mathbb{Z}} (-1)^k \sigma(n - \omega(k)) \right] q^n.$$

The assertion follows by comparing coefficients in (+). □

There appear to be no known elementary proofs of the recursion formula for $\sigma(n)$. The function $\sigma(n)$ can be expressed recursively by means of the function $p(n)$. For all $n \geq 1$,

$$\begin{aligned} \sigma(n) &= p(n-1) + 2p(n-2) - 5p(n-5) - 7p(n-7) + \\ &\quad + 12p(n-12) + 15p(n-15) - \cdots \\ &= \sum_{k \geq 1} (-1)^{k-1} [\omega(k) p(n - \omega(k)) + \omega(-k) p(n - \omega(-k))]. \end{aligned}$$

This was observed in 1884 by C. Zeller [Z]. We note another formula that can be derived by means of the pentagonal number theorem:

$$p(n) = \frac{1}{n} \sum_{\nu=1}^{n} \sigma(\nu) p(n - \nu).$$

3. Series expansion of $\prod_{\nu=1}^{\infty}(1 + q^\nu z)$ in powers of z. Although the power series expansion of this function in powers of q is known only for special values of z (cf. Subsections 1 and 2), its expansion in powers of z can be found easily. If we set $Q(z,q) := \prod_{\nu \geq 1}(1 + q^\nu z)$, it follows at once that

$$(1) \qquad (1 + qz) Q(qz, q) = Q(z, q);$$

[8]Series of the type $\sum_{\nu=1}^{\infty} a_\nu q^\nu/(1-q^\nu)$ are called *Lambert series.* Since $q^\nu \cdot (1-q^\nu)^{-1} = \sum_{\mu=1}^{\infty} q^{\mu\nu}$, the following is immediate (cf. also [Kn], p. 450).

If the Lambert series $\sum_{\nu=1}^{\infty} a_\nu q^\nu/(1-q^\nu)$ converges normally in $\mathbb{E}$, then

$$\sum_{\nu=1}^{\infty} a_\nu \frac{q^\nu}{1-q^\nu} = \sum_{\nu=1}^{\infty} A_\nu q^\nu, \; q \in \mathbb{E}, \text{ where } A_\nu := \sum_{d|\nu} a_d.$$

for $(q, z) \in \mathbb{E} \times \mathbb{C}$, this functional equation immediately gives

$$(2) \qquad \prod_{\nu=1}^{\infty}(1+q^\nu z) = 1 + \sum_{\nu=1}^{\infty} \frac{q^{\frac{1}{2}\nu(\nu+1)}}{(1-q)(1-q^2)\dots(1-q^\nu)} z^\nu.$$

Proof. For fixed $q \in \mathbb{E}$, let $\sum_{\nu \geq 0} a_\nu z^\nu$ be the Taylor series for $Q(z, q)$. Then $a_0 = 1$, and (1) gives the recursion formula

$$a_\nu q^\nu + a_{\nu-1} q^\nu = a_\nu; \quad \text{i.e.} \quad a_\nu = \frac{q^\nu}{1-q^\nu} a_{\nu-1} \quad \text{for } \nu \geq 1.$$

It follows from this (by induction, for example) that $a_\nu = q^{\frac{1}{2}\nu(\nu+1)}[(1-q) \cdot \dots \cdot (1-q^\nu)]^{-1}$. □

For $z := 1$, we see that

$$(3) \qquad \begin{aligned} (1+q)(1+q^2)(1+q^3)\dots \quad &= \quad 1 + \frac{q}{1-q} + \frac{q^3}{(1-q)(1-q^2)} \\ &\quad + \frac{q^6}{(1-q)(1-q^2)(1-q^3)} + \cdots . \end{aligned}$$

If we write q^2 instead of q in (2) and set $z := q^{-1}$, we obtain

$$\prod_{\nu=1}^{\infty}(1+q^{2\nu-1}) = 1 + \sum_{\nu=1}^{\infty} \frac{q^{\nu^2}}{(1-q^2)(1-q^4)\dots(1-q^{2\nu})},$$

or, written out,

$$\begin{aligned} (1+q)(1+q^3)(1+q^5)\dots \quad &= \quad 1 + \frac{q}{1-q^2} + \frac{q^4}{(1-q^2)(1-q^4)} \\ &\quad + \frac{q^9}{(1-q^2)(1-q^4)(1-q^6)} + \cdots . \end{aligned}$$

This derivation and more can be found in [I], p. 251 ff. □

The product $Q(z, q)$ is simpler than the sine product. Not only does normal convergence already follow because of the geometric series, but the functional equation (1), which replaces the duplication formula for $s(z)$, follows easily and is also more fruitful.

Exercises. Show that the following hold for all $(q, z) \in \mathbb{E} \times \mathbb{C}$:

a) $\displaystyle\prod_{\nu=1}^{\infty} \frac{1}{1-q^\nu z} = 1 + \sum_{\nu=1}^{\infty} \frac{q^\nu}{(1-q)(1-q^2) \cdot \dots \cdot (1-q^\nu)} z^\nu,$

b) $$\prod_{\nu=1}^{\infty} \frac{1}{1-q^\nu z} = 1 + \sum_{\nu=1}^{\infty} \frac{q^{\nu^2}}{(1-q)(1-q^2)\cdot\ldots\cdot(1-q^\nu)} \times \frac{z^\nu}{(1-qz)(1-q^2z)\cdot\ldots\cdot(1-q^\nu z)}.$$

Compare the results for $z = 1$.

Hint. For a), first consider $\prod_{\nu=1}^{n} \frac{1}{1-q^\nu z}$, $1 \le n < \infty$. Find functional equations in each case and imitate the proof of (2); for a), conclude by letting $n \to \infty$. Equation b) can be found, for example, in the *Fundamenta* ([Ja$_1$], pp. 232–233).

4. On the history of partitions and the pentagonal number theorem. As early as 1699, G. W. Leibniz asked Joh. Bernoulli in a letter whether he had studied the function $p(n)$; he commented that this problem was important though not easy (*Math. Schriften*, ed. Gerhardt, vol. III-2, p. 601). Euler was asked by P. Naudé, a Berlin mathematician of French origin, in how many ways a given natural number n could be represented as a sum of s distinct natural numbers. Euler repeatedly considered these and related questions and thus became the father of a new area of analysis, which he called "partitio numerorum." In April 1741, shortly before his departure for Berlin, he had already submitted his first results to the Petersburg Academy ([Eu], I-2, pp. 163–193). At the end of this work he stated the pentagonal number theorem, after he had determined the initial terms of the pentagonal number series up to the summand q^{51} by multiplying out the first 51 factors of $\prod(1-q^\nu)$ (loc. cit., pp. 191–192). But almost 10 years passed before he could prove the theorem (letter to Goldbach, 9 June 1750; [C], vol. 1, pp. 522–524). In its introduction, Chapter 16 deals thoroughly "with the decomposition of numbers into parts"; the pentagonal number theorem is mentioned and applied (p. 269).

The recursion formula for the function $p(n)$ first appears in 1750, in the treatise *De Partitione Numerorum* ([Eu], I-2, p. 281). It was used in 1918 by P. A. Macmahon to compute $p(n)$ up to $n = 200$; he found that $p(200) = 3{,}972{,}999{,}029{,}388$ (*Proc. London Math. Soc.* (2) 17, 1918; pp. 114–115 in particular).

In 1741, Euler had already verified the recursion formula for $\sigma(n)$ numerically for all $n < 300$ (letter to Goldbach, 1 April 1741; [C], vol. 1, pp. 407–410). In that letter, he called his discovery *"a very surprising pattern in the numbers"* and wrote that he *"would have [no] rigorous proof. But even if I had none at all, no one could doubt its truth, since this rule is always valid up to over 300."* He then informed Goldbach of the derivation of the recursion formula from the (then still unproved) pentagonal number theorem. He gave a complete statement with a proof in 1751, in "Découverte d'une loi tout extraordinaire des nombres par rapport à la somme de leurs diviseurs" ([Eu], I-2, pp. 241–253). — The reader can find further historical information and commentary in [Weil], pp. 276–281. Not until almost eighty years later could Jacobi give the complete explanation of the Euler identities with his theory of theta functions. We examine this a bit more closely in the next section.

§5*. Jacobi's Product Representation of the Series $J(z,q) := \sum_{\nu=-\infty}^{\infty} q^{\nu^2} z^\nu$

The Laurent series $\sum_{\nu=-\infty}^{\infty} q^{\nu^2} z^\nu = 1 + \sum_{\nu=1}^{\infty} q^{\nu^2}(z^\nu + z^{-\nu})$ converges for every $q \in \mathbb{E}$; thus $J(z,q) \in \mathcal{O}(\mathbb{C}^\times)$ for all $q \in \mathbb{E}$. Readers familiar with the theta function will immediately observe that

$$\vartheta(z,\tau) = J(e^{2\pi i z}, e^{-\pi\tau}) \quad \text{(cf. I.12.4);}$$

this relation, however, plays no role in what follows. It is immediate that

$$(1) \qquad J(i,q) = J(-1,q^4), \quad q \in \mathbb{E}.$$

Jacobi saw in 1829 that his series $J(z,q)$ coincided with the product

$$A(z,q) := \prod_{\nu=1}^{\infty} [(1-q^{2\nu})(1+q^{2\nu-1}z)(1+q^{2\nu-1}z^{-1})],$$

which had been studied by Abel. $A(z,q) \in \mathcal{O}(\mathbb{C}^\times)$ for every $q \in \mathbb{E}$, since the product converges normally in $\mathbb{C}^\times$ for each q. The following relation holds between the Euler product $Q(z,q)$ of 4.3 and $A(z,q)$:

$$A(z,q) = \prod_{\nu=1}^{\infty} (1-q^{2\nu}) \cdot Q(q^{-1}z, q^2) \cdot Q(q^{-1}z^{-1}, q^2).$$

The identity $J(z,q) = A(z,q)$, called Jacobi's triple product identity, is one of many deep formulas that appear in Jacobi's *Fundamenta Nova.* We obtain it in Subsection 1 with the aid of the functional equations

$$(2) \quad A(q^2 z, q) = (qz)^{-1} A(z,q), \quad A(z^{-1}, q) = A(z,q), \quad (z,q) \in \mathbb{C}^\times \times \mathbb{E}^\times,$$

$$(3) \qquad A(i,q) = A(-1,q^4), \quad q \in \mathbb{E},$$

all of which can easily be deduced from the definition of A; in the proof of (3), we observe that

$$\prod_{\nu=1}^{\infty}(1-q^{2\nu}) = \prod_{\nu=1}^{\infty}[(1-q^{4\nu})(1-q^{4\nu-2})], \quad (1+q^{2\nu-1}i)(1-q^{2\nu-1}i) = 1+q^{4\nu-2}.$$

Fascinating identities, some of which go back to Euler, result from considering special cases of the equation $J(z,q) = A(z,q)$; we give samples in Subsection 2.

1. Jacobi's theorem. *For all* $(q,z) \in \mathbb{E} \times \mathbb{C}^\times$,

$$\text{(J)} \qquad \sum_{\nu=-\infty}^{\infty} q^{\nu^2} z^\nu = \prod_{\nu=1}^{\infty} [(1-q^{2\nu})(1+q^{2\nu-1}z)(1+q^{2\nu-1}z^{-1})].$$

Proof (cf. [HW], pp. 282–283). For every $q \in \mathbb{E}$, the product $A(z,q)$ has a Laurent expansion $\sum_{-\infty}^{\infty} a_\nu z^\nu$ about 0 in $\mathbb{C}^\times$, with coefficients a_ν that depend on q. Equations (2) of the introduction imply that $a_{-\nu} = a_\nu$ and $a_\nu = q^{2\nu-1} a_{\nu-1}$ for all $\nu \in \mathbb{Z}$. From this it follows (first inductively for $\nu > 0$ and then in general) that $a_\nu = q^{\nu^2} a_0$ for all $\nu \in \mathbb{Z}$. It is thus already clear, if we write $a(q)$ for a_0, that

$$A(z,q) = a(q) J(z,q) \quad \text{with} \quad a(0) = 1.$$

$A(1,q)$ and $J(1,q)$ are holomorphic in $\mathbb{E}$ as functions of q and $J(1,0) = 1$; hence $a(q)$ is holomorphic in a neighborhood of zero. From equations (1) and (3) of the introduction it follows, because $J(i,q) \not\equiv 0$, that

$$a(q) = a(q^4) \quad \text{and hence} \quad a(q) = a(q^{4^n}), \quad n \geq 1, \quad \text{for all } q \in \mathbb{E}.$$

The continuity of $a(q)$ at 0 forces $a(q) = \lim_{n\to\infty} a(q^{4^n}) = a(0) = 1$ for all $q \in \mathbb{E}$. □

The idea of this elegant proof is said to date back to Jacobi (cf. [HW], p. 296). The reader is advised to look at Kronecker's proof ([Kr], pp. 182–186). With $z := e^{2iw}$, (J) can be written in the form

$$\sum_{\nu=-\infty}^{\infty} q^{\nu^2} e^{2i\nu w} = \prod_{\nu=1}^{\infty} [(1 - q^{2\nu})(1 + 2q^{2\nu-1} \cos 2w + q^{4\nu-2})].$$

The identity (J) is occasionally also written as

$$(\mathrm{J}') \quad \sum_{\nu=-\infty}^{\infty} (-1)^\nu q^{\frac{1}{2}\nu(\nu+1)} z^\nu = (1 - z^{-1}) \prod_{\nu=1}^{\infty} [(1 - q^\nu)(1 - q^\nu z)(1 - q^\nu z^{-1})].$$

(J′) follows from (J) by substituting $-qz$ for z, rearranging the resulting product, and finally writing q instead of q^2.

2. Discussion of Jacobi's theorem. For $z := 1$, (J) gives the product representation of the classical theta series

$$(1) \qquad \sum_{\nu=-\infty}^{\infty} q^{\nu^2} = 1 + 2\sum_{\nu=1}^{\infty} q^{\nu^2} = \prod_{\nu=1}^{\infty} [(1 + q^{2\nu-1})^2 (1 - q^{2\nu})],$$

which converges in $\mathbb{E}$. We also note:

Suppose that $k, l \in \mathbb{N}\backslash\{0\}$ *are both even or both odd. Then, for all* $(z,q) \in \mathbb{C}^\times \times \mathbb{E}$,

$$(2) \quad \sum_{-\infty}^{\infty} q^{\frac{1}{2}\nu(k\nu+l)} z^\nu = \prod_{\nu=1}^{\infty} [(1 - q^{k\nu})(1 + q^{k\nu - \frac{1}{2}(k-l)} z)(1 + q^{k\nu - \frac{1}{2}(k+l)} z^{-1})].$$

Proof. First let $0 < q < 1$. Then $q^{\frac{1}{2}k}$, $q^{\frac{1}{2}l} \in (0,1)$ are uniquely determined, and substituting $q^{\frac{1}{2}k}$ for q and $q^{\frac{1}{2}l}z$ for z turns (J) into (2). By the hypothesis on k and l, all the exponents in (2) are integers (!); hence the left- and right-hand sides of (2) are holomorphic functions in $q \in \mathbb{E}$ for fixed z. The assertion follows from the identity theorem. □

For $k = l = 1$ and $z = 1$, (2) becomes

$$(3) \quad \sum_{-\infty}^{\infty} q^{\frac{1}{2}\nu(\nu+1)} = 2 + 2\sum_{\nu=1}^{\infty} q^{\frac{1}{2}\nu(\nu+1)} = \prod_{\nu=1}^{\infty}[(1-q^{2\nu})(1+q^{\nu-1})];$$

this identity, due to Euler, was written by Gauss in 1808 as follows ([Ga], p. 20):

$$(3') \quad 1 + q + q^3 + q^6 + q^{10} + \text{etc.} = \frac{1-qq}{1-q} \cdot \frac{1-q^4}{1-q^3} \cdot \frac{1-q^6}{1-q^5} \cdot \frac{1-q^8}{1-q^7} \cdot \text{etc.}$$

(to prove this, use Exercise 2) of 2.1). □

For $k = 3$, $l = 1$, and $z = -1$, equation (2) says that

$$\prod_{\nu=1}^{\infty}[(1-q^{3\nu})(1-q^{3\nu-1})(1-q^{3\nu-2})] = \sum_{\nu=-\infty}^{\infty}(-1)^\nu q^{\frac{1}{2}\nu(3\nu+1)}.$$

Since each factor $1 - q^\nu$, $\nu \geq 1$, appears here on the left-hand side exactly once, this yields the pentagonal number theorem

$$(4) \qquad \prod_{\nu=1}^{\infty}(1-q^\nu) = 1 + \sum_{\nu=1}^{\infty}(-1)^\nu[q^{\frac{1}{2}(3\nu-1)} + q^{\frac{1}{2}(3\nu+1)}], \quad q \in \mathbb{E},$$

as announced in 4.2. Written out, this becomes

$$(4') \qquad \begin{aligned}&(1-q)(1-q^2)(1-q^3)\dots \\ &\quad = 1 - q - q^2 + q^5 + q^7 - q^{12} - q^{15} + \cdots.\end{aligned}$$ □

Now, in principle, the power series about 0 of $\prod(1+q^\nu)$ can also be computed. Since $\prod(1-q^\nu) \cdot \prod(1+q^\nu) = \prod(1-q^{2\nu})$, we use (4′) to obtain

$$\begin{aligned}\prod_{\nu=1}^{\infty}(1+q^\nu) &= \frac{1-q^2-q^4+q^{10}+q^{14}-\cdots}{1-q-q^2+q^5+q^7-\cdots} \\ &= 1+q+q^2+2q^3+2q^4+3q^5+4q^6+5q^7+\cdots.\end{aligned}$$

The first coefficients on the right-hand side were already given by Euler; no simple explicit representation of all the coefficients is known. Number-theoretic interpretations of the formulas above, as well as further identities, can be found in [HW]. □

We conclude this discussion by noting Jacobi's famous formula for the cube of the Euler product (cf. [Ja$_1$], p. 237, and [JF], p. 60):

$$\prod_{\nu=1}^{\infty}(1-q^\nu)^3 = \sum_{\nu=0}^{\infty}(-1)^\nu(2\nu+1)q^{\frac{1}{2}\nu(\nu+1)}. \tag{5}$$

To prove this, Jacobi differentiates the identity (J′) of Section 1 with respect to z, then sets $z := 1$ (the reader should carry out the details, grouping the terms in the series with index ν and $-\nu-1$). In 1848, referring to identity (5), Jacobi wrote ([JF], p. 60): "Dies mag wohl in der Analysis das einzige Beispiel sein, daß eine Potenz einer Reihe, deren Exponenten eine arithmetische Reihe zweiter Ordnung [= quadratischer Form an^2+bn+c] bilden, wieder eine solche Reihe giebt." (This may well be the only example in analysis where a power of a series whose exponents form an arithmetic series of second order [= quadratic form an^2+bn+c] again gives such a series.)

3. On the history of Jacobi's identity. Jacobi proved the triple product identity in 1829, in his great work *Fundamenta Nova Theoriae Functionum Ellipticarum*; at that time he wrote ([Ja$_1$], p. 232):

> Aequationem identicam, quam antecedentibus comprobatum ivimus:
>
> $$\begin{aligned}&(1-2q\cos 2x+q^2)(1-2q^3\cos 2x+q^6)(1-2q^5\cos 2x+q^{10})\ldots\\ &\quad= \frac{1-2q\cos 2x+2q^4\cos 4x-2q^9\cos 6x+2q^{16}\cos 8x-\cdots}{(1-q^2)(1-q^4)(1-q^6)(1-q^8)\ldots}.\end{aligned}$$

In a paper published in 1848, Jacobi systematically exploited his equation and wrote ([Ja$_2$], p. 221):

> Die sämmtlichen diesen Untersuchungen zum Grunde gelegten Entwicklungen sind particuläre Fälle einer Fundamentalformel der Theorie der elliptischen Functionen, welche in der Gleichung
>
> $$\begin{aligned}&(1-q^2)(1-q^4)(1-q^6)(1-q^8)\ldots\\ &\times(1-qz)(1-q^3z)(1-q^5z)(1-q^7z)\ldots\\ &\times(1-qz^{-1})(1-q^3z^{-1})(1-q^5z^{-1})(1-q^7z^{-1})\ldots\\ &=1-q(z+z^{-1})+q^4(z^2+z^{-2})-q^9(z^3+z^{-3})+\cdots\end{aligned}$$
>
> enthalten ist. (All the developments that underlie these investigations are special cases of a fundamental formula of the theory of elliptic functions, which is contained in the equation ....)

The preliminary work for the Jacobi formula was carried out by Euler through his pentagonal number theorem. In 1848 Jacobi wrote to the secretary of the Petersburg Academy, P. H. von Fuss (1797–1855) (cf. [JF], p.60): "Ich möchte mir bei dieser Gelegenheit noch erlauben, Ihnen zu sagen, warum ich mich so für diese Eulersche Entdeckung interessiere. Sie ist nämlich der erste Fall gewesen, in welchem Reihen aufgetreten sind, deren Exponenten eine arithmetische Reihe *zweiter* Ordnung bilden, und auf diese Reihen ist durch mich die Theorie der elliptischen Transcendenten gegründet worden. Die Eulersche Formel ist ein specieller Fall einer Formel, welche wohl das wichtigste und fruchtbarste ist, was ich in reiner Mathematik erfunden habe." (I would also like to take this opportunity to tell you why I am so interested in Euler's discovery. It was, you see, the first case where series appeared whose exponents form an arithmetic series of *second* order, and these series, through my work, form the basis of the theory of elliptic transcendental functions. The Euler formula is a special case of a formula that is probably the most important and fruitful I have discovered in pure mathematics.)

Jacobi did not know that, long before Euler, Jacob Bernoulli and Leibniz had already come across series whose exponents form a series of second order. In 1685 Jacob Bernoulli, in the *Journal des Scavans*, posed a problem in probability theory whose solution he gave in 1690 in *Acta Eruditorum*: series appear there whose exponents are explicitly asserted to be arithmetic series of second order. Shortly after Bernoulli, Leibniz — in *Acta Eruditorum* — also solved the problem; he considered the question especially interesting because it might lead to series that had not yet been thoroughly studied (ad series tamen non satis adhuc examinatas ducit). For further details, see the article [En] of G. E. Eneström.

In his *Ars Conjectandi*, Bernoulli returned to the problem; the series

$$1 - m + m^3 - m^6 + m^{10} - m^{15} + m^{21} - m^{28} + m^{36} - m^{45} + \cdots$$

appears in [B] (p. 142). Bernoulli says that he cannot sum the series but that one can easily "compute approximate values to arbitrarily prescribed accuracy" (from R. Haussner's German translation of [B], p. 59). Bernoulli gives the approximation 0.52393, which is accurate up to a unit in the last decimal place.

Gauss informed Jacobi that he had already known this formula by about 1808; cf. the first letter from Jacobi to Legendre ([JL], p. 394). Legendre, bitter toward Gauss because of the reciprocity law and the method of least squares, writes to Jacobi on the subject ([JL], p. 398): "Comment se fait-il que M. Gauss ait osé vous faire dire que la plupart de vos théorèmes lui était connus et qu'il en avait fait la découverte dès 1808? Cet excès d'impudence n'est pas croyable de la part d'un homme qui a assez de mérite personnel pour n'avoir besoin de s'approprier les découvertes des autres ... ." (How could Mr. Gauss have dared inform you that most of your theorems were known to him and that he had discovered them as early as 1808? Such outrageous impudence is incredible in a man with enough ability of his own that he shouldn't have to take credit for other people's discoveries ... .)

But Gauss was right: Jacobi's fundamental formula and more were found in the papers he left behind. Gauss's manuscripts were printed in 1876, in the third volume of his *Werke*; on page 440 (without any statements about convergence) is the formula

$$(1+xy)(1+x^3y)(1+x^5y)\ldots\left(1+\tfrac{x}{y}\right)\left(1+\tfrac{x^3}{y}\right)\left(1+\tfrac{x^5}{y}\right)\ldots$$
$$=\tfrac{1}{[xx]}\left\{1+x\left(y+\tfrac{1}{y}\right)+x^4\left(yy+\tfrac{1}{yy}\right)+x^9\left(y^3+\tfrac{1}{y^3}\right)+\cdots\right\},$$

where $[xx]$ stands for $(1-x^2)(1-x^4)(1-x^6)\ldots$. This does in fact give Jacobi's result (J). Schering, the editor of this volume, declares on page 494 that this research of Gauss *probably belongs to the year 1808.*

Kronecker, generally sparing of praise, paid tribute to the triple product identity as follows ([Kr], p. 186): "Hierin besteht die ungeheure Entdeckung Jacobi's; die Umwandlung der Reihe in das Produkt war sehr schwierig. Abel hat auch das Produkt, aber nicht die Reihe. Deshalb wollte Dirichlet sie auch als Jacobi'sche Reihe bezeichnen." (Jacobi's tremendous discovery consists of this: the transformation of the series into the product was very difficult. Abel too had the product, but not the series. This is why Dirichlet also wanted it to be called the Jacobi series.)

The Jacobi formulas are only the tip of an iceberg of fascinating identities. In 1929, G. N. Watson ([Wa], pp. 44–45) discovered the

Quintuple product identity. *For all* $(q,z)\in\mathbb{E}\times\mathbb{C}^\times$,

$$\sum_{\nu=-\infty}^{\infty} q^{3\nu^2-2\nu}(z^{3\nu}+z^{-3\nu}-z^{3\nu-2}-z^{-3\nu+2})$$
$$=\prod_{\nu=1}^{\infty}(1-q^{2\nu})(1-q^{2\nu-1}z)(1-q^{2\nu-1}z^{-1})(1-q^{4\nu-4}z^2)(1-q^{4\nu-4}z^{-2}).$$

Many additional formulas come from considering special cases; see also [Go] and [Ew]. For some years there has been a renaissance of the Jacobi identities in the theory of affine root systems. As a result, identities have been discovered that were unknown in the classical theory. E. Neher, in [N], gives an introduction with many references to the literature.

Bibliography

[B] BERNOULLI, J.: *Ars Conjectandi, Die Werke von Jakob Bernoulli*, vol. 3, 107–286; Birkhäuser, 1975; German translation by R. HAUSSNER, Oswald's Klassiker 107 (1895).

[C] Von FUSS, P. J. (ed.): *Correspondance mathématique et physique de quelques célèbres géomètres du XVIIIième siècle*, St. Petersburg, 1843, 2 vols.; reprinted in 1968 by Johnson Reprint Corp.

[D] DICKSON, L. E.: *History of the Theory of Numbers*, vol. 2, Chelsea Publ. Co., New York, 1952.

[Ei] EISENSTEIN, F. G. M.: Genaue Untersuchung der undendlichen Doppelproducte, aus welchen die elliptischen Functionen als Quotienten zusammengesetzt sind, und der mit ihnen zusammenhängenden Doppelreihen, *Journ. reine angew. Math.* 35, 153–274 (1847); *Math. Werke* 1, 357–478.

[En] ENESTRÖM, G. E.: Jacob Bernoulli und die Jacobischen Thetafunction, *Bibl. Math.* 9, 3. Folge, 206–210 (1908–1909).

[Eu] Leonhardi EULERI *Opera omnia*, sub auspiciis societatis scientarium naturalium Helveticae, Series I–IV A, 1911–.

[Ew] EWELL, J. A.: Consequences of Watson's quintuple-product identity, *Fibonacci Quarterly* 20(3), 256–262 (1982).

[Ga] GAUSS, C. F.: Summatio quarundam serierum singularium, *Werke* 2, 9–45.

[Go] GORDON, B.: Some identities in combinatorial analysis, *The Quart. Journ. Math.* (Oxford) 12, 285–290 (1961).

[HW] HARDY, G. H. and E. M. WRIGHT: *An Introduction to the Theory of Numbers*, 4th ed., Oxford, Clarendon Press, 1960.

[I] EULER, L.: *Introductio in Analysin Infinitorum*, vol. 1, Lausanne 1748; in [Eu], I-8; German translation, *Einleitung in der Analysis des Unendlichen*, 1885, published by Julius Springer; reprinted by Springer, 1983.

[Ja_1] JACOBI, C. G. J.: Fundamenta nova theoriae functionum ellipticarum, *Ges. Werke* 1, 49–239.

[Ja_2] JACOBI, C. G. J.: Ueber unendliche Reihen, deren Exponenten zugleich in zwei verschiedenen quadratischen Formen enthalten sind, *Journ. reine angew. Math.* 37, 61–94 and 221–254 (1848); *Ges. Werke* 2, 217–288.

[JF] JACOBI, C. G. J. and P. H. von FUSS: *Briefwechsel zwischen C. G. J. Jacobi und P. H. von Fuss über die Herausgabe der Werke Leonhard Eulers*, ed. P. STÄCKEL and W. AHRENS, Teubner, Leipzig, 1908.

[JL] JACOBI, C. G. J.: Correspondance mathématique avec Legendre, *Ges. Werke* 1, 385–461.

[Kn] KNOPP, K.: *Theory and Application of Infinite Series*, Heffner Publishing Co., New York, 1947 (trans. R. C. H. YOUNG).

[Kr] KRONECKER, L: *Theorie der einfachen und der vielfachen Integrale*, ed. E. NETTO, Teubner, Leipzig, 1894.

[M] MOORE, E. H.: Concerning the definition by a system of functional properties of the function $f(z) = (\sin \pi z)/\pi$, *Ann. Math.* 9, 1st ser., 43–49 (1894).

[N] NEHER, E.: Jacobi's Tripelprodukt Identität und η-Identitäten in der Theorie affiner Lie-Algebren, *Jber. DMV* 87, 164–181 (1985).

[Nu] *Numbers*, Springer-Verlag, New York, 1993; ed. J. H. EWING; trans. H. L. S. ORDE.

[P] PRINGSHEIM, A.: Über die Convergenz unendlicher Produkte, *Math. Ann.* 33, 119–154 (1889).

[R] RITT, J. F.: Representation of analytic functions as infinite products, *Math. Zeitschr.* 32, 1–3 (1930).

[V] VALIRON, G.: *Théorie des fonctions*, 2nd ed., Masson, Paris, 1948.

[Wa] WATSON, G. N.: Theorems stated by Ramanujan (VII): Theorems on continued fractions, *Journ. London Math. Soc.* 4, 39–48 (1929).

[Wei] WEIERSTRASS, K.: Über die Theorie der analytischen Facultäten, *Journ. reine angew. Math.* 51, 1–60 (1856); *Math. Werke* 1, 153–221.

[Weil] WEIL, A.: *Number Theory: An Approach through History, from Hammurapi to Legendre*, Birkhäuser, 1984.

[Z] ZELLER,, C.: Zu Eulers Recursionsformel für die Divisorensummen, *Acta Math.* 4, 415–416 (1884).

2

The Gamma Function

> Also das Product $1 \cdot 2 \cdot 3 \ldots x$ ist die Function, die meiner Meinung nach in der Analyse eingeführt werden muss. (Thus the product $1 \cdot 2 \cdot 3 \ldots x$ is the function that, in my opinion, must be introduced into analysis.)
>
> — C. F. Gauss to F. W. Bessel, 21 November 1811

1. The problem of extending the function $n!$ to real arguments and finding the simplest possible "factorial function" with value $n!$ at $n \in \mathbb{N}$ led Euler in 1729 to the Γ-function. He gave the infinite product

$$\Gamma(z+1) := \frac{1 \cdot 2^z}{1+z} \cdot \frac{2^{1-z}3^z}{2+z} \cdot \frac{3^{1-z}4^z}{3+z} \cdots = \prod_{\nu=1}^{\infty} \left(1+\frac{1}{\nu}\right)^z \left(1+\frac{z}{\nu}\right)^{-1}$$

as a solution.[1] Euler considered only real arguments; Gauss, in 1811, admitted complex numbers as well. On 21 November 1811, he wrote to Bessel (1784-1846), who was also concerned with the problem of general factorials, "Will man sich aber nicht ... zahllosen Paralogismen und Paradoxen und Widersprüchen blossstellen, so muss $1 \cdot 2 \cdot 3 \ldots x$ nicht als Definition von $\prod x$ gebraucht werden, da eine solche nur, wenn x eine ganze Zahl ist, einen bestimmten Sinn hat, sondern man muss von einer höheren allgemein, selbst

[1] Precise references to Euler can be found in the appropriate sections of this chapter; we rely to a large extent on the article "Übersicht über die Bände 17, 18, 19 der ersten Serie" of A. Krazer and G. Faber in [Eu], I-19, pp. XLVII–LXV in particular.

auf imaginäre Werthe von x anwendbaren, Definition ausgehen, wovon ... jene als specieller Fall erscheint. Ich habe folgenden gewählt

$$\Pi x = \frac{1.2.3\ldots k.k^x}{x+1.x+2.x+3\ldots x+k},$$

wenn k unendlich wird." (But if one doesn't want ... countless fallacies and paradoxes and contradictions to be exposed, $1 \cdot 2 \cdot 3 \ldots x$ must not be used as the definition of $\prod x$, since such a definition has a precise meaning only when x is an integer; rather, one must start with a definition of greater generality, applicable even to imaginary values of x, of which that one occurs as a special case. I have chosen the following ... when k becomes infinite.) (Cf. [G_1], pp. 362–363.) We will understand in §2.1 why, in fact, Gauss had no other choice.

The functions of Euler and Gauss are linked by the equations

$$\Gamma(z+1) = \Pi(z), \quad \Gamma(n+1) = \Pi(n) = n! \quad \text{for } n = 1, 2, 3, \ldots.$$

The Γ-function is *meromorphic in* $\mathbb{C}$; all its poles are of *first* order and occur at the points $-n$, $n \in \mathbb{N}$. This function has the value $n!$ at $n+1$ (rather than n) for purely historical reasons. Gauss's notation Πz did not last. Legendre introduced the now-standard notation $\Gamma(z)$ in place of $\Pi(z-1)$ (cf. [L_1], vol. 2, p. 5); since then, one speaks of the gamma function.

2. In 1854, Weierstrass made the reciprocal

$$Fc(z) := \frac{1}{\Gamma(z)} := z\prod_1^\infty \left(\frac{\nu}{\nu+1}\right)^z \left(1+\frac{z}{\nu}\right) = z\prod_1^\infty \left(1+\frac{z}{\nu}\right) e^{-z\log\left(\frac{\nu+1}{\nu}\right)}$$

of the Euler product the starting point for the theory; $Fc(z)$, in contrast to $\Gamma(z)$, is holomorphic everywhere in $\mathbb{C}$. Weierstrass says of his product ([We_1], p. 161): "Ich möchte für dasselbe die Benennung 'Factorielle von u' und die Bezeichnung $Fc(u)$ vorschlagen, indem die Anwendung dieser Function in der Theorie der Facultäten dem Gebrauch der Γ-Function deshalb vorzuziehen sein dürfte, weil sie für keinen Wert von u eine Unterbrechung der Stetigkeit erleidet und überhaupt ... im Wesentlichen den Charakter einer rationalen ganzen Function besitzt." (I would like to propose the name "Factorielle of u" and the notation $Fc(u)$ for it, since the application of this function in the theory of factorials is surely preferable to the use of the Γ-function because it suffers no break in continuity for any value of u and, overall, ... essentially has the character of a rational entire function.) Moreover, Weierstrass almost apologized for his interest in the function $Fc(u)$; he writes (p. 158) "daß die Theorie der analytischen Facultäten in meinen Augen durchaus nicht die Wichtigkeit hat, die ihr in früherer Zeit viele Mathematiker beimassen" (that the theory of analytic factorials, in my opinion, does not by any means have the importance that many mathematicians used to attribute to it).

Weierstrass's "Factorielle" Fc is now usually written in the form

$$ze^{\gamma z}\prod_{1}^{\infty}\left(1+\frac{z}{\nu}\right)e^{-\frac{z}{\nu}}, \quad \gamma := \lim_{n\to\infty}\left(\sum_{1}^{n}\frac{1}{\nu}-\log n\right) = \text{Euler's constant.}$$

We set $\Delta := Fc$ and compile a list of the most important properties of Δ in Section 1. The Γ-function is studied in Section 2. Wieland's uniqueness theorem, which, for example, immediately yields Gauss's multiplication formula, is central.

3. A theory of the gamma function is incomplete without classical integral formulas and Stirling's formula. Euler was familiar with integral representations from the outset: the equation

$$n! = \int_0^1 (-\log x)^n dx, \quad n \in \mathbb{N},$$

appears in his first work on the Γ-function, in 1729.

For a long time, Euler's identity

$$\Gamma(z) = \int_0^\infty t^{z-1}e^{-t}dt \quad \text{for} \quad z \in \mathbb{C}, \quad \operatorname{Re} z > 0,$$

has played the central role; in Section 3 we derive it and Hankel's formulas by using Wielandt's theorem. In Section 4 Stirling's formula, with a *universal estimate of the error function,* is also derived by means of Wielandt's theorem; at the same time, following the example of Stieltjes (1889), the error function is defined by an improper integral. In Section 5, again using the uniqueness theorem, we prove that

$$\mathsf{B}(w,z) = \int_0^1 t^{w-1}(1-t)^{z-1}dt = \frac{\Gamma(w)\Gamma(z)}{\Gamma(w+z)}.$$

Textbooks on the Γ-function:

[A] ARTIN, E.: *The Gamma Function*, trans. M. BUTLER, Holt, Rinehart and Winston, New York, 1964.

[Li] LINDELÖF, E.: *Le calcul des résidus*, Paris, 1905; reprinted 1947 by Chelsea Publ. Co., New York; Chap. IV in particular.

[Ni] NIELSEN, N.: *Handbuch der Theorie der Gammafunktion*, first printing Leipzig, 1906; reprinted 1965 by Chelsea Publ. Co., New York.

[WW] WHITTAKER, E. T. and G. N. WATSON: *A Course of Modern Analysis*, 4th ed., Cambridge Univ. Press, 1927, Chap. XII in particular.

Another instructive reference is an encyclopedic article by J. L. W. V. JENSEN: An elementary exposition of the theory of the gamma function, *Ann. Math.* 17 (2nd ser.), 1915–16, 124–166. For the reader's convenience, these references are listed again in the bibliography at the end of this chapter.

§1. The Weierstrass Function $\Delta(z) = ze^{\gamma z}\prod_{\nu\geq 1}(1+z/\nu)e^{-z/\nu}$

In this section we collect basic properties of the function Δ, including

$$\Delta \in \mathcal{O}(\mathbb{C}), \quad \Delta(z) = z\Delta(z+1), \quad \pi\Delta(1-z) = \sin \pi z.$$

1. The auxiliary function $H(z) := z\prod_{\nu=1}^{\infty}(1+z/\nu)e^{-z/\nu}$**.** The next result is fundamental.

(1) *The product* $\prod_{\nu\geq 1}(1+z/\nu)e^{-z/\nu}$ *converges normally in* $\mathbb{C}$.

Proof. Let $B_n := B_n(0)$, $n \in \mathbb{N}\backslash\{0\}$. It suffices to show that

$$\sum_{\nu\geq 1}\left|1-\left(1+\frac{z}{\nu}\right)e^{-z/\nu}\right|_{B_n} < \infty \quad \text{for all } n \geq 1.$$

In the identity

$$1-(1-w)e^w = w^2\left[\left(1-\frac{1}{2!}\right)+\left(\frac{1}{2!}-\frac{1}{3!}\right)w+\cdots + \left(\frac{1}{\nu!}-\frac{1}{(\nu+1)!}\right)w^{\nu-1}+\cdots\right],$$

all expressions in parentheses on the right-hand side $(\dots)$ are positive. Hence

$$|1-(1-w)e^w| \leq |w|^2\sum_1^\infty\left(\frac{1}{\nu!}-\frac{1}{(\nu+1)!}\right) = |w|^2 \quad \text{whenever } |w| \leq 1.$$

For $w = -z/\nu$, it follows that $|1-(1+z/\nu)e^{-z/\nu}| \leq |z|^2/\nu^2$ if $|z| \leq \nu$; thus

$$\sum_{\nu\geq n}|1-(1+z/\nu)e^{-z/\nu}|_{B_n} \leq n^2\sum_{\nu\geq n}\frac{1}{\nu^2} < \infty. \qquad \square$$

Convergence is produced in the preceding expression by inserting the exponential factor $\exp(-z/\nu)$ into the divergent product $\prod_{\nu\geq 1}(1+z/\nu)$. Weierstrass was the first to recognize the importance of this trick. He developed a general theory from it; see Chapter 3.

Because of (1), $H(z) := z\prod(1+z/\nu)e^{-z/\nu}$ is an *entire* function. By 1.2.2, H has zeros, each of first order, precisely at the points $-n$, $n \in \mathbb{N}$. The identity

$$(2) \qquad -H(z)H(-z) = z^2\prod_{\nu\geq 1}(1-z^2/\nu^2) = \pi^{-1}z\sin \pi z$$

follows immediately; it says that $H(z)$ consists essentially of "half the factors of the sine product." Furthermore,

$$(3)\quad H(1) = e^{-\gamma}, \quad \text{with} \quad \gamma := \lim_{n\to\infty}\left(1+\frac{1}{2}+\frac{1}{3}+\cdots+\frac{1}{n}-\log n\right) \in \mathbb{R}.$$

Proof. Since $\prod_{\nu=1}^{n}\left(1+\frac{1}{\nu}\right) = n+1$, we have

$$H(1) = \lim_{n\to\infty}\prod_{\nu=1}^{n}\left(1+\frac{1}{\nu}\right)\exp\left(-\frac{1}{\nu}\right) = \lim_{n\to\infty}\exp\left(\log(n+1)-\sum_{\nu=1}^{n}\frac{1}{\nu}\right).$$

Clearly $H(1)>0$; hence $\gamma := -\log H(1) = \lim_{n\to\infty}\left(\sum_{\nu=1}^{n}\frac{1}{\nu}-\log(n+1)\right) \in \mathbb{R}$. Since $\log(n+1)-\log n = \log\left(1+\frac{1}{n}\right)$ and $\lim_{n\to\infty}\log\left(1+\frac{1}{n}\right)=0$, the assertion follows. □

The real number γ is called *Euler's constant*; $\gamma = 0.5772156\ldots$.

Euler introduced this number in 1734 and computed it to 6 decimal places ([Eu], I-14, p. 94); in 1781 he gave it to 16 decimal places ([Eu], I-15, p. 115), of which the first 15 are correct. *It is not known whether γ is rational or irrational*, nor has anyone yet succeeded in finding a representation for γ with simple arithmetic formation rules like those known, for example, for e and π.

With $n^z := e^{z\log n}$, we have

$$z\prod_{\nu=1}^{n}(1+z/\nu)e^{-z/\nu} = \frac{z(z+1)\ldots(z+n)}{n!n^z}\exp\left[z\left(\log n - \sum_{\nu=1}^{n}\frac{1}{\nu}\right)\right];$$

thus H can also be written as follows:

$$(4)\qquad H(z) = e^{-\gamma z}\lim_{n\to\infty}\frac{z(z+1)\ldots(z+n)}{n!n^z}.$$

In the next subsection, the annoying factor $e^{-\gamma z}$ is interwoven with the product.

Exercise (Pringsheim 1915). Let $p,q\in\mathbb{N}\backslash\{0\}$. Prove that

$$\lim_{n\to\infty}\left[\prod_{\nu=1}^{pn}\left(1-\frac{z}{\nu}\right)\prod_{\nu=1}^{qn}\left(1+\frac{z}{\nu}\right)\right] = \left[\exp\left(z\log\frac{q}{p}\right)\right]\cdot\frac{\sin\pi z}{\pi z}, \quad z\in\mathbb{C}^\times.$$

Hint. Prove, among other things, that $\lim_{n\to\infty}\sum_{pn+1}^{qn}\frac{1}{\nu} = \log\frac{q}{p}$ for $q>p$.

2. The entire function $\Delta(z) := e^{\gamma z}H(z)$ has zeros, all of first order, precisely at the points $-n$, $n\in\mathbb{N}$. We have

$$\overline{\Delta(z)} = \Delta(\overline{z}), \quad \Delta(x)>0 \quad \text{for every } x\in\mathbb{R},\ x>0.$$

It follows from 1(3) and 1(4) that

(1) $$\Delta(1) = 1, \quad \Delta(z) = \lim_{n\to\infty} \frac{z(z+1)\dots(z+n)}{n!n^z}.$$

From this, since $\lim(z+n+1)/n = 1$, we immediately obtain the

Functional equation. $\Delta(z) = z\Delta(z+1)$.

The sine function and the function Δ are linked by the equation

(2) $$\pi\Delta(z)\Delta(1-z) = \sin \pi z.$$

Proof. This is clear by 1(2), since $\Delta(z)\Delta(1-z) = -z^{-1}\Delta(z)\Delta(-z) = -z^{-1}H(z)H(-z)$. □

In 2.5 we will need the multiplication formula

(3) $$(2\pi)^{\frac{1}{2}(k-1)}\Delta\left(\frac{1}{k}\right)\Delta\left(\frac{2}{k}\right)\dots\Delta\left(\frac{k-1}{k}\right) = \sqrt{k} \quad \text{for } k = 2, 3, \dots.$$

Proof. We use the well-known equation

(*) $$2^{k-1}\prod_{\kappa=1}^{k-1} \sin\frac{\kappa}{k}\pi = k.$$

(The quickest way to see this is to observe that $\sin z = (2i)^{-1}e^{iz}(1-e^{-2iz})$ and $\prod_{\kappa=1}^{k-1} e^{i\pi\kappa/k} = e^{i\pi(k-1)/2} = i^{k-1}$, write the sine product in (*) in the form

$$(2i)^{1-k}i^{k-1}\prod_{\kappa=1}^{k-1}(1-e^{-2i\pi\kappa/k}),$$

and use the identity $1+w+\dots+w^{k-1} = (w^k-1)/(w-1) = \prod_{\kappa=1}^{k-1}(w-e^{-2i\pi\kappa/k})$ for $w := 1$.)

Since $\prod_{\kappa=1}^{k-1}\Delta(\kappa/k) = \prod_{\kappa=1}^{k-1}\Delta(1-\kappa/k)$ holds trivially, (2) and (*) yield

$$\prod_{\kappa=1}^{k-1}\Delta\left(\frac{\kappa}{k}\right)^2 = \prod_{\kappa=1}^{k-1}\Delta\left(\frac{\kappa}{k}\right)\Delta\left(1-\frac{\kappa}{k}\right) = \prod_{\kappa=1}^{k-1}\frac{1}{\pi}\sin\frac{\kappa}{k}\pi = \frac{k}{(2\pi)^{k-1}}.$$

Since $\Delta(x) > 0$ for $x > 0$, the assertion follows by taking roots.

Exercise (Weierstrass, 1876). Show that

$$\Delta(z) = z\prod_{\nu\geq 1}\left(\frac{\nu}{\nu+1}\right)^z\left(1+\frac{z}{\nu}\right).$$

§2. The Gamma Function

We define

$$\Gamma(z) := 1/\Delta(z)$$

and translate the results of the preceding section into statements about the gamma function, thus giving its theory a purely multiplicative foundation.

1. Properties of the Γ-function. Our first result is immediate.

$\Gamma(z)$ is holomorphic and nonvanishing in $\mathbb{C}\setminus\{0,-1,-2,\ldots\}$; every point $-n$, $n \in \mathbb{N}$, is a first-order pole of $\Gamma(z)$. Moreover,

(F) $\quad \Gamma(z+1) = z\Gamma(z)$, *with* $\Gamma(1) = 1$ *(functional equation).*

The functional equation (F) is central to the whole broader theory. For instance, if $\Gamma(z)$ is known in the strip $0 < \operatorname{Re} z \leq 1$, then (F) can immediately be used to find its values in the adjacent strip $1 < \operatorname{Re} z \leq 2$, and so on. In general, it follows inductively from (F), for $n \in \mathbb{N}\setminus\{0\}$, that

(1) $\quad \Gamma(z+n) = z(z+1)\ldots(z+n-1)\Gamma(z), \quad \Gamma(n) = (n-1)!.$

We immediately determine the residues of the gamma function:

(2) $$\operatorname{res}_{-n}\Gamma = \frac{(-1)^n}{n!}, \quad n \in \mathbb{N}.$$

Proof. Since $-n$ is a *first*-order pole of Γ, we know that $\operatorname{res}_{-n}\Gamma = \lim_{z\to -n}(z+n)\Gamma(z)$ (see, for example, I.13.1.2). By (1),

$$\begin{aligned}\operatorname{res}_{-n}\Gamma &= \lim_{z\to -n}\frac{\Gamma(z+n+1)}{z(z+1)\ldots(z+n-1)}\\ &= \frac{\Gamma(1)}{(-n)(-n+1)\ldots(-1)} = \frac{(-1)^n}{n!}.\end{aligned}$$ □

Remark. Every function $h(z) \in \mathcal{M}(\mathbb{C})$ that satisfies the equation $h(z+1) = zh(z)$ with $h(1) \in \mathbb{C}^\times$ has a first-order pole at each $-n$, $n \in \mathbb{N}$, with residue $(-1)^n h(1)(n!)^{-1}$.

The formula 1.2(1) for $\Delta(z)$ becomes *Gauss's product representation:*

(G) $$\boxed{\Gamma(z) = \lim_{n\to\infty}\frac{n!n^z}{z(z+1)\ldots(z+n)}.}$$

Plausibility argument that (G) *is the "only" equation for functions f that satisfy* (F): By (F), for all z, $n \in \mathbb{N}$,

$$\begin{aligned}f(z+n) &= (n-1)!n(n+1)\cdot\ldots\cdot(n+z-1)\\ &= (n-1)!n^z\left(1+\frac{1}{n}\right)\left(1+\frac{2}{n}\right)\cdot\ldots\cdot\left(1+\frac{z-1}{n}\right).\end{aligned}$$

Clearly $f(z+n) \sim (n-1)!n^z$ for large n; more precisely, $\lim_{n\to\infty} f(z+n)/((n-1)!n^z) = 1$. If *one postulates* this asymptotic behavior for arbitrary z, then (1) forces

$$f(z) = \lim_{n\to\infty} \frac{f(z+n)}{z(z+1)\dots(z+n-1)} = \lim_{n\to\infty} \frac{(n-1)!n^z}{z(z+1)\dots(z+n-1)},$$

which, since $\lim n/(z+n) = 1$, is just Gauss's equation (G). See also Subsection 4.

It follows immediately from (1) and (G) that

$$\lim_{n\to\infty} \frac{\Gamma(z+n)}{\Gamma(n)n^z} = 1. \tag{3}$$

Formula 1.2(2) can be rewritten as *Euler's supplement*:

$$\boxed{\Gamma(z)\Gamma(1-z) = \frac{\pi}{\sin \pi z}.} \tag{E}$$

It follows immediately from the definition of $\Gamma(z)$ that

$$\overline{\Gamma(z)} = \Gamma(\overline{z}) \quad \text{and} \quad \Gamma(x) > 0 \text{ for } x > 0.$$

Since $|n^z| = n^x$ and $|z+\nu| \geq x+\nu$ for all z with $x = \operatorname{Re} z > 0$, (G) implies that

$$|\Gamma(z)| \leq \Gamma(x) \quad \text{for all } z \in \mathbb{C} \text{ with } x = \operatorname{Re} z > 0. \tag{4}$$

In particular, $\Gamma(z)$ is *bounded* in every strip $\{z \in \mathbb{C} : r \leq x \leq s\}$ with $0 < r < s < \infty$; this is needed in the proof of the uniqueness theorem 2.4.

We note some consequences of (E).

1) $\Gamma\left(\frac{1}{2}\right) = \sqrt{\pi}$; *more generally*, $\Gamma\left(n+\frac{1}{2}\right) = \dfrac{(2n)!}{4^n n!}\sqrt{\pi}$, $n \in \mathbb{N}$.

2) $\Gamma\left(\frac{1}{2}+z\right)\Gamma\left(\frac{1}{2}-z\right) = \dfrac{\pi}{\cos \pi z}$, $\quad \Gamma(z)\Gamma(-z) = -\dfrac{\pi}{z \sin \pi z}$.

3) $|\Gamma(iy)|^2 = \dfrac{\pi}{y \sinh \pi y}$, $\quad \left|\Gamma\left(\frac{1}{2}+iy\right)\right|^2 = \dfrac{\pi}{\cosh \pi y}$.

4) $\displaystyle\int_0^1 \log \Gamma(t)dt = \log \sqrt{2\pi}$ (Raabe, 1843, *Crelle* 25 and 28).

Proof. ad 1) and 2). These follow from (E).

ad 3). This follows from 2) by observing that $\overline{\Gamma(z)} = \Gamma(\overline{z})$, $\sinh t = -i \sin it$, and $\cosh t = \cos it$.

ad 4). The supplement (E) yields

$$\int_0^1 \log \Gamma(t)dt + \int_0^1 \log \Gamma(1-t)dt = \log \pi - \int_0^1 \log \sin \pi t dt.$$

4) follows immediately from this, by using 1.3.2(1) and the footnote there. □

Exercises. 1) For all $z \in \mathbb{C}\backslash\{1,-2,3,-4,\ldots\}$,

$$(1-z)\left(1+\frac{z}{2}\right)\left(1-\frac{z}{3}\right)\left(1+\frac{z}{4}\right)\cdot\ldots = \frac{\sqrt{\pi}}{\Gamma(1+\frac{1}{2}z)\Gamma(\frac{1}{2}-\frac{1}{2}z)}.$$

2) For all $z \in \mathbb{C}$, $\sin \pi z = \pi z(1-z)\prod_{n=1}^{\infty}\left(1+\frac{z(1-z)}{n(n+1)}\right).$

Hint. Use the factorization $n^2+n+z(1-z)=(n+z)(n+1-z)$ and (E).

2. Historical notes. Euler had discovered the relation 1(E) by 1749 at the latest; cf. [Eu], I-15, p. 82. In 1812, Gauss made the product 1(G) the starting point of the theory ([G$_2$], p. 145). Gauss seems not to have known that Euler had already anticipated the formula 1(G) in 1776 ([Eu], I-16, p. 144); Weierstrass too, as late as 1876, gave Gauss credit for the discovery ([We$_2$], p. 91).

It has become customary (cf., for example, [WW], p. 236) to call

$$\text{(W)} \qquad z\Gamma(z) = e^{-\gamma z}\prod_{\nu\geq 1}\frac{e^{z/\nu}}{1+z/\nu}$$

the "Weierstrass product." But it does not appear in this form in his work; in [We$_2$], p. 91, however, the product $\prod_{n=1}^{\infty}\left\{\left(1+\frac{x}{n}\right)e^{-x\log[(n+1)/n]}\right\}$ does appear for the "Factorielle" $1/\Gamma(x)$. The formula (W) was very much admired in the last century. Hermite writes on 31 December 1878 to Lipschitz: "...son [Weierstrass's] théorème concernant $1/\Gamma(z)$ aurait dû occuper une place d'honneur qu'il est bien singulier qu'on ne lui ait pas donné" (...his [Weierstrass's] theorem about $1/\Gamma(z)$ should have held a place of honor that very strangely wasn't given to it); cf. [Scha], p. 140.[2] — The equation (W) had already appeared in an 1843 paper of O. Schlömilch and an 1848 paper of F. W. Newman (cf. [Schl], p. 171, and [Ne], p. 57).

Since $e^{z/\nu} = (1+1/\nu)^z \exp z\,[1/\nu + \log\nu - \log(\nu+1)]$ and

$$\lim_{n\to\infty}\left(\sum_{\nu\geq 1}^{n}\frac{1}{\nu} - \log(n+1)\right) = \gamma,$$

it follows immediately from (W) that

$$z\Gamma(z) = \prod_{\nu\geq 1}\frac{\nu^{1-z}\cdot(\nu+1)^z}{\nu+z} = \prod_{\nu\geq 1}\left(1+\frac{1}{\nu}\right)^z\left(1+\frac{z}{\nu}\right)^{-1} \quad \text{(Euler, 1729)}.$$

[2]Letters of praise were hardly unusual at that time; for example, there was the "Société d'admiration mutuelle," as the astronomer H. Glydën called the group consisting of Hermite, Kovalevskaya, Mittag-Leffler, Picard, and Weierstrass.

For Euler, this product was the solution of the problem of interpolating the sequence of factorials 1, 2, 6, 24, 120, ...; cf. [Eu], I-14, pp. 1–24. Weierstrass makes no reference to the Euler product.

Writing u/v instead of z in 1(1) gives

$$u(u+v)(u+2v)\dots(u+(n-1)v) = v^n \frac{\Gamma\left(\frac{u}{v}+n\right)}{\Gamma\left(\frac{u}{v}\right)}.$$

The *finite* product on the left-hand side was studied intensively in the first half of the nineteenth century, under the name "*analytic factorial.*" This function of three variables had even been given a symbol of its own, $u^{n|v}$. Gauss opposed this nonsense in 1812 with the words, "Sed consultius videtur, functionem *unius* variabilis in analysin introducere, quam functionem trium variabilium, praesertim quum hanc ad illam reducere liceat." (It seems, however, more advisable to introduce a function of *one* variable into analysis than a function of three variables, especially since the latter can be reduced to the former.) ([G2], p. 147) The theory of analytic factorials continued to flourish despite such criticism, e.g. in the work of Bessel, Crelle, and Raabe. It was Weierstrass who, with his 1856 paper [We2], finally brought this activity to an end.

3. The logarithmic derivative $\psi := \Gamma'/\Gamma \in \mathcal{M}(\mathbb{C})$ satisfies the equations

$$(1) \qquad \psi(z+1) = \psi(z) + z^{-1}, \quad \psi(1-z) - \psi(z) = \pi \cot \pi z.$$

These formulas can be read off from the following series expansion.

Proposition *(Partial fraction representation of $\psi(z)$).*

$$\psi(z) = -\gamma - \frac{1}{z} - \sum_{\nu=1}^{\infty}\left(\frac{1}{z+\nu} - \frac{1}{\nu}\right),$$

where the series converges normally in $\mathbb{C}$.

Proof. Since $\Gamma = 1/\Delta$, we have $\psi = -\Delta'/\Delta$. Hence the assertion follows from Theorem 1.2.3 by logarithmic differentiation of $\Delta(z) = ze^{\gamma z}\prod(1+z/\nu)e^{-z/\nu}$. □

Corollary 1. $\Gamma'(1) = \psi(1) = -\gamma$; $\psi(k) = 1 + \frac{1}{2} + \dots + \frac{1}{k-1} - \gamma$ *for* $k = 2, 3, \dots$.

Proof. $\Gamma'(1) = \psi(1) = -\gamma - 1 - \sum_{\nu\geq 1}(1/(\nu+1) - 1/\nu) = -\gamma - 1 + 1 = -\gamma$. The assertion for $\psi(k)$ then follows inductively by (1). □

Corollary 2 *(Partial fraction representation of $\psi'(z)$).*

$$\psi'(z) = \sum_{\nu=0}^{\infty} \frac{1}{(z+\nu)^2},$$

where the series converges normally in $\mathbb{C}$.

Proof. This is clear, since (by I.11.1.2, for instance) normally convergent series of meromorphic functions can be differentiated term by term. □

Note that the series for ψ and ψ' are essentially "half" the partial fraction series for $\pi \cot \pi z$ and $\pi^2/\sin^2 \pi z$, respectively (cf. I.11.2.1 and 2.3).

The first equation in (1) makes possible an *additive* approach to the gamma function. This path was chosen by N. Nielsen in 1906 in his manual [Ni]. One can also proceed from the functional equation

$$g(z+1) = g(z) - z^{-2},$$

which is satisfied by ψ': for every solution $g \in \mathcal{M}(\mathbb{C})$ of this equation,

$$g(z) = \sum_{\nu=0}^{n} \frac{1}{(z+\nu)^2} + g(z+n+1)$$

(proof by induction); the partial fraction series for ψ' is thus no surprise.

Exercise. Show that $\psi(1) - \psi(\frac{1}{2}) = 2\log 2$.

4. The uniqueness problem. The exponential function is the only function $F : \mathbb{C} \to \mathbb{C}$ holomorphic at 0 and with $F'(0) = 1$ that satisfies the functional equation $F(w+z) = F(w)F(z)$. Can the Γ-function also be characterized by its functional equation $F(z+1) = zF(z)$? To begin with, this equation is satisfied by all functions $F := g\Gamma$, where $g \in \mathcal{M}(\mathbb{C})$ has period 1. The following theorem was proved by H. Wielandt in 1939.

Uniqueness theorem. *Let F be holomorphic in the right half-plane $\mathbb{T} := \{z \in \mathbb{C} : \operatorname{Re} z > 0\}$. Suppose that $F(z+1) = zF(z)$ and also that F is bounded in the strip $S := \{z \in \mathbb{C} : 1 \leq \operatorname{Re} z < 2\}$. Then $F = a\Gamma$ in $\mathbb{T}$, where $a := F(1)$.*

Proof (Demonstratio fere pulchrior theoremate). The equation $v(z+1) = zv(z)$ also holds for $v := F - a\Gamma \in \mathcal{O}(\mathbb{T})$. Hence v has a meromorphic extension to $\mathbb{C}$. Its poles, if any, can occur only at $0, -1, -2, \ldots$. Since $v(1) = 0$, it follows that $\lim_{z \to 0} zv(z) = 0$; thus v continues holomorphically to 0. Since $v(z+1) = zv(z)$, v can also be continued holomorphically to every point $-n$, $n \in \mathbb{N}$.

Since $\Gamma|S$ is bounded — see 1(4) — so is $v|S$. But then v is also bounded in the strip $S_0 := \{z \in \mathbb{C} : 0 \leq \operatorname{Re} z \leq 1\}$ (for $z \in S_0$ with $|\operatorname{Im} z| \leq 1$, this follows from continuity; for $|\operatorname{Im} z| > 1$ it follows, since $v(z) = v(z+1)/z$, from the boundedness of $v|S$). Since $v(1-z)$ and $v(z)$ assume the same values in S_0, $q(z) := v(z)v(1-z) \in \mathcal{O}(\mathbb{C})$ is bounded in S_0. It follows from Liouville that $q(z) \equiv q(1) = v(1)v(0) = 0$. Thus $v \equiv 0$, i.e. $F = a\Gamma$. □

We will encounter five compelling applications of the uniqueness theorem. In the next subsection, it gives Gauss's multiplication formula in a few lines; in Section 3 it makes possible short proofs of Euler's and Hankel's product representations of $\Gamma(z)$; in Section 4 it leads quickly to Stirling's formula; and in Section 5 it immediately yields Euler's identity for the beta integral.

The following elementary characterization of the *real* Γ-function by means of the concept of *logarithmic convexity* — without differentiability conditions — can be found in E. Artin's little book [A], which appeared in 1931.

Uniqueness theorem (H. Bohr and J. Mollerup, 1922; cf. [BM], p. 149 ff). *Let* $F : (0, \infty) \to (0, \infty)$ *be a function with the following properties:*

a) $F(x+1) = xF(x)$ *for all* $x > 0$ *and* $F(1) = 1$.
b) F *is logarithmically convex (i.e.* $\log F$ *is convex) in* $(0, \infty)$.

Then $F = \Gamma|(0, \infty)$.

$\Gamma(x)$ satisfies property b), since by 2.3

$$(\log \Gamma(x))'' = \psi'(x) = \sum \frac{1}{(x+\nu)^2} > 0 \quad \text{for } x > 0.$$

Historical remark. Weierstrass observed in 1854 ([We$_1$], pp. 193-194) that the Γ-function is the only solution of the functional equation $F(z+1) = zF(z)$ with the normalization $F(1) = 1$ that also satisfies the limit condition

$$\lim_{n \to \infty} \frac{F(z+n)}{n^z F(n)} = 1.$$

(This is trivial: the first two assertions imply that

$$F(z) = \frac{(n-1)!}{z(z+1)\dots(z+n-1)} \cdot \frac{F(z+n)}{F(n)};$$

with the third condition, this becomes Gauss's product.)

Hermann Hankel (1839–1873, a student of Riemann), in his 1836 *Habilitationsdissertation* (Leipzig, published by L. Voss), sought tractable conditions "on the behavior of the function for infinite values of x [$= z$]." He was dissatisfied with his result: "Überhaupt scheint es, als ob die Definition von $\Gamma(x)$ durch ein System von Bedingungen, ohne Voraussetzung einer explicirten Darstellung derselben, nur in der Weise gegeben werden kann, daß man das Verhalten von $\Gamma(x)$ für $x = \infty$ in dieselbe aufnimmt. Die Brauchbarkeit einer solchen Definition ist aber sehr gering, insofern es nur in den seltensten Fällen möglich ist, ohne grosse Weitläufigkeiten und selbst Schwierigkeiten den asymptotischen Werth einer Function zu bestimmen." (In fact, it seems as if the definition of $\Gamma(x)$ by a system of conditions, without assuming an explicit representation for it, could be given only by including the behavior of $\Gamma(x)$ for $x = \infty$ in the definition. The usefulness of such a definition is, however, very modest, in that it is possible only in the rarest cases to determine the asymptotic value of a function without great tediousness and even difficulties.) ([H], p. 5)

It was not until 1922 that Bohr and Mollerup succeeded in characterizing the real Γ-function by means of logarithmic convexity. But this — despite the immediately compelling applications (see [A]) — was not the kind of characterization that Hankel had had in mind. Such a characterization was first given in 1939 by H. Wielandt. His theorem can hardly be found in the literature, although K. Knopp promptly included it in 1941 in his *Funktionentheorie II*, Sammlung Göschen 703, 47–49.

In his paper "Note on the gamma function," *Bull. Amer. Math. Soc.* 20, 1–10 (1914), G. D. Birkhoff had already derived Euler's theorem (p. 51) and Euler's identity (p. 68) by using Liouville's theorem. He first investigates the quotients of functions in the closed strip $\{z \in \mathbb{C} : 1 \leq \operatorname{Re} z \leq 2\}$, then shows that they are bounded entire functions and therefore constant (loc. cit., p. 8 and p. 10). Was he perhaps already thinking of a uniqueness theorem à la Wielandt?

5. Multiplication formulas. The gamma function satisfies the equations

$$\text{(1)} \quad \begin{aligned} &\Gamma(z)\Gamma\left(z+\frac{1}{k}\right)\Gamma\left(z+\frac{2}{k}\right)\dots\Gamma\left(z+\frac{k-1}{k}\right) \\ &= (2\pi)^{\frac{1}{2}(k-1)}k^{\frac{1}{2}-kz}\Gamma(kz), \quad k=2,3,\dots. \end{aligned}$$

Proof. Set $F(z) := \Gamma\left(\frac{z}{k}\right)\Gamma\left(\frac{z+1}{k}\right)\dots\Gamma\left(\frac{z+k-1}{k}\right)\Big/(2\pi)^{\frac{1}{2}(k-1)}k^{\frac{1}{2}-z}$. Then $F(z) \in \mathcal{O}(\mathbb{C}^-)$, where $\mathbb{C}^- = \mathbb{C}\backslash(-\infty, 0]$. We have

$$F(z+1) = k\Gamma\left(\frac{z}{k}\right)^{-1} F(z)\cdot\Gamma\left(\frac{z+k}{k}\right) = zF(z);$$

moreover, it follows immediately from 1.2(3) that $F(1) = 1$. Since $|k^z| = k^x$ and $|\Gamma(z)| \leq \Gamma(x)$ whenever $x = \operatorname{Re} z > 0$ (cf. 1(4)), F is bounded in $\{z \in \mathbb{C} : 1 \leq \operatorname{Re} z < 2\}$. By the uniqueness theorem of the preceding subsection, it follows that $F = \Gamma$; hence $F(kz) = \Gamma(kz)$, i.e. (1). □

Historical note. By about 1776, Euler already knew the formulas

$$\text{(1')} \quad \sqrt{k}\Gamma\left(\frac{1}{k}\right)\Gamma\left(\frac{2}{k}\right)\dots\Gamma\left(\frac{k-1}{k}\right) = (2\pi)^{\frac{1}{2}(k-1)}$$

([Eu], I-19, p. 483); they generalize the equation $\Gamma(\frac{1}{2}) = \sqrt{\pi}$. The equations (1) were proved by Gauss in 1812 ([G$_2$], p. 150); E. E. Kummer gave another proof in 1847 [Ku]. □

Logarithmic differentiation turns (1) into the convenient

Summation formula: $\psi(kz) = \log k + \frac{1}{k}\sum_{\kappa=0}^{k-1}\psi\left(z+\frac{\kappa}{k}\right)$, $k = 2, 3, \dots$.

For $k = 2$, (1) becomes the

Duplication formula: $\sqrt{\pi}\Gamma(2z) = 2^{2z-1}\Gamma(z)\Gamma(z+\frac{1}{2})$,

which was already stated by Legendre in 1811 ([L$_1$], vol. 1, p. 284).

The identities (1) contain

Multiplication formulas for $\sin \pi z$. *For all $k \in \mathbb{N}$, $k \geq 2$,*

$$\sin k\pi z = 2^{k-1} \sin \pi z \sin \pi \left(z + \frac{1}{k}\right) \sin \pi \left(z + \frac{2}{k}\right) \ldots \sin \pi \left(z + \frac{k-1}{k}\right).$$

Proof. Since $1 - kz = k(-z + 1/k)$, Euler's formulas 1(E) and (1) yield

$$\begin{aligned} \pi(\sin k\pi z)^{-1} &= \Gamma(kz)\Gamma(k(-z+1/k)) \\ &= (2\pi)^{1-k} \prod_{\kappa=0}^{k-1} \left[\Gamma\left(z + \frac{\kappa}{k}\right) \Gamma\left(-z + \frac{1+\kappa}{k}\right)\right]. \end{aligned}$$

It is clear that $\prod_{\kappa=0}^{k-2} \Gamma\left(-z + \frac{1+\kappa}{k}\right) = \prod_{\kappa=1}^{k-1} \Gamma\left(1 - z - \frac{\kappa}{k}\right)$; hence

$$\begin{aligned} \pi(\sin k\pi z)^{-1} &= (2\pi)^{1-k}\Gamma(z)\Gamma(1-z) \prod_{\kappa=1}^{k-1} \left[\Gamma\left(z + \frac{\kappa}{k}\right) \Gamma\left(1 - \left(z + \frac{\kappa}{k}\right)\right)\right] \\ &= (2\pi)^{1-k} \pi(\sin \pi z)^{-1} \prod_{\kappa=1}^{k-1} \left[\pi \left(\sin \pi \left(z + \frac{\kappa}{k}\right)\right)^{-1}\right]. \end{aligned}$$ □

The duplication formula leads to another

Uniqueness theorem. *Suppose that $F \in \mathcal{M}(\mathbb{C})$ is positive in $(0, \infty)$ and satisfies*

$$F(z+1) = zF(z) \quad \text{and} \quad \sqrt{\pi}F(2z) = 2^{2z-1}F(z)F(z+\tfrac{1}{2}).$$

Then $F = \Gamma$.

Proof. For $g := F/\Gamma \in \mathcal{M}(\mathbb{C})$, we have $g(2z) = g(z)g(z+\frac{1}{2})$ and $g(z+1) = g(z)$. Therefore $g(x) > 0$ for all $x \in \mathbb{R}$. Hence, by Lemma 1.3.2, $g(z) = ae^{bz}$, where b is now real. Since g has period 1, it follows that $b = 0$; hence $g(z) \equiv 1$, i.e. $F = \Gamma$.

Exercises. Prove the following:

1) $\int_0^1 \log \Gamma(t)dt = \log \sqrt{2\pi}$ directly, using the duplication formula (cf. 1.4).

2) $\int_0^1 \log \Gamma(\zeta + z)d\zeta = \log \sqrt{2\pi} + z \log z - z$ for $z \in \mathbb{C}\backslash(-\infty, 0)$ (Raabe's function).

3) $1 + \frac{1}{2} + \frac{1}{3} + \cdots + \frac{1}{k-1} - \gamma = \frac{1}{k} \sum_{\kappa=0}^{k-1} \psi\left(1 + \frac{\kappa}{k}\right)$, $k = 2, 3, \ldots$.

6*. Hölder's theorem. One can ask whether the Γ-function — by analogy with the functions $\exp z$ and $\cos z$, $\sin z$ — satisfies a simple differential equation. O. Hölder proved that this is not the case ([Hö], 1886).

Hölder's theorem. *The Γ-function does not satisfy any algebraic differential equation. In other words, there is no polynomial* $F(X, X_0, X_1, \ldots, X_n) \neq 0$ *in finitely many indeterminates over* $\mathbb{C}$ *such that*

$$F(z, \Gamma(z), \Gamma'(z), \ldots, \Gamma^{(n)}(z)) \equiv 0.$$

Weierstrass assigned the proof of this theorem as an exercise. There is a series of proofs, for example those of Moore (1897), F. Hausdorff (1925), and A. Ostrowski (1919 and 1925) (cf. [M], [Ha], and [O]). Ostrowski's 1925 proof is considered especially simple; it can be found, among other places, in [Bi], pp. 356–359. All the proofs construct a contradiction between the functional equation $\Gamma(z+1) = z\Gamma(z)$ and the hypothesized differential equation.

7*. The logarithm of the Γ-function. Since $\Gamma(z)$ has no zeros in the star-shaped domain $\mathbb{C}^-$, the function $\psi(z) = \Gamma'(z)/\Gamma(z)$ is holomorphic there and

$$(1) \qquad l(z) := \int_{[1,z]} \psi(\zeta)\, d\zeta, \quad z \in \mathbb{C}^-, \text{ with } l(1) = 0,$$

as an antiderivative of the logarithmic derivative of $\Gamma(z)$, is a logarithm of $\Gamma(z)$ (that is, $e^{l(z)} = \Gamma(z)$; cf. I.9.3). We write $\log \Gamma(z)$ for the function $l(z)$; this notation, however, does not mean that $l(z)$ is obtained in $\mathbb{C}^-$ by substituting $\Gamma(z)$ into the function $\log z$. It follows easily from (1) and Proposition 3 that

$$(1') \qquad \log \Gamma(z) = -\gamma z - \log z + \sum_{\nu=1}^{\infty} \left[\frac{z}{\nu} - \log\left(1 + \frac{z}{\nu}\right)\right], \quad z \in \mathbb{C}^-.$$

Proof. Since the partial fraction series $-\gamma - 1/\zeta - \sum_{\nu=1}^{\infty} [1/(\zeta+\nu) - 1/\nu]$ converges normally in $\mathbb{C}^-$, it can be integrated term by term; cf. I.8.4.4. For $z \in \mathbb{C}^-$ and $\nu \geq 1$, $1 + z/\nu \in \mathbb{C}^-$ and hence $\log(z+\nu) - \log(1+\nu) = \log(1 + z/\nu) - \log(1 + 1/\nu)$ (!). Thus

$$\begin{aligned} \log \Gamma(z) &= -\gamma z + \gamma - \log z - \sum_{\nu=1}^{\infty} \left[\log(z+\nu) - \log(1+\nu) - \frac{z}{\nu} + \frac{1}{\nu}\right] \\ &= -\gamma z + \gamma + \sum_{\nu=1}^{\infty} \left[\frac{z}{\nu} - \log\left(1 + \frac{z}{\nu}\right) + \log\left(1 + \frac{1}{\nu}\right) - \frac{1}{\nu}\right]. \end{aligned}$$

Since $\sum_{\nu=1}^{n} [1/\nu - \log(1 + 1/\nu)] = \sum_{\nu=1}^{n} 1/\nu - \log(n+1)$ tends to γ, (1′) follows. □

We now consider the function $\log\Gamma(z+1)$ in $\mathbb{E}$. Its Taylor series about 0 has radius of convergence 1; we claim that

$$(2)\quad \log\Gamma(z+1) = -\gamma z + \sum_{n=2}^{\infty} \frac{(-1)^n}{n}\zeta(n) z^n, \quad \text{where} \quad \zeta(n) := \sum_{\nu=1}^{\infty} \frac{1}{\nu^n}.$$

Proof. Since

$$\frac{1}{z+\nu} - \frac{1}{\nu} = \frac{1}{\nu}\left[\frac{1}{1+z/\nu} - 1\right] = \sum_{n=1}^{\infty} \frac{(-1)^n}{\nu}\left(\frac{z}{\nu}\right)^n \quad \text{if } |z| < \nu,$$

it follows from 3(1) and Proposition 3 that

$$\psi(z+1) = -\gamma - \sum_{\nu=1}^{\infty}\left(\sum_{n=1}^{\infty} \frac{(-1)^n}{\nu^{n+1}} z^n\right) = -\gamma + \sum_{n=2}^{\infty} (-1)^n \zeta(n) z^{n-1}, \quad z \in \mathbb{E}.$$

But $(\log\Gamma(z+1))' = \psi(z+1)$ and $\log\Gamma(1) = 0$; this gives (2). □

For $z = 1$, the series (2) gives the formula

$$(3)\quad \gamma = \sum_{n=2}^{\infty} \frac{(-1)^n}{n}\zeta(n) \quad \text{(Euler, 1769)}.$$

Proof. Since $\zeta(n+1) < \zeta(n)$, the terms of the alternating series on the right-hand side tend monotonically to 0; hence the series is convergent. Abel's limit theorem can be applied to (2):

$$\sum_{n=2}^{\infty} \frac{(-1)^n}{n}\zeta(n) = \lim_{x\nearrow 1} \sum_{n=2}^{\infty} \frac{(-1)^n}{n}\zeta(n) x^n = \gamma + \log\Gamma(2) = \gamma.$$

Historical note. Rapidly convergent series for $\log\Gamma(z+1)$ can be obtained from the series (2); cf., for example, [Ni], p. 38. These give enough information to tabulate the initial values $\zeta(n)$ of the logarithm of the gamma function. Legendre established the first such table: it contains the values of $\log\Gamma(x+1)$ from $x = 0$ to $x = 0.5$, with increment 0.005, up to seven decimal places. Legendre later published tables from $x = 0$ to $x = 1$ with increment 0.001, correct to seven decimal places ([L_1], vol. 1, pp. 302–306); in 1817 he improved these tables to twelve decimal places ([L_1], vol. 2, pp. 85–95). Gauss, in 1812, gave the functional values of $\psi(1+x)$ and $\log\Gamma(1+x)$ from $x = 0$ to $x = 1$, with increment 0.01, up to twenty decimal places ([G_2], pp. 161–162). Euler announced equation (3) in 1769 ([Eu], I-15, p. 119).

§3. Euler's and Hankel's Integral Representations of $\Gamma(z)$

Euler observed as early as 1729 — in his first work ([Eu], I-14, pp. 1-24) on the gamma function — that the sequence of factorials 1, 2, 6, 24, ... is given by the integral

$$n! = \int_0^1 (-\log \tau)^n d\tau, \quad n \in \mathbb{N}$$

(loc. cit., p. 12). In general,

$$\Gamma(z+1) = \int_0^1 (-\log \tau)^z d\tau \quad \text{whenever} \quad \operatorname{Re} z > -1;$$

with z instead of $z+1$ and $t := -\log \tau$, this yields the equation

$$(1) \qquad \Gamma(z) = \int_0^\infty t^{z-1} e^{-t} dt, \quad z \in \mathbb{T} := \{z \in \mathbb{C} : \operatorname{Re} z > 0\}.$$

The *improper* integral on the right-hand side of (1) was called *Euler's integral of the second kind* by Legendre in 1811 ([L_1], vol. 1, p. 221). Its existence is not obvious; we prove in Subsection 1 that it converges and is holomorphic. The identity (1) is a cornerstone of the theory of the gamma function; we prove it in Subsection 2, using the uniqueness theorem 2.4. In Subsection 4 we use the uniqueness theorem to obtain Hankel's formulas for $\Gamma(z)$.

Integral representations of the Γ-function have repeatedly attracted the interest of mathematicians since Euler. R. Dedekind obtained his doctorate in 1852 with a paper entitled "Über die Elemente der Theorie der Eulerschen Integrale" (cf. his *Ges. Math. Werke* 1, pp. 1–31), and H. Hankel qualified as a university lecturer in 1863 in Leipzig with a paper called "Die Eulerschen Integrale bei unbeschränkter Variabilität des Argumentes"; cf. [H].

1. Convergence of Euler's integral. We recall the following result.

Majorant criterion. *Let $g : D \times [a, \infty) \to \mathbb{C}$ be continuous, where $D \subset \mathbb{C}$ is a region and $a \in \mathbb{R}$. Suppose there exists a function $M(t)$ on $[a, \infty)$ such that*

$$|g(z,t)| \le M(t) \text{ for all } z \in D,\ t \ge a, \quad \text{and} \quad \int_a^\infty M(t)dt \in \mathbb{R} \text{ exists.}$$

Then $\int_a^\infty g(z,t)dt$ converges uniformly and absolutely in D. If $g(z,t) \in \mathcal{O}(D)$ for every $t \ge a$, then $\int_a^\infty g(z,t)dt$ is holomorphic in D.

Proof. Let $\varepsilon > 0$. Choose $b \geq a$ such that $\int_b^\infty M(t)dt \leq \varepsilon$. Then

$$\left|\int_b^c g(z,t)dt\right| \leq \int_b^c |g(z,t)|dt \leq \int_b^c M(t)dt \leq \varepsilon \quad \text{for all } z \in D \text{ and } c \geq b.$$

The uniform and absolute convergence of the integral in D follows from Cauchy's convergence criterion. If g is always holomorphic in D for fixed t, then $\int_r^s g(z,t)dt \in \mathcal{O}(D)$ for all r and s such that $a < r < s < \infty$ (cf. I.8.2.2). Then we also have $\int_a^\infty g(z,t)dt \in \mathcal{O}(D)$. (Incidentally, it is easier to show that this integral is holomorphic by using Vitali's theorem; cf. 7.4.2.) □

For $r \in \mathbb{R}$, let S_r^+ (resp., S_r^-) denote the *right* half-plane $\operatorname{Re} z \geq r$ (resp., the *left* half-plane $\operatorname{Re} z \leq r$). For brevity, we set

$$u(z) := \int_0^1 t^{z-1}e^{-t}dt, \quad v(z) := \int_1^\infty t^{z-1}e^{-t}dt.$$

Convergence theorem *The integral $v(z)$ converges uniformly and absolutely in S_r^- for every $r \in \mathbb{R}$; moreover, $v(z) \in \mathcal{O}(\mathbb{C})$.*

The integral $u(z)$ converges uniformly and absolutely in S_r^+ for every $r > 0$. Moreover, $u(z) \in \mathcal{O}(\mathbb{T})$ and

(1) $$u(z) = \sum_{\nu=0}^\infty \frac{(-1)^\nu}{\nu!}\frac{1}{z+\nu} \quad \textit{for every } z \in \mathbb{T}.$$

Proof. a) For all $z \in S_r^-$, we have $|t^{z-1}| \leq t^{r-1}$. Since $\lim_{t\to\infty} t^{r-1}e^{-\frac{1}{2}t} = 0$, there exists an $M > 0$ such that $|t^{z-1}e^{-t}| \leq Me^{-\frac{1}{2}t}$ for all $z \in S_r^-$, $t \geq 1$. Since $\int_1^\infty e^{-\frac{1}{2}t}dt = 2/\sqrt{e}$ and $t^{z-1}e^{-t} \in \mathcal{O}(\mathbb{C})$ for all $t \geq 1$, the claims about v follow from the majorant criterion.

b) Set $s := 1/t$; then $u(z) = \int_1^\infty e^{-1/s}s^{-z-1}ds$. If $r > 0$, then $|e^{-1/s}s^{-z-1}| \leq s^{-r-1}$ for all $z \in S_r^+$, and moreover $\int_1^\infty s^{-r-1}ds = r^{-1}$. The majorant criterion now gives all the claims about u except for equation (1). This follows from the identity

$$\begin{aligned}\int_\delta^1 t^{z-1}e^{-t}\,dt &= \sum_{\nu=0}^\infty \frac{(-1)^\nu}{\nu!}\int_\delta^1 t^{z+\nu-1}\,dt \\ &= \sum_{\nu=0}^\infty \frac{(-1)^\nu}{\nu!}\frac{1}{z+\nu} - \delta^z \sum_0^\infty \frac{(-1)^\nu}{\nu!}\frac{\delta^\nu}{z+\nu},\end{aligned}$$

which holds for all $\delta \in (0,1)$ (theorem on interchanging the order of integration and summation; cf. I.6.2.3), since $\operatorname{Re} z > 0$ and the last summand therefore tends to 0 as $\delta \to 0$. □

The integrals $u(r)$, $r < 0$, diverge. Since $t^{r-1}e^{-t} \geq e^{-1}t^{r-1}$ in $(0,1)$,

$$\int_\delta^1 t^{r-1}e^{-t}dt \geq e^{-1}\int_\delta^1 t^{r-1}dt = e^{-1}r^{-1}(1-\delta^r); \quad \text{thus } \lim_{\delta\to 0}\int_\delta^1 t^{r-1}e^{-t}dt = \infty.$$

2. Euler's theorem. *The integral* $\int_0^\infty t^{z-1}e^{-t}dt$ *converges uniformly and absolutely to* $\Gamma(z)$ *in every strip* $\{z \in \mathbb{C} : a \leq \operatorname{Re} z \leq b\}$, $0 < a < b < \infty$:

$$\boxed{\Gamma(z) = \int_0^\infty t^{z-1}e^{-t}dt \text{ for } z \in \mathbb{T}.}$$

Proof. Convergence follows from the convergence theorem 1, since the integral coincides with $F := u + v$ in $\mathbb{T}$. For $F \in \mathcal{O}(\mathbb{T})$, it is immediate that

$$F(z+1) = zF(z), \quad F(1) = 1, \quad |F(z)| \leq |F(\operatorname{Re} z)|, \quad \text{for all } z \in \mathbb{T}.$$

In particular, F is bounded if $1 \leq \operatorname{Re} z < 2$. That $F = \Gamma$ follows from Theorem 2.4. □

Of course, there are also direct proofs of the equation $F = \Gamma$. The reader may consult, for instance, [A], where the logarithmic convexity of $F(x)$, $x > 0$, is proved, or [WW], where Gauss's proof is given. One verifies the equations

$$\frac{n!n^z}{z(z+1)\ldots(z+n)} = \int_0^n t^{z-1}(1-t/n)dt, \quad z \in \mathbb{T},\ n = 1, 2, \ldots$$

by induction, then proves that the sequence on the right-hand side converges to $F(z)$. □

The Γ-integral can be used to determine a number of integrals. The Gaussian error integral, discussed at length in Volume I, is a special Γ-value:

$$\int_0^\infty e^{-x^\alpha}dx = \alpha^{-1}\Gamma(\alpha^{-1}) \text{ for } \alpha > 0; \quad \text{in particular, } \int_0^\infty e^{-x^2}dx = \frac{1}{2}\sqrt{\pi}.$$

Proof. For $t := x^\alpha$, we have $t^{\alpha^{-1}-1} = x^{1-\alpha}$ and $dt = \alpha x^{\alpha-1}dx$; thus

$$\Gamma(\alpha^{-1}) = \int_0^\infty t^{\alpha^{-1}-1}e^{-t}dt = \int_0^\infty x^{1-\alpha}e^{-x^\alpha}\alpha x^{\alpha-1}dx = \alpha\int_0^\infty e^{-x^\alpha}dx.$$

The last equality is clear, since $\Gamma(\frac{1}{2}) = \sqrt{\pi}$ by 2.1.1). □

An inductive argument using integration by parts yields

$$\int_0^\infty x^{2n}e^{-x^2}dx = \tfrac{1}{2}\Gamma(n+\tfrac{1}{2}), \quad n \in \mathbb{N}.$$

(Cf. also I.12.4.6(3).)

The Fresnel integrals, already determined in I.7.1.6*, can also be derived from the Γ-integral; for more on this, see Subsection 3.

We mention in addition the representation given by F. E. Prym ([Pr], 1876).

Partial fraction representation of the Γ-function. *The identity*

$$\Gamma(z) = \sum_{\nu=0}^{\infty} \frac{(-1)^\nu}{\nu!} \frac{1}{z+\nu} + \int_1^\infty t^{z-1}e^{-t}dt$$

holds for all $z \in \mathbb{C}\backslash\{0, -1, -2 \dots\}$.

Proof. The assertion is true for $z \in \mathbb{T}$. Since the functions that appear are holomorphic in $\mathbb{C}\backslash\{0, -1, -2, \dots\}$, the general case follows from the identity theorem. □

3*. The equation $\int_0^\infty t^{z-1}e^{-it}dt = e^{-\pi iz/2}\Gamma(z)$, $0 < \operatorname{Re} z < 1$. To prove this, let

a) $g(\zeta) := \zeta^{z-1}e^{-\zeta}$; then $|g(\zeta)| \le e^{\pi|y|}r^{x-1}e^{-r\cos\varphi}$, where $z = x + iy \in \mathbb{C}$, $\zeta = re^{i\varphi} \in \mathbb{C}^-$. Since $g \in \mathcal{O}(\mathbb{C}^-)$, we have by Cauchy (see Figure 2.1):

b) $\int_{\gamma'+\gamma_R} g\,d\zeta = \int_{\gamma_\delta+\gamma} g\,d\zeta$.

If we show that, for $0 < \operatorname{Re} z < 1$,

c) $\lim\limits_{\delta\to 0} \int_{\gamma_\delta} g\,d\zeta = \lim\limits_{R\to\infty} \int_{\gamma_R} g\,d\zeta = 0$,

the assertion will follow from b) by taking limits, since γ is the path $\zeta(t) := it$, $\delta \le t \le R$. By a),

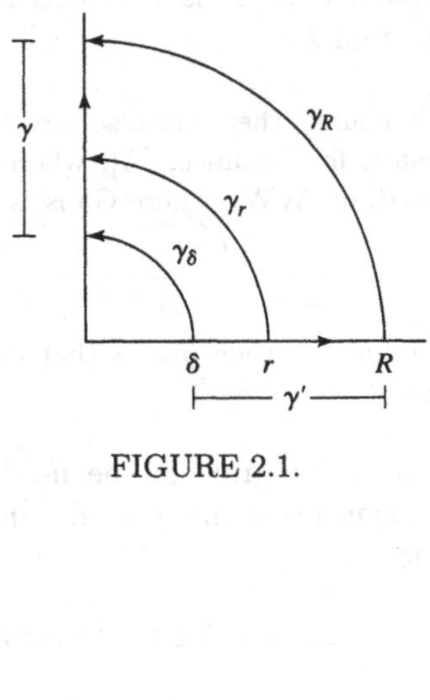

FIGURE 2.1.

$$\int_{\gamma_r} g\,d\zeta = \int_0^{\pi/2} g(re^{i\varphi})ire^{i\varphi}d\varphi;$$

thus

d) $$\left|\int_{\gamma_r} g\,d\zeta\right| \le e^{\pi|y|}r^x \int_0^{\pi/2} e^{-r\cos\varphi}d\varphi$$

for all $r \in (0, \infty)$. To verify the second equality in c), we observe that $\cos\varphi \ge 1 - \frac{2}{\pi}\varphi$ for all $\varphi \in [0, \frac{1}{2}\pi]$ (concavity of the cosine). Hence

$$\int_0^{\pi/2} e^{-R\cos\varphi}d\varphi \le \int_0^{\pi/2} \exp(2R\pi^{-1}\varphi - R)d\varphi = e^{-R}\frac{\pi}{2R}e^{2R\pi^{-1}\varphi}\Big|_0^{\frac{1}{2}\pi} < \frac{\pi}{2R}.$$

By d),

$$\left|\int_{\gamma_R} g\,d\zeta\right| \le \frac{1}{2}\pi e^{\pi|y|}R^{x-1}; \quad \text{thus} \quad \lim_{R\to\infty}\int_{\gamma_R} g\,d\zeta = 0 \quad \text{if } x < 1. \qquad \square$$

For $z := x \in (0, 1)$, splitting into real and imaginary parts gives

$$(1) \qquad \int_0^\infty t^{x-1} \cos t\, dt = \cos(\tfrac{1}{2}\pi x)\Gamma(x), \qquad \int_0^\infty t^{x-1} \sin t\, dt = \sin(\tfrac{1}{2}\pi x)\Gamma(x).$$

For $x := \frac{1}{2}$ and $\tau^2 := t$, these are the Fresnel formulas (cf. I.7.1.6*):

$$\int_0^\infty \cos^2 \tau d\tau = \int_0^\infty \sin^2 \tau d\tau = \frac{1}{2}\sqrt{\frac{1}{2}\pi}.$$

The equations (1), combined with Euler's summation formula, give an extremely simple proof of the functional equation for the Riemann ζ-function; cf. [T], p. 15.

Exercises. 1) Argue as above to prove that

$$\int_0^\infty t^{z-1} e^{-wt} dt = w^{-z}\Gamma(z) \quad \text{for } w,\, z \in \mathbb{T}.$$

(We had $w = i$ above. The concavity of $\cos\varphi$ is no longer needed.)

2) Prove that the following holds for the ζ-function $\zeta(z) := \sum_{n=1}^\infty n^{-z}$:

$$\zeta(z)\Gamma(z) = \int_0^\infty \frac{t^{z-1}}{e^t - 1} dt \quad \text{for all } z \in \mathbb{T}.$$

(This formula can be used to obtain the functional equation

$$\zeta(1-z) = 2(2\pi)^{-z} \cos \tfrac{1}{2}\pi z \Gamma(z)\zeta(z).)$$

Historical note. Euler knew the formulas (1) in 1781. In [Eu] (I-19, p. 225), by taking real and imaginary parts of $w^x \int_0^\infty t^{x-1} e^{-wt} dt = \Gamma(x)$, with $w = p + iq$, he obtained the equations

$$\begin{aligned} \int_0^\infty t^{x-1} e^{-pt} \cos qt\; dt &= \Gamma(x) \cdot f^{-x} \cos x\theta \quad \text{and} \\ \int_0^\infty t^{x-1} e^{-pt} \sin qt\; dt &= \Gamma(x) \cdot f^{-x} \sin x\theta, \end{aligned}$$

where $\theta := \arctan(q/p)$ and $f := |w| = \sqrt{p^2 + q^2}$. Euler did not worry about the region in which his identities were valid; (1) follows from setting $p = 0$, $q = 1$ (cf. also I.7.1.6*).

4*. Hankel's loop integral. Euler's integral represents $\Gamma(z)$ only in the right half-plane. We now introduce an integral, with integrand $w^{-z}e^w$, which represents $\Gamma(z)$ in all of $\mathbb{C}\backslash(-\mathbb{N})$; we will "make a detour" around the annoying singularity of $w^{-z}e^w$ at 0. Clearly

$$(*) \qquad |w^{-z}e^w| \le e^{\pi|y|}|w|^{-x}e^{\operatorname{Re} w} \quad \text{for } z = x + iy \in \mathbb{C},\ w \in \mathbb{C}^-.$$

Now let $s \in (0, \infty)$ and $c \in \partial B_s(0)$, $c \ne \pm s$, be chosen and fixed. We denote by γ the "improper loop path" $\gamma_1 + \delta + \gamma_2$ (Figure 2.2) and by

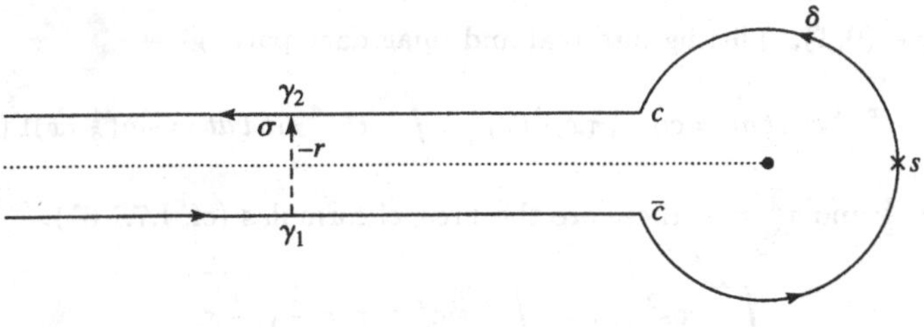

FIGURE 2.2.

S a strip $[a,b] \times i\mathbb{R}$, $a < b$. The next statement follows from $(*)$, since $\lim_{t\to\infty} |t-c|^q e^{-\frac{1}{2}t} = 0$ for every $q \in \mathbb{R}$.

(1) *There exists a* t_0 *such that* $\max_{z\in S} |w^{-z}e^w| \le e^{\pi|y|}e^{-\frac{1}{2}t}$ *for* $w = \gamma_2(t) = c - t$, $t \ge t_0$.

We claim that the following holds.

Lemma. *The "loop integral"* $\frac{1}{2\pi i}\int_\gamma w^{-z}e^w dw$ *converges compactly and absolutely in* $\mathbb{C}$ *to an entire function* h *satisfying* $h(1) = 1$ *and* $h(-n) = 0$, $n \in \mathbb{N}$. *Moreover,* $h(z)e^{-\pi|y|}$ *is bounded in every strip* S.

Proof. Since $e^{\pi|y|}$ is bounded on every compact set $K \subset \mathbb{C}$, the integral converges, by (1), uniformly and absolutely along γ_2 (majorant criterion). As the same holds for the integral along γ_1, the claim about convergence follows.

For every $m \in \mathbb{Z}$, we have $\lim_{r\to\infty} \int_\sigma w^{-m}e^w dw = 0$ (Figure 2.2). Hence $h(m) = \text{res}_0(w^{-m}e^w)$ for $m \in \mathbb{Z}$. It follows that $h(1) = 1$ and $h(-\mathbb{N}) = 0$. (1) also shows that $h(z)e^{-\pi|y|}$ is bounded in S. □

Hankel's formulas now follow quickly:

$$\frac{1}{\Gamma(z)} = \frac{1}{2\pi i}\int_\gamma w^{-z}e^w dw, \quad z \in \mathbb{C};$$
$$\Gamma(z) = \frac{1}{2i\sin \pi z}\int_\gamma w^{z-1}e^w dw, \quad z \in \mathbb{C}\backslash(-\mathbb{N}).$$

Proof. We denote the functions on the right-hand side by h and F, respectively. Then

$$(*) \qquad F(z) = \pi\frac{h(1-z)}{\sin \pi z}, \quad z \in \mathbb{C}\backslash\mathbb{Z}.$$

Since $h(-\mathbb{N}) = 0$ (by the lemma), it follows that $F \in \mathcal{O}(\mathbb{T})$. Integration by parts in the integral for h gives $h(z) = zh(z+1)$, whence $F(z+1) = zF(z)$. Since $|2\sin z| \ge e^{|y|} - e^{-|y|}$, it follows from the lemma that

$$|F(z)| = \pi\frac{|h(1-z)|}{|\sin \pi z|} \le \frac{A}{1 - e^{-2\pi|y|}} \quad \text{for } 1 \le \text{Re}\, z < 2,\ y \ne 0,$$

with a constant $A > 0$. Thus F is *bounded* for $1 \le \operatorname{Re} z < 2$. Hence, by the uniqueness theorem 2.4, $F = a\Gamma$, $a \in \mathbb{C}$. The supplement $\Gamma(z)\Gamma(1-z)\sin \pi z = \pi$ and (∗) then give $h = a/\Gamma$. Since $a = h(1) = 1$ (by the lemma), Hankel's formulas are proved. □

Historical note. Hankel discovered his formulas in 1863; cf. [H], p. 7. The proof presented here follows an idea of H. Wielandt.

"By varying the path of integration in [his] generally valid integral," Hänkel "easily" obtains "the forms of the integral $\Gamma(x)$ or the quotient $1 : \Gamma(x)$ that have been familiar so far"; thus, for example, the equations 3*(1) (cf. [H], top of p.10). The reader may also consult [WW], p. 246.

Hankel's formulas remain valid for $c = -s$ if in the integrals along γ_1 (from $-\infty$ to $-s$), resp. γ_2 (from $-s$ to $-\infty$), we substitute for the integrands the limiting values of $w^{-z}e^w$, resp. $w^{z-1}e^w$, as $(-\infty, -s)$ is approached from the *lower*, resp. *upper* half-plane. Thus we have $e^{\mp i\pi(z-1)}e^t|t|^{z-1}$, $-\infty < t < -s$, in the second formula. If we now assume in addition that $z \in \mathbb{T}$, we may also let s approach 0. Thus integrating along the degenerate loop path (from $-\infty$ to 0 and back) gives, for all $z \in \mathbb{T}$,

$$\begin{aligned} 2i\Gamma(z)\sin \pi z &= e^{-i\pi(z-1)}\int_{-\infty}^{0} |t|^{z-1}e^t dt + e^{i\pi(z-1)}\int_{0}^{-\infty} |t|^{z-1}e^t dt \\ &= 2i \sin \pi z \int_{-\infty}^{0} |t|^{z-1}e^t dt. \end{aligned}$$

Here the final integral on the right-hand side is $\int_0^\infty t^{z-1}e^t dt$. We have proved that

Euler's formula for $\Gamma(z)$, $z \in \mathbb{T}$, follows from Hankel's second formula.

Euler's formula is thus a degenerate case of Hankel's. Conversely, Hankel's formulas can be recovered from this *degenerate case* (cf., for example, [H], pp. 6–8; [K], pp. 198–199; or [WW], pp. 244–245).

§4. Stirling's Formula and Gudermann's Series

Invenire summam quotcunque Logarithmorum, quorum numeri sint in progressione Arithmetica.

— J. Stirling, 1730, *Methodus Differentialis*

For applications — not just numerical applications — the growth of the function $\Gamma(z)$ must be known: For large z, we would like to approximate $\Gamma(z)$ in the slit plane $\mathbb{C}^- := \mathbb{C}\backslash(-\infty, 0]$ by "simpler" functions (we omit the half-line $(-\infty, 0]$ since $\Gamma(z)$ has poles at $-\mathbb{N}$). We are guided in this search by the growth of the sequence $n!$, which is described by Stirling's classical formula

$$\text{(ST)} \qquad n! = \sqrt{2\pi} n^{n+\frac{1}{2}} e^{-n} e^{a_n}, \quad \text{with } \lim a_n = 0;$$

cf. [W], pp. 351–353. This formula suggests looking for an "error function" $\mu \in \mathcal{O}(\mathbb{C}^-)$ for the Γ-function such that an equation

$$\Gamma(z) = \sqrt{2\pi} z^{z-\frac{1}{2}} e^{-z} e^{\mu(z)}, \quad \text{with } \lim_{z\to\infty} \mu(z) = 0,$$

holds in all of $\mathbb{C}^-$, where $z^{z-\frac{1}{2}} = \exp[(z - \frac{1}{2})\log(z)]$; (ST) would be contained in this since $n\Gamma(n) = n!$. We will see that

$$\mu(z) = \log \Gamma(z) - (z - \tfrac{1}{2}) \log z + z - \tfrac{1}{2} \log 2\pi$$

is an ideal error function. *It even tends to zero like $1/z$ as the distance from z to the negative real axis tends to infinity.* Thus $\sqrt{2\pi} z^{z-\frac{1}{2}} e^{-z}$ is a "simpler" function that approximates $\Gamma(z)$ in $\mathbb{C}^-$.

The equation given for $\mu(z)$ is hardly suitable as a definition. We define $\mu(z)$ in Subsection 1 by an improper integral that makes the main properties of this function obvious and leads immediately, in Subsection 2, to Stirling's formula with solid estimates for $\mu(z)$. These estimates are further improved in Subsection 4. In Subsections 5 and 6, Stirling's formula is generalized to Stirling's series with estimates for the remainder.

To estimate integrands with powers of $z+t$ in the denominator, we always use the following inequality:

$$(*) \qquad |z + t| \geq (|z| + t) \cos \tfrac{1}{2}\varphi \quad \text{for } z = |z|e^{i\varphi} \text{ and } t \geq 0.$$

Proof. Let $r := |z|$. Since $\cos\varphi = 1 - 2\sin^2 \frac{1}{2}\varphi$ and $(r+t)^2 \geq 4rt$, it follows that

$$|z+t|^2 = r^2 + 2rt\cos\varphi + t^2 = (r+t)^2 - 4rt\sin^2 \tfrac{1}{2}\varphi \geq (r+t)^2 \cos^2 \tfrac{1}{2}\varphi. \qquad \square$$

One consequence of this is a "uniform" estimate in angular sectors.

(**) *Let $0 < \delta \leq \pi$ and $t \geq 0$. Then (since $\cos \frac{1}{2}\varphi \geq \sin \frac{1}{2}\delta$)*

$$|z + t| \geq (|z| + t) \sin \tfrac{1}{2}\delta \quad \textit{for all } z = |z|e^{i\varphi} \textit{ with } |\varphi| \leq \pi - \delta.$$

1. Stieltjes's definition of the function $\mu(z)$. The real functions

$$(1) \qquad P_1(t) := t - [t] - \tfrac{1}{2} \quad \text{and} \quad Q(t) := \tfrac{1}{2}(t - [t] - (t - [t])^2),$$

where $[t]$ denotes the greatest integer $\leq t$, are continuous in $\mathbb{R}\backslash\mathbb{Z}$ and have

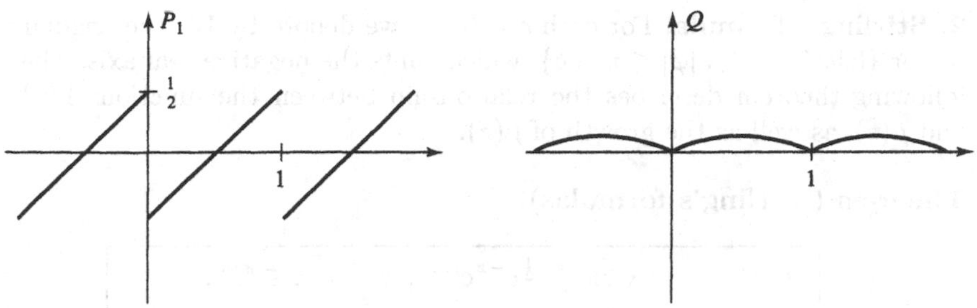

FIGURE 2.3.

period 1 (see Figure 2.3); $P_1(t)$ is the "sawtooth function." The function $Q(t)$ is an antiderivative of $-P_1(t)$ in $\mathbb{R}\backslash\mathbb{Z}$; we have $0 \le Q(t) \le \frac{1}{8}$. Moreover, Q is continuous on all of $\mathbb{R}$. The starting point for all further considerations is the following definition:

$$\text{(2)} \qquad \boxed{\mu(z) := -\int_0^\infty \frac{P_1(t)}{z+t}dt = \int_0^\infty \frac{Q(t)}{(z+t)^2}dt \in \mathcal{O}(\mathbb{C}^-).}$$

This definition is certainly legitimate once we prove that the integrals in (2) converge locally uniformly in $\mathbb{C}^-$ to the same function. Let $\delta \in (0, \pi]$ and $\varepsilon > 0$. For all $t \ge 0$, we then have (by (**) of the introduction)

$$\left|\frac{Q(t)}{(z+t)^2}\right| \le \frac{1}{8}\frac{1}{\sin^2\frac{1}{2}\delta} \cdot \frac{1}{(\varepsilon+t)^2},$$

if $z = |z|e^{i\varphi}$ with $|z| \ge \varepsilon$ and $|\varphi| \le \pi - \delta$. The second integral thus converges locally uniformly in $\mathbb{C}^-$ by the majorant criterion 3.1. The first integral also converges locally uniformly in $\mathbb{C}^-$ to the same limit function since

$$-\int_r^s \frac{P_1(t)}{z+t}dt = \left.\frac{Q(t)}{z+t}\right|_r^s + \int_r^s \frac{Q(t)}{(z+t)^2}dt \quad \text{for } 0 < r < s < \infty.$$

(Integration by parts is permissible because Q is continuous.)

We immediately obtain a *functional equation* for the μ-function:

$$\text{(3)} \quad \mu(z) - \mu(z+1) = \int_0^1 \frac{\frac{1}{2} - t}{z+t}dt = (z + \tfrac{1}{2})\log(1 + \tfrac{1}{z}) - 1, \quad z \in \mathbb{C}^-.$$

Proof. Observe that $P_1(t+1) = P_1(t)$ and write

$$\mu(z+1) = -\int_0^\infty \frac{P_1(t+1)}{z+t+1}dt = -\int_1^\infty \frac{P_1(t)}{z+t}dt = \mu(z) - \int_0^1 \frac{\frac{1}{2}-t}{z+t}dt.$$

The integrand on the right-hand side has $(z+\frac{1}{2})\log(z+t) - t$ as an antiderivative; (3) follows since $\log(z+1) - \log z = \log(1 + \frac{1}{z})$ in all of $\mathbb{C}^-$.

2. Stirling's formula. For each $\delta \in (0, \pi]$, we denote by W_δ the angular sector $\{|z|e^{i\varphi} \in \mathbb{C}^\times : |\varphi| \le \pi - \delta\}$, which omits the negative real axis. The following theorem describes the relationship between the functions $\Gamma(z)$ and $\mu(z)$, as well as the growth of $\mu(z)$.

Theorem (Stirling's formulas).

(ST)
$$\boxed{\begin{aligned} \Gamma(z) &= \sqrt{2\pi} z^{z-\frac{1}{2}} e^{-z} e^{\mu(z)}, && z \in \mathbb{C}^-, \\ |\mu(z)| &\le \frac{1}{8} \frac{1}{\cos^2 \frac{1}{2}\varphi} \frac{1}{|z|}, && z = |z|e^{i\varphi} \in \mathbb{C}^-, \\ |\mu(z)| &\le \frac{1}{8} \frac{1}{\sin^2 \frac{1}{2}\delta} \frac{1}{|z|}, && z \in W_\delta,\ 0 < \delta \le \pi. \end{aligned}}$$

Proof. Since $Q(t) \le \frac{1}{8}$ and $|z+t| \ge |z| \cos \frac{1}{2}\varphi \ge |z| \sin \frac{1}{2}\delta$ (see the introduction to this section), the inequalities follow from 1(2). We show, moreover, that $F(z) := z^{z-\frac{1}{2}} e^{-z} e^{\mu(z)} \in \mathcal{O}(\mathbb{C}^-)$ satisfies the hypotheses of the uniqueness theorem 2.4. The functional equation 1(3) for $\mu(z)$ immediately gives

$$F(z+1) = (z+\tfrac{1}{2})^{z+\frac{1}{2}} e^{-z-1} e^{\mu(z)-(z+\frac{1}{2})\log(1+\frac{1}{z})+1} = z^{z+\frac{1}{2}} e^{-z} e^{\mu(z)} = zF(z).$$

Furthermore, F is bounded in the strip $S = \{z \in \mathbb{C} : 1 \le \operatorname{Re} z < 2\}$: Certainly $e^{\mu(z)}$ is bounded there. For all $z = x + iy = |z|e^{i\varphi} \in \mathbb{C}^-$, we have $|z^{z-\frac{1}{2}} e^{-z}| = |z|^{x-\frac{1}{2}} e^{-y\varphi}$. If $z \in S$ and $|y| \ge 2$, then $x - \frac{1}{2} \le 2$, $|z| \le 2y$, and $-y\varphi \le -\frac{1}{2}\pi|y|$; for such z, it follows that $|z^{z-\frac{1}{2}} e^{-z}| \le 4y^2 e^{-\frac{1}{2}\pi|y|}$. Since $\lim_{|y|\to\infty} y^2 e^{-\frac{1}{2}\pi|y|} = 0$, F is bounded in S.

That $\Gamma(z) = az^{z-\frac{1}{2}} e^{-z} e^{\mu(z)}$ now follows from Theorem 2.4. In order to show that $a = \sqrt{2\pi}$, we substitute the right-hand side into the Legendre duplication formula of 2.5. After simplifying, we obtain

$$\sqrt{2\pi e}\ e^{\mu(2z)-\mu(z)-\mu(z+\frac{1}{2})} = a(1 + \tfrac{1}{2z})^z.$$

Since $\lim_{x\to\infty} \mu(x) = 0$ and $\lim_{x\to\infty}(1 + \frac{1}{2x})^x = \sqrt{e}$, it follows that $a = \sqrt{2\pi}$. □

The equation (ST) shows — as claimed in the introduction — that the following holds:

(ST′)
$$\boxed{\log \Gamma(z) = \tfrac{1}{2} \log 2\pi + (z - \tfrac{1}{2}) \log z - z + \mu(z).}$$

For real numbers, (ST) can be written as

(ST*)
$$\boxed{\Gamma(x+1) = \sqrt{2\pi x}\ x^x e^{-x+\theta(x)/(8x)}, \quad x > 0, \quad 0 < \theta < 1;}$$

for $x := n$, this is a more precise version of the equation (ST) of the introduction.

The great power of this theorem lies in the estimates for $\mu(z)$. They are actually seldom used with this precision. Usually it suffices to know that, in every angular sector W_δ, $\mu(z)$ tends uniformly to zero like $1/z$ as z tends to ∞.

The statements of (ST) are easily summarized in the "asymptotic equation"

$$\boxed{\Gamma(z) \sim \sqrt{2\pi} z^{z-\frac{1}{2}} e^{-z}, \quad \text{or} \quad \Gamma(z+1) \sim \sqrt{2\pi z}\left(\frac{z}{e}\right)^z,}$$

where the symbol $\sim$ means that the quotient of the left- and right-hand sides converges *uniformly* to 1 as $z \to \infty$ in every angular sector W_δ punctured at 0. One consequence is that

$$\Gamma(z+a) \sim z^a \Gamma(z) \quad \text{for fixed } a \in \mathbb{C}\backslash\{-1,-2,-3,\ldots\}.$$

The inequalities in (ST) can immediately be sharpened through better estimates for the integral defining $\mu(z)$. We first note that

$$|\mu(z)| \le \frac{1}{8}\int_0^\infty \frac{dt}{|z+t|^2} = \frac{1}{8|z|}\int_0^\infty \frac{ds}{(s+\cos\varphi)^2+\sin^2\varphi}.$$

If we now observe that $\arctan x = \frac{1}{2}\pi - \operatorname{arccot} x$ is an antiderivative of $(x^2+1)^{-1}$, it follows immediately (with $\varphi/\sin\varphi := 1$ for $\varphi = 0$) that

$$(1) \qquad |\mu(z)| \le \frac{1}{8}\frac{\varphi}{\sin\varphi}\frac{1}{|z|} \quad \text{for } z = |z|e^{i\varphi} \in \mathbb{C}^-.$$

Since $\varphi/\sin\varphi$ is monotone increasing in $[0, \pi)$, this contains the inequality

$$(2) \qquad |\mu(z)| \le \frac{1}{8}\frac{\pi-\delta}{\sin\delta}\frac{1}{|z|}, \quad z \in W_\delta, \quad 0 < \delta \le \pi.$$

The bounds in (1) and (2) are *better* than the old ones in (ST) when $\varphi \ne 0$ or $\delta \ne \pi$; for then $|\varphi| < 2\tan\frac{1}{2}|\varphi|$, or $\pi-\delta < 2\cot\frac{1}{2}\delta$, whence it follows immediately that $\varphi/\sin\varphi < (\cos\frac{1}{2}\varphi)^{-2}$ (resp. $(\pi-\delta)/\sin\delta < (\sin\frac{1}{2}\delta)^{-2}$).

Historical note. For the *slit plane* $\mathbb{C}^-$, Stirling's formula was first proved in 1889 by T.-J. Stieltjes; cf. [St]. Until then, the formula had been known to hold only in the *right* half-plane. Stieltjes *systematically* used the definition of the μ-function by means of $P_1(t)$ given in Subsection 1 ([St], p. 428 ff.). It has the advantage over older formulas of Binet and Gauss of holding in all of $\mathbb{C}^-$, not just in the right half-plane $\mathbb{T}$. This formula for $\mu(z)$ was published in 1875 by P. Gilbert in "Recherches sur le développement de la fonction Γ et sur certaines intégrales définies qui en dépendent," *Mém. de l'Acad. de Belgique* 41, 1–60, especially p. 12. However, I know of no compelling applications for large angular sectors.

3. Growth of $|\Gamma(x+iy)|$ for $|y| \to \infty$. An elementary consequence of Stirling's formulas is that $|\Gamma(x+iy)|$ tends *exponentially* to zero as y increases. As early as 1889, S. Pincherle observed ([Pi], p. 234):

(1) *The following holds uniformly as* $|y| \to \infty$, *for* x *in a compact subset of* $\mathbb{R}$:

$$|\Gamma(x+iy)| \sim \sqrt{2\pi}|y|^{x-\frac{1}{2}}e^{-\frac{1}{2}\pi|y|}.$$

Proof. $\Gamma(z)| \sim \sqrt{2\pi}|z|^{z-\frac{1}{2}}|e^{-z}|$ by 2(ST). Since $|z^{z-\frac{1}{2}}| = |z|^{x-\frac{1}{2}}e^{-y\varphi}$ for $z = x+iy = |z|e^{i\varphi}$, $\varphi \in (-\pi, \pi)$, it follows that

$$(*) \qquad |\Gamma(x+iy)| \sim \sqrt{2\pi}|z|^{x-\frac{1}{2}}e^{-x-y\varphi} \quad \text{uniformly as } |y| \to \infty,$$

for x in a compact subset of $\mathbb{R}$. Since $|z| \sim |y|$ as $|y| \to \infty$,

$$(**) \qquad |z|^{x-\frac{1}{2}} \sim |y|^{x-\frac{1}{2}} \quad \text{uniformly as } |y| \to \infty,$$

for x in a compact subset of $\mathbb{R}$. To deal with $\exp(-x-y\varphi)$ asymptotically, we may restrict to the case $y \to +\infty$ (because $\Gamma(\overline{z}) = \overline{\Gamma(z)}$). Since $\tan(\frac{1}{2}\pi - \varphi) = xy^{-1}$,

$$\varphi = \frac{1}{2}\pi - \arctan xy^{-1}, \text{ where } \arctan w = w - \tfrac{1}{3}w^3 + \tfrac{1}{5}w^5 - + \cdots, \ |w| < 1.$$

Since $\lim_{y\to\infty} y \arctan xy^{-1} = x$ uniformly for x in a compact subset of $\mathbb{R}$, we see that $e^{-x-\varphi y} \sim e^{-\frac{1}{2}\pi y}$ as $y \to \infty$. (1) now follows from $(*)$ and $(**)$. □

4*. Gudermann's series. Equation 1(3) yields

$$\sum_{\nu=0}^{n}\left[\left(z+\nu+\frac{1}{2}\right)\log\left(1+\frac{1}{z+\nu}\right)-1\right] = -\int_0^n \frac{P_1(t)}{z+t}dt.$$

This and 1(2) give Gudermann's series representation:

$$(1) \qquad \mu(z) = \sum_{\nu=0}^{\infty}\left[\left(z+\nu+\frac{1}{2}\right)\log\left(1+\frac{1}{z+\nu}\right)-1\right] \quad \text{in } \mathbb{C}^-.$$

The series (1) can be used to improve the factor 1/8 in 2(ST) to 1/12. We write $(z+\frac{1}{2})\log(1+\frac{1}{z})-1$ as $\lambda(z)$ for brevity, and begin by proving the following:

$$(\text{o}) \qquad \lambda(z) = \frac{1}{2}\int_0^1 \frac{t(1-t)}{(z+t)^2}dt = 2\int_0^{\frac{1}{2}} \frac{(\frac{1}{2}-t)^2}{(z+t)(z+1-t)}dt, \quad z \in \mathbb{C}^-;$$

$$(\text{oo}) \qquad |\lambda(z)| \le \frac{1}{12}\frac{1}{\cos^2\frac{1}{2}\varphi}\left(\frac{1}{|z|}-\frac{1}{|z+1|}\right).$$

Proof. (o) Integration by parts in 1(3) gives the first integral for $\lambda(z)$. The second integral results from integrating from 0 to 1/2 and 1/2 to 1 in 1(3), then substituting $1-t$ for t in the second integrand.

(oo) Since $t(1-t) \geq 0$ in $[0,1]$, $(*)$ of the introduction and (o) give

$$|\lambda(z)| \leq \frac{1}{2}\int_0^1 \frac{t(1-t)}{|z+t|^2}dt \leq \left(\cos\frac{1}{2}\varphi\right)^{-2}\lambda(|z|).$$

Since $(r+t)(r+1-t) \geq r(r+1)$ for $t \in [0,1]$, the second integral in (o) gives

$$\lambda(r) \leq \frac{2}{r(r+1)}\int_0^{\frac{1}{2}}\left(\frac{1}{2}-t\right)^2 dt = \frac{1}{12}\left(\frac{1}{r}-\frac{1}{r+1}\right) \quad \text{for all } r > 0. \quad \square$$

One can also estimate $\lambda(r)$ by means of the power series for $\log\frac{1+r}{1-r}$ (cf. [A], p. 21).

The following theorem is now immediate.

Theorem. *Gudermann's series converges normally in* $\mathbb{C}^-$, *and*

$$|\mu(z)| \leq \frac{1}{12}\frac{1}{\cos^2\frac{1}{2}\varphi}\frac{1}{|z|} \quad \textit{for } z = |z|e^{i\varphi} \in \mathbb{C}^-.$$

Proof. By (1) and (oo),

$$|\mu(z)| \leq \sum_{\nu=0}^{\infty}|\lambda(z+\nu)| \leq \frac{1}{12}\left(\cos\frac{1}{2}\varphi\right)^{-2}\sum_{\nu=0}^{\infty}\left(\frac{1}{|z+\nu|}-\frac{1}{|z+\nu+1|}\right). \quad \square$$

2(ST*) now holds with 1/12 instead of 1/8 in the exponent. Moreover, it follows at once that

$$|\mu(z)| \leq \frac{1}{12}\frac{1}{\operatorname{Re} z} \text{ when } \operatorname{Re} z > 0 \quad \text{and} \quad |\mu(iy)| \leq \frac{1}{6}\frac{1}{|y|} \text{ for } y \in \mathbb{R}.$$

The bounds for $\mu(z)$ are better than the bounds in 2(1) and 2(2) whenever $\tan\frac{1}{2}\varphi < \frac{3}{4}\varphi$ and $\cot\frac{1}{2}\delta > \frac{3}{4}(\pi-\delta)$, i.e. for $\varphi < 110.8°$ and $\delta > 69.2°$. We will see in §5.3 that $|\mu(z)| \leq \frac{1}{12}|z|^{-1}$ in the angular sector $|\varphi| \leq \frac{1}{4}\pi$.

Historical note. C. Gudermann discovered the series $\mu(z)$ in 1845 [G]; it has sinced been named after him. The inequality with the classical initial factor $\frac{1}{12}$ instead of $\frac{1}{8}$ is due to Stieltjes ([St], p. 443).

5*. Stirling's series. We seek an *asymptotic* expansion of the function $\mu(z)$ in powers of z^{-1}. We work with the Bernoulli polynomials

$$B_k(w) = \sum_{\kappa=0}^{k}\binom{k}{\kappa}B_\kappa^{k-\kappa} = w^k - \frac{1}{2}kw^{k-1} + \cdots + B_k, \quad k \geq 1,$$

where $B_\kappa := B_\kappa(0)$ is the κth Bernoulli number. The following identities hold (cf. I.7.5.4):

$$(*)\quad B'_{k+1}(w) = (k+1)B_k(w), \quad k \in \mathbb{N}, \quad \text{and} \quad B_k(0) = B_k(1) \text{ for } k \geq 2.$$

To every polynomial $B_n(t)$, we assign a *periodic* function $P_n : \mathbb{R} \to \mathbb{R}$ defined by

$$(1) \qquad P_n(t) := B_n(t) \text{ for } 0 \leq t < 1, \quad P_n(t) \text{ has period } 1.$$

Then $P_1(t)$ is the sawtooth function. We now set

$$(2) \qquad \mu_k(z) := -\frac{1}{k}\int_0^\infty \frac{P_k(t)}{(z+t)^k}dt, \quad k \geq 1.$$

All the functions μ_k are holomorphic in $\mathbb{C}^-$; furthermore,

$$(3) \qquad \begin{aligned} &\mu_1(z) = \mu(z),\ \mu_k(z) = \frac{B_{k+1}}{k(k+1)}\frac{1}{z^k} + \mu_{k+1}(z),\ \mu_{2n}(z) = \mu_{2n+1}(z) \\ &\mu(z) = \sum_{\nu=1}^{n} \frac{B_{2\nu}}{(2\nu-1)2\nu}\frac{1}{z^{2\nu-1}} + \mu_{2n+1}(z). \end{aligned}$$

Proof of (3). The recursion formulas follow from integrating by parts in (2); the equations $\mu_{2n} = \mu_{2n+1}$ hold because $B_3 = B_5 = \cdots = 0$. □

The series in (3) is called Stirling's series with remainder term μ_{2n+1}. For $n = 1$,

$$(3') \qquad \begin{aligned} \mu(z) &= \frac{1}{12}\cdot\frac{1}{z} - \frac{1}{3}\int_0^\infty \frac{P_3(t)}{(z+t)^3}, \quad \text{where} \\ P_3(t) &= t^3 - \frac{3}{2}t^2 + \frac{1}{2}t \quad \text{for } t \in [0,1]. \end{aligned}$$

Since $|P_3(t)| < 1/20$ (the maximum occurs at $1/2 \pm \sqrt{3}/6$) and $\int_0^\infty |z+t|^{-3}dt = |z|^{-1}(|z| + \operatorname{Re} z)^{-1}$ (because $a^{-2}x(x^2+a^2)^{-1/2}$ is an antiderivative of $(x^2+a^2)^{-3/2}$), it follows that

$$(3'') \qquad \left|\mu(z) - \frac{1}{12}\frac{1}{z}\right| < \frac{1}{60}\frac{1}{|z|}\frac{1}{|z| + \operatorname{Re} z}, \quad z \in \mathbb{C}^-.$$

Stirling's series (3) does *not* give a Laurent expansion for μ as $n \to \infty$ (since μ does not have an isolated singularity at 0). In fact:

For every $z \in \mathbb{C}^\times$, the sequence $\dfrac{B_{2\nu}}{(2\nu-1)2\nu}\cdot\dfrac{1}{z^{2\nu-1}}$ *is unbounded.*

This follows since $|B_{2\nu}| > 2(2\nu)!/(2\pi)^{2\nu}$ (cf. I.11.3.2) and $\lim n!/r^n = \infty$ for $r > 0$. □

The full significance of Stirling's series is realized only through useful estimates for the complicated remainder term. By (3),

$$\mu_{2n-1}(z) = \frac{1}{2n}\int_0^\infty \frac{B_{2n} - P_{2n}(t)}{(z+t)^{2n}}dt \tag{4}$$
$$\left(\text{note that } \frac{1}{k}\frac{1}{z^k} = \int_0^\infty \frac{dt}{(z+t)^{k+1}}\right).$$

For $n \geq 1$, we set $M_n := \sup_{t\geq b}|B_{2n} - P_{2n}(t)| \in \mathbb{R}$ and $z = |z|e^{i\varphi} \in \mathbb{C}^-$. It follows immediately that

$$|\mu_{2n-1}(z)| \leq \frac{M_n}{(2n-1)2n} \cdot \frac{1}{\cos^{2n}\frac{1}{2}\varphi} \cdot \frac{1}{|z|^{2n-1}}. \tag{5}$$ □

A direct estimate for $\mu_{2n-1}(z)$ without the detour via (4) would have given, instead of (5), only the power $|z|^{2n-2}$ in the denominator. (2) and (5) immediately give

$$\lim_{z\in W_\delta, z\to\infty} z^{2n-1}\mu_{2n-1}(z) = \frac{B_{2n}}{(2n-1)2n}.$$

From (3) and (5) we obtain the following limit equation for every angular sector W_δ:

$$\lim_{z\in W_\delta, z\to\infty}\left|\mu(z) - \sum_{\nu=1}^{n}\frac{B_{2\nu}}{(2\nu-1)2\nu}\frac{1}{z^{2\nu-1}}\right||z|^{2n} = 0, \quad n \geq 1. \tag{6}$$

Stirling's series is thus an *asymptotic expansion of* $\mu(z)$ (at ∞ — see I.9.6.1). If z is large compared to n, this gives a very good approximation for $\mu(z)$, but making the index n large for fixed z yields nothing. For $n = 3$, for example,

$$\log\Gamma(z) = (z-\frac{1}{2})\log z - z + \log\sqrt{2\pi} + \frac{1}{12}\frac{1}{z} - \frac{1}{360}\frac{1}{z^3} + \frac{1}{1260}\frac{1}{z^5} - \text{error term}.$$

6*. Delicate estimates for the remainder term. Whoever is ambitious looks for good numerical values for the bounds M_n in 5(5). Stieltjes already proved ([St], pp. 434–436):

$$|\mu_{2n-1}(z)| \leq \frac{|B_{2n}|}{(2n-1)2n} \cdot \frac{1}{\cos^{2n}\frac{1}{2}\varphi} \cdot \frac{1}{|z|^{2n-1}}, \tag{1}$$
$$z = |z|e^{i\varphi} \in \mathbb{C}, \ n \geq 1.$$

For $n = 1$, this is the inequality of Theorem 4. The proof of (1) uses the following unobvious property of the sign of the function $P_{2n}(t)$, $t \geq 0$:

(S) $\quad B_{2n} - P_{2n}(t)$ *always has the sign* $(-1)^{n-1}$, $n \geq 1$.

(1) follows quickly from (S): Since $B_{2n} - P_{2n}(t)$ never changes sign, 5(4) (with $z = re^{i\varphi}$) immediately gives

$$\begin{aligned} \cos^{2n}\frac{1}{2}\varphi|\mu_{2n-1}(z)| &\le \frac{1}{2n}\int_0^\infty \frac{|B_{2n} - P_{2n}(t)|}{(r+t)^{2n}}dt \\ &= \frac{1}{2n}\left|\int_0^\infty \frac{B_{2n} - P_{2n}(t)}{(r+t)^{2n}}dt\right| = |\mu_{2n-1}(r)|. \end{aligned}$$

To estimate $\mu_{2n-1}(r)$, we use 5(3). We have

$$\mu_{2n-1}(r) - \mu_{2n+1}(r) = \frac{B_{2n}}{(2n-1)2n}\frac{1}{r^{2n-1}}.$$

Since, by 5(4) and (S), $\mu_{2n-1}(r)$ and $-\mu_{2n+1}(r)$ have the same sign for all $r > 0$, it follows that

$$|\mu_{2n-1}(r)| \le |\mu_{2n-1}(r) - \mu_{2n+1}(r)| \le \frac{|B_{2n}|}{(2n-1)2n}\frac{1}{r^{2n-1}}.$$

(1) is then proved by applying (S). To prove (S), we exploit the Fourier series of $P_{2n}(t)$ (cf. I.14.3.4):

$$P_{2n}(t) = (-1)^{n-1}\frac{2(2n)!}{(2\pi)^{2n}}\sum_{\nu=1}^{\infty}\frac{\cos 2\pi\nu t}{\nu^{2n}}, \quad t \ge 0,\ n \ge 1.$$

Since $P_{2n}(0) = B_{2n}$ (Euler's formula),

$$B_{2n} - P_{2n}(t) = (-1)^{n-1}\frac{2(2n)!}{(2\pi)^{2n}}\sum_{\nu=1}^{\infty}\frac{1-\cos 2\pi\nu t}{\nu^{2n}}.$$

Since no summand on the right-hand side is negative, it is clear that (S) holds.

7*. Binet's integral. There are other interesting representations of the μ-function besides the Stieltjes integral formula and Gudermann's series, but these are valid only in the right half-plane. J. M. Binet proved in 1839 that

$$(1) \qquad \mu(z) = 2\int_0^\infty \frac{\arctan(t/z)}{e^{2\pi t} - 1}dt \quad \text{if } \operatorname{Re} z > 0$$

([B], p. 243). The following formula is convenient for the proof.

Plana's summation formula. *Let f be holomorphic in a neighborhood of the closed half-plane $\{z \in \mathbb{C} : \operatorname{Re} z \ge 0\}$ and have the following properties:*

1. $\sum_0^\infty f(\nu)$ *and* $\int_0^\infty f(x)dx$ *exist;*
2. $\lim_{|t|\to\infty} f(x+it)e^{-2\pi|t|} = 0$ *uniformly for $x \in [0,s]$, $s > 0$ arbitrary;*
3. $\lim_{s\to\infty}\int_{-\infty}^{\infty}|f(s+iy)|e^{-2\pi|y|}dy = 0.$

Then

$$\boxed{\sum_{\nu=0}^{\infty} f(\nu) = \frac{1}{2} f(0) + \int_0^\infty f(x)dx + i\int_0^\infty \frac{f(iy)-f(-iy)}{e^{2\pi y}-1}dy.}$$

A proof of this formula can be found in [SG], pp. 438–440. Plana gave his formula (which he described as "remarquable") in 1820 in [Pl]. Abel arrived at this formula three years later. Cauchy, in 1826, gave one of the first correct proofs; Kronecker treated this and related questions in 1889. For further details, see [Li], pp. 68–69.

Clearly the function $h(w) := (z+w)^{-2}$ satisfies the hypotheses of Plana's summation formula for every fixed number $z \in \mathbb{T}$. We have (cf. Corollary 2.3)

$$\sum_{\nu=0}^{\infty} h(\nu) = (\log\Gamma)''(z),\ h(0) = \frac{1}{z^2},\ \int_0^\infty h(x)dx = \frac{1}{z},\ h(it) - h(-it) = \frac{-4izt}{(z^2+t^2)^2}.$$

It follows that

$$(2) \qquad (\log\Gamma)''(z) = \frac{1}{2z^2} + \frac{1}{z} + \int_0^\infty \frac{4zt}{(z^2+t^2)^2}\frac{dt}{e^{2\pi t}-1}, \qquad \operatorname{Re} z > 0.$$

Integrating under the integral sign in this equation yields

$$(3) \qquad \boxed{\begin{aligned} \mu'(z) &= -\int_0^\infty \frac{2t}{z^2+t^2}\frac{dt}{e^{2\pi t}-1} \\ \mu(z) &= \int_0^\infty \frac{2\arctan t/z}{e^{2\pi t}-1}dt \end{aligned} \quad ,\ \operatorname{Re} z > 0.}$$

Proof. Integrating (2) once gives

$$(\text{o}) \qquad (\log\Gamma)'(z) = c_1 - \frac{1}{2z} + \log z - \int_0^\infty \frac{2t}{z^2+t^2}\frac{dt}{e^{2\pi t}-1}, \qquad c_1 := \text{constant}.$$

Since $[(z-\frac{1}{2})\log z - z]' = \log z - 1/2z$, integrating again gives

$$(\text{oo}) \qquad \log\Gamma(z) = c_0 + c_1 z + \left(z - \frac{1}{2}\right)\log z - z + \int_0^\infty \frac{2\arctan t/z}{e^{2\pi t}-1}dt.$$

Comparing (oo) with 4.2(ST′) leads to the equation

$$\mu(z) = c_0 - \frac{1}{2}\log 2\pi + c_1 z + \int_0^\infty \frac{2\arctan t/z}{e^{2\pi t}-1}dt.$$

For the integral $I(z)$ on the right-hand side, since $0 \le \arctan t/x \le t/x$ for $x > 0$,

$$0 \le I(x) \le \frac{2}{x}\int_0^\infty \frac{t}{e^{2\pi t}-1}dt; \quad \text{thus } \lim_{x\to\infty} I(x) = 0.$$

Since $\lim_{x\to\infty}\mu(x) = 0$ as well, it follows that $c_0 = \frac{1}{2}\log 2\pi$ and $c_1 = 0$. □

The equations (3) are called *Binet's integrals for μ' and μ.* There are other integral representations for $\mu(z)$; for example,

$$\mu(z) = -\frac{z}{\pi}\int_0^\infty \frac{\log(1-e^{-2\pi t})}{z^2+t^2}dt, \quad \operatorname{Re} z > 0.$$

(Integration by parts gives (3).) All these formulas — except for the Stieltjes formula — are valid only in the right half-plane.

8*. Lindelöf's estimate. A series expansion of $2t/(z^2+t^2)$ in Binet's integral formula yields formulas for the Bernoulli numbers B_{2n} and the functions $\mu'_{2n-1}(z)$:

$$\text{(1)}\qquad \boxed{\begin{aligned} &\int_0^\infty \frac{t^{2n-1}}{e^{2\pi t}-1}dt = \frac{(-1)^{n-1}}{4n}\cdot B_{2n}, \quad n \geq 1 \\ &\mu'_{2n-1}(z) = 2\frac{(-1)^{n-1}}{z^{2n-1}}\int_0^\infty \frac{t^{2n-1}}{z^2+t^2}\frac{dt}{e^{2\pi t}-1}, \quad n \geq 1 \text{ and } \operatorname{Re} z > 0. \end{aligned}}$$

Proof. Since $1/(1+q) = \sum_1^{n-1}(-1)^{\nu-1}q^{\nu-1} + (-q)^{n-1}/(1+q)$, we have (with $q := t^2/z^2$)

$$\frac{2t}{z^2+t^2} = 2\sum_{\nu=1}^{n-1}(-1)^{\nu-1}\frac{t^{2\nu-1}}{z^{2\nu}} + 2(-1)^{n-1}\frac{t^{2n-1}}{z^2+t^2}\cdot\frac{1}{z^{2n-2}}.$$

Hence, for Re $z > 0$, 7(3) gives the series

$$\text{(o)}\qquad \begin{aligned} \mu'(z) = &-\sum_{\nu=1}^{n-1} 2(-1)^{\nu-1}\frac{1}{z^{2\nu}}\int_0^\infty \frac{t^{2\nu-1}}{e^{2\pi t}-1}dt \\ &+ \frac{2(-1)^{n-1}}{z^{2n-2}}\int_0^\infty \frac{t^{2n-1}}{z^2+t^2}\frac{dt}{e^{2\pi t}-1}. \end{aligned}$$

On the other hand, differentiating the series in 4.5(3) gives

$$\text{(oo)}\qquad \mu'(z) = -\sum_{\nu=1}^{n-1}\frac{B_{2\nu}}{2\nu}\frac{1}{z^{2\nu}} + \mu'_{2n-1}(z).$$

For fixed n and large z, the last terms in (o) and (oo) tend to zero like z^{-2n}: In (o) this follows directly; in (oo) one estimates the equation

$$\mu'_{2n-1}(z) = -\int_0^\infty \frac{B_{2n}-P_{2n}(t)}{(z+t)^{2n+1}}dt, \quad \operatorname{Re} z > 0,$$

which comes from differentiating under the integral sign in 5(4). Thus (o) and (oo) are "asymptotic Laurent expansions" for $\mu'(z)$. The uniqueness of such expansions follows as for power series (cf. I.9.6.1, p. 294). Comparison of coefficients gives (1). □

We now estimate $\mu'_{2n-1}(z)$. For the function defined in the interval $(-\frac{1}{2}\pi, \frac{1}{2}\pi)$ by

$$c(\varphi) := 1 \text{ for } |\varphi| \le \tfrac{1}{4}\pi \quad \text{and} \quad c(\varphi) := |\sin 2\varphi|^{-1} \text{ for } \tfrac{1}{4}\pi < |\varphi| < \tfrac{1}{2}\pi,$$

we have $|z^2 + t^2| \ge |z|^2/c(\varphi)$ if $z = |z|e^{i\varphi}$ and $|\varphi| < \frac{1}{2}\pi$. It follows by (1) that

$$(2) \qquad |\mu'_{2n-1}(z)| \le \frac{|B_{2n}|}{2n}\frac{c(\varphi)}{|z|^{2n}} \quad \text{if } \operatorname{Re} z > 0.$$

Since $\lim_{t\to\infty} \mu'(zt) = 0$ for $z \in \mathbb{C}^-$, integration along $\zeta(t) = zt$, $t \ge 1$, gives

$$\mu_{2n-1}(z) = -\int_1^\infty z\mu'_{2n-1}(zt)dt.$$

Thus, for all z with Re $z > 0$,

$$|\mu_{2n-1}(z)| \le |z| \int_1^\infty |\mu'_{2n-1}(zt)|dt \le \frac{|B_{2n}|}{2n}\frac{c(\varphi)}{|z|^{2n-1}}\int_1^\infty \frac{dt}{t^{2n}}.$$

An immediate result of this is *Lindelöf's estimates* ([Li], p. 99):

$$\text{(L)} \qquad \boxed{|\mu_{2n-1}(z)| \le \frac{|B_{2n}|}{(2n-1)2n}\frac{c(\varphi)}{|z|^{2n-1}}, \quad \text{for } z = |z|e^{i\varphi},\ |\varphi| < \tfrac{1}{2}\pi;\ n \ge 1.}$$

In the angular sector $|\varphi| \le \frac{1}{4}\pi$, these inequalities are better than 4.6(1); for example, it follows that

$$|\mu(z)| \le \frac{1}{12}\frac{1}{|z|} \quad \text{for all } z = |z|e^{i\varphi} \text{ with } |\varphi| \le \tfrac{1}{4}\pi.$$

Lindelöf's bound $c(\varphi) = 1$ cannot be improved for $|\varphi| \le \frac{1}{4}\pi$ because of 4.5(6). For $|\varphi| > \frac{1}{4}\pi$, (L) is better than 4.6(1) as long as $|\sin 2\varphi| > \cos^{2n+2} \frac{1}{2}\varphi$.

Interest in delicate estimates like (L) is still alive today. In their recent article [Sch F] in the venerable *Crelle's Journal*, W. Schäfke and A. Finsterer showed that $|\sin 2\varphi|^{-1}$, in the angular sector $\frac{1}{4}\pi < |\varphi| < \frac{1}{2}\pi$, is *the best of n independent bounds* for which (L) holds. For each individual n, however, there exists a better bound $c_n(\varphi) < c(\varphi)$ (cf. [Sch S]).

§5. The Beta Function

The improper integral

$$(1) \qquad \mathsf{B}(w, z) := \int_0^1 t^{w-1}(1-t)^{z-1}dt$$

converges *compactly* and *absolutely* in the quadrant $\mathbb{T} \times \mathbb{T} = \{(w, z) \in \mathbb{C}^2 : \operatorname{Re} z > 0, \operatorname{Re} w > 0\}$, and is therefore holomorphic in $z \in \mathbb{T}$ (resp., $w \in \mathbb{T}$) for fixed $w \in \mathbb{T}$ (resp. $z \in \mathbb{T}$); the proof, like that for the Γ-integral, uses a majorant test (cf. also 7.4.2). The function $\mathsf{B}(w, z)$ is called the (Euler) *beta function*; Legendre referred in 1811 to *Euler's integral of the first kind* ([L_1], vol. 1, p. 221). The main result of the theory of the beta function is

Euler's identity: $\mathsf{B}(w,z) = \dfrac{\Gamma(w)\Gamma(z)}{\Gamma(w+z)}$ *for all* w, $z \in \mathbb{T}$.

It will be derived in Subsection 1 by means of the uniqueness theorem 2.4.

1. Proof of Euler's identity. We need the following results:

a) $\mathsf{B}(w,1) = w^{-1}$, $\mathsf{B}(w,z+1) = \frac{z}{w+z}\mathsf{B}(w,z)$,

b) $|\mathsf{B}(w,z)| \le \mathsf{B}(\operatorname{Re} w, \operatorname{Re} z)$.

Proof. a) The first formula is trivial; the second is proved as follows:

$$\begin{aligned}
&(w+z)\mathsf{B}(w,z+1) - z\mathsf{B}(w,z)\\
&= (w+z)\int_0^1 t^{w-1}(1-t)^z dt - z\int_0^1 t^{w-1}(1-t)^{z-1}dt\\
&= \int_0^1 \{wt^{w-1}(1-t)^z - t^w z(1-t)^{z-1}\}dt\\
&= [t^w(1-t)^z]_0^1 = 0.
\end{aligned}$$

b) This is clear, since $|(1-t)^{w-1}t^{z-1}| \le (1-t)^{\operatorname{Re} w-1}t^{\operatorname{Re} z-1}$. □

To prove Euler's identity, we now fix $w \in \mathbb{T}$ and set $F(z) := \mathsf{B}(w,z)\Gamma(w+z) \in \mathcal{O}(\mathbb{T})$. By a), $F(1) = \Gamma(w)$ and $F(z+1) = zF(z)$. Since $|\Gamma(w+z)| \le \Gamma(\operatorname{Re}(w+z))$, b) shows that F is bounded in the strip $\{1 \le \operatorname{Re} z < 2\}$. By the uniqueness theorem 2.4, $F(z) = \Gamma(w)\Gamma(z)$. □

A proof of Euler's identity for real arguments, using the logarithmic convexity of the product $\mathsf{B}(x,y)\Gamma(x+y)$, can be found in Artin's book ([A], pp. 18–19).

Because of the formula $\mathsf{B}(w,z) = \Gamma(w)\Gamma(z)/\Gamma(w+z)$, the beta function is not interesting in its own right. Despite this, it survived for quite a while alongside the gamma function as a separate function: a profusion of relations between beta functions was derived, especially by means of the identities a); these often reduce to trivialities as soon as the Euler identities are applied. See, for example, the classical works of Legendre ([$L_{1,2}$], passim) and Binet [Bin], or even those of Euler himself (and also [Ni], p. 15).

The following integral formulas, valid for all w, $z \in \mathbb{T}$, are useful:

$$(1)\qquad \mathsf{B}(w,z) = 2\int_0^{\frac{1}{2}\pi} (\sin\varphi)^{2w-1}(\cos\varphi)^{2z-1}d\varphi = \int_0^\infty \frac{s^{w-1}}{(1+s)^{w+z}}ds.$$

Proof. In (1) of the introduction, substitute $t = \sin^2\varphi$ (resp., $s = \tan^2\varphi$); thus $(1+s)^{-1} = \cos^2\varphi$ and $ds = 2\tan\varphi(\cos\varphi)^{-2}d\varphi$. □

Historical note. Euler, in 1766, systematically studied the integral

$$(*) \qquad \int_0^1 x^{p-1}(1-x^n)^{\frac{q}{n}-1}dx = \int_0^1 \frac{x^{p-1}dx}{\sqrt[n]{(1-x^n)^{n-q}}}$$

([Eu], I–17, pp. 268–287); he writes $\left(\frac{p}{q}\right)$ for his integral. Substituting $y := x^n$ yields

$$\left(\frac{p}{q}\right) = \frac{1}{n}\int_0^1 y^{\frac{p}{n}-1}(1-y)^{\frac{q}{n}-1}dy = \frac{1}{n}\mathsf{B}\left(\frac{p}{n}, \frac{q}{n}\right).$$

Integrals of the type $(*)$ already occur in Euler's "De productis ex infinitis factoribus ortis," which was submitted to the Petersburg Academy on 12 January 1739 but not published until 1750 ([Eu], I-14, pp. 260–290).

Euler knew by 1771 at the latest that the beta function could be reduced to the gamma function (cf. [Eu], I-17, p. 355).

2. Classical proofs of Euler's identity. Because B is holomorphic in $\mathbb{T}$, it suffices to verify the formula for real numbers $w > 0$, $z > 0$ (identity theorem).

Dirichlet's proof (1839, [D], p. 398). First, we have

$$(1+s)^{-z}\Gamma(z) = \int_0^\infty t^{z-1}e^{-(1+s)t}dt, \quad \mathrm{Re}\, s > -1, \quad z > 0 \quad (\text{even } z \in \mathbb{T}).$$

Substituting $w + z$ for z and using 1(1), we find that

$$\Gamma(w+z)\mathsf{B}(w,z) = \int_0^\infty s^{w-1}\left[\int_0^\infty t^{w+z-1}e^{-(1+s)t}dt\right]ds, \quad w > 0, \quad z > 0.$$

By theorems of real analysis, reversing the order of integration is legitimate here for all real $w > 0$, $z > 0$ (!); thus

$$\Gamma(w+z)\mathsf{B}(w,z) = \int_0^\infty \left[\int_0^\infty s^{w-1}e^{-ts}ds\right] t^{w+z-1}e^{-t}dt.$$

The inner integral equals $\Gamma(w)t^{-w}$. Hence

$$\Gamma(w+z)\mathsf{B}(w,z) = \Gamma(w)\int_0^\infty t^{z-1}e^{-t}dt = \Gamma(w)\Gamma(z). \qquad \square$$

Dirichlet carefully examined the theorem used to reverse the order of integration. Jacobi, in 1833, argued concisely as follows [J]:

Demonstratio formulae

$$\int_0^1 w^{a-1}(1-w)^{b-1}\partial w = \frac{\int_0^\infty e^{-x}x^{a-1}\partial x . \int_0^\infty e^{-x}x^{b-1}\partial x}{\int_0^\infty e^{-x}x^{a+b-1}\partial x} = \frac{\Gamma a\,\Gamma b}{\Gamma(a+b)}.$$

(Auct. Dr. *C. G. J. Jacobi*, prof. math. Regiom.)

Quoties variabilibus x, y valores omnes positivi tribuuntur inde a 0 usque ad $+\infty$, posito

$$x+y=r, \quad x=rw,$$

variabili novae z valores conveniunt omnes positivi a 0 usque ad $+\infty$, variabili w valores omnes positivi a 0 usque ad $+1$. Fit simul

$$\partial x\,\partial y = r\partial r\,\partial w.$$

Sit iam e notatione nota:

$$\Gamma(a) = \int_0^\infty e^{-x}x^{a-1}\partial x,$$

habetur

$$\Gamma(a)\Gamma(b) = \iint e^{-x-y}x^{a-1}y^{b-1}\partial x\,\partial y,$$

variabilibus x, y tributis valoribus omnibus positivis a 0 usque ad $+\infty$. Posito autem:

$$x+y=r, \quad x=rw,$$

integrale duplex propositum ex antecedentibus altero quoque modo in duos factores discerpitur:

$$\Gamma(a)\Gamma(b) = \int_0^\infty e^{-r}r^{a+b-1}\partial r\int_0^1 w^{a-1}(1-w)^{b-1}\partial w,$$

unde

$$\int_0^1 w^{a-1}(1-w)^{b-1}\partial w = \frac{\Gamma(a)\Gamma(b)}{\Gamma(a+b)}.$$

Quod est theorema fundamentale, quo integralium Eulerianorum, quae ill. Legendre vocavit, altera species per alteram exhibetur.

23. Aug. 1833.

Exercises. Prove the following identities.

1) $\int_0^{\pi/2}(\cos\varphi)^{2m-1}(\sin\varphi)^{2n-1}d\varphi = \frac{1}{2}\frac{(m-1)!(n-1)!}{(m+n-1)!}$ for $m, n \in \mathbb{N}\backslash\{0\}$.

2) $\frac{\pi}{\sin\pi z} = \int_0^\infty \frac{t^{z-1}}{(1-t)^z}dt = 2\int_0^\infty(\tan\varphi)^{2z-1}d\varphi = \int_0^\infty \frac{s^{z-1}}{1+s}ds$ for $0 < \mathrm{Re}\, z < 1$.

Bibliography

[A] ARTIN, E.: *The Gamma Function*, trans. M. BUTLER, Holt, Rinehart and Winston, New York, 1964.

[Bi] BIEBERBACH, L.: *Theorie der gewöhnlichen Differentialgleichungen*, Grdl. Math. Wiss. 66, 2nd ed., Springer, 1965.

[B] BINET, M. J.: Mémoire sur les intégrales définies Eulériennes, *Journ. de l'École Roy. Polyt.* 16, 123–343 (1839).

[BM] BOHR, H. and J. MOLLERUP: *Laerebog i matematisk Analyse*, vol. 3, Copenhagen, 1922.

[D] DIRICHLET, P. G.: Über eine neue Methode zur Bestimmung vielfacher Integrale, *Abh. Königl. Preuss. Akad. Wiss.* 1839, 61–79; *Werke* 1, 393–410.

[Eu] Leonhardi EULERI Opera omnia, sub auspiciis societatis scientarium naturalium Helveticae, Series I-IV A, 1911–.

[G_1] GAUSS, C. F.: Letter to Bessel, *Werke* 10, Part 1, 362–365.

[G_2] GAUSS, C. F.: Disquisitiones generales circa seriam infinitam

$$1+\frac{\alpha\cdot\beta}{1\cdot\gamma}x+\frac{\alpha(\alpha+1)\beta(\beta+1)}{1\cdot 2\cdot\gamma(\gamma+1)}xx+\frac{\alpha(\alpha+1)(\alpha+2)\beta(\beta+1)(\beta+2)}{1\cdot 2\cdot 3\cdot\gamma(\gamma+1)(\gamma+2)}x^3+\text{etc.},$$

Werke 3, 123–162.

[Gu] GUDERMANN, C.: Additamentum ad functionis $\Gamma(a) = \int_0^\infty e^{-x}\cdot x^{a-1}\partial x$ theoriam, *Journ. reine angew. Math.* 29, 209–212 (1845).

[H] HANKEL, H.: Die Eulerschen Integrale bei unbeschränkter Variabilität des Argumentes, *Zeitschr. Math. Phys.* 9, 1–21 (1864).

[Ha] HAUSDORFF, F.: Zum Hölderschen Satz über $\Gamma(x)$, *Math. Ann.* 94, 244–247 (1925).

[Hö] HÖLDER, O.: Ueber die Eigenschaft der Gammafunktion keiner algebraischen Differentialgleichungen zu genügen, *Math. Ann.* 28, 1–13 (1887).

[J] JACOBI, C. G. J.: Demonstratio formulae $\int_0^1 w^{a-1}(1-w)^{b-1}dw = \ldots$, *Journ. reine angew. Math.* 11, p. 307 (1833); *Ges. Werke* 6, 62–63.

[Je] JENSEN: An elementary exposition of the theory of the gamma function, *Ann. Math.* 17 (2nd ser.), 1915–16, 124–166.

[Kn] KNESER, H.: *Funktionentheorie*, Vandenhoeck & Ruprecht 1958.

[Ku] KUMMER, E. E.: Beitrag zur Theorie der Function $\Gamma(x) = \int_0^\infty e^{-\eta}\eta^{x-1}d\eta$, *Journ. reine angew. Math.* 35, 1–4 (1847): *Coll. Papers* II, 325–328.

[L_1] LEGENDRE, A. M.: *Exercices de calcul intégral*, 3 vols., Paris 1811, 1816, and 1817.

[L_2] LEGENDRE, A. M.: *Traité des fonctions elliptiques et des intégrales Eulériennes*, 3 vols., Paris 1825, 1826, and 1828.

[Li] LINDELÖF, E.: *Le calcul des résidus*, Paris 1905; reprinted 1947 by Chelsea Publ. Co., New York; Chap. IV in particular.

[M] MOORE, E. H.: Concerning transcendentally transcendental functions, *Math. Ann.* 48, 49–74 (1897).

[Ne] NEWMAN, F. W.: On $\Gamma(a)$ especially when a is negative, *Cambridge and Dublin Math. Journ.* 3, 57–60 (1848).

[Ni] NIELSEN, N.: *Handbuch der Theorie der Gammafunktion*, first printing Leipzig 1906, reprinted 1965 by Chelsea Publ. Co., New York.

[O] OSTROWSKI, A.: Zum Hölderschen Satz über $\Gamma(x)$, *Math. Ann.* 94, 248-251 (1925); *Coll. Math. Pap.* 4, 29–32.

[Pi] PINCHERLE, S.: Sulla funzioni ipergeometriche generalizzate, II, *Atti. Rend. Reale Accad. Lincei* (4), 792–799 (1889), *Opere scelte* I, 231–238.

[Pl] PLANA, J.: Note sur une nouvelle expression analytique des nombres Bernoulliens, propre à exprimer en termes finis la formule générale pour la sommation des suites, *Mem. Reale Accad. Sci. Torino* 25, 403–418 (1820).

[Pr] PRYM, F. E.: Zur Theorie der Gammafunction, *Journ. reine angew. Math.* 82, 165–172 (1876).

[Re$_1$] REMMERT, R.: Wielandt's theorem about the Γ-function, *Amer. Math. Monthly* 103, 214–220 (1996).

[Re$_2$] REMMERT, R.: Wielandt's characterization of the Γ-function, in *Math. Works of Helmut Wielandt*, vol. 2, 265–268 (1996).

[Scha] SCHARLAU, W. (ed.): *Rudolph Lipschitz, Briefwechsel mit Cantor, Dedekind, Helmholtz, Kronecker, Weierstrass und anderen*, Dok. Gesch. Math. 2, DMV, Vieweg u. Sohn, 1986.

[Sch F] SCHÄFKE, F. W. and A. FINSTERER: On Lindelöf's error bound for Stirling's series, *Journ. reine angew. Math.* 404, 135–139 (1990).

[Sch S] SCHÄFKE, F. W. and A. SATTLER: Restgliedabschätzungen für die Stirlingsche Reihe, *Note. Mat.* X, Suppl. no. 2, 453–470 (1990).

[Schl] SCHLÖMILCH, O.: Einiges über die Eulerischen Integrale der zweiten Art, *Arch. Math. Phys.* 4, 167–174 (1843).

[SG] SANSONE, G. and J. GERRETSEN: *Lectures on the Theory of Functions of a Complex Variable*, vol. I, P. Noordhoff-Groningen, 1960.

[St] STIELTJES, T.-J.: Sur le développement de $\log\Gamma(a)$, *Journ. Math. Pur. Appl.* (4)5, 425–444 (1889); *Œuvres* 2, Noordhoff, 1918, 211–230; 2nd ed., Springer, 1993, 215–234.

[T] TITCHMARSH, E. C.: *The Theory of the Riemann Zeta-Function*, Oxford University Press, 1951.

[W] WALTER, W.: *Analysis I*, Grundwissen 3, Springer, 1985, 2nd ed. 1990.

[We$_1$] WEIERSTRASS, K.: Über die Theorie der analytischen Facultäten, *Journ. reine angew. Math.* 51, 1–60 (1856), Math. Werke 1, 153–221.

[We$_2$] WEIERSTRASS, K.: Zur Theorie der eindeutigen analytischen Functionen, *Math. Werke* 2, 77–124.

[WW] WHITTAKER, E. T. and G. N. WATSON: *A Course of Modern Analysis*, 4th ed., Cambridge Univ. Press, 1927, Chap. XII in particular.

3

Entire Functions with Prescribed Zeros

> Es ist also stets möglich, eine ganze eindeutige Function $G(x)$ mit vorgeschriebenen Null-Stellen a_1, a_2, a_3, ... zu bilden, wofern nur die nothwendige Bedingung $\text{Lim}_{n=\infty}|a_n| = \infty$ erfüllt ist. (It is therefore always possible to construct a single-valued entire function $G(x)$ with prescribed zeros a_1, a_2, $a_3, \ldots$, provided only that the necessary condition $\text{Lim}_{n=\infty} |a_n| = \infty$ is satisfied.)
>
> — Weierstrass, *Math. Werke* 2, p. 97

If $f \neq 0$ is a holomorphic function on a domain G, its zero set $Z(f)$ is *locally finite* in G by the identity theorem (cf. I.8.1.3). It is natural to pose the following problem:

Let T be any locally finite subset of G, and let every point $d \in T$ be assigned a natural number $\mathfrak{d}(d) \geq 1$ in some way. Construct functions holomorphic in G which each have zero set T and, moreover, whose zeros at each point $d \in T$ have order $\mathfrak{d}(d)$.

It is not at all clear that such functions exist. Of course, if T is finite, the polynomials

$$\prod_{d\in T}(z-d)^{\mathfrak{d}(d)} \quad \text{or} \quad z^{\mathfrak{d}(0)} \prod_{d\in T\setminus\{0\}} \left(1-\frac{z}{d}\right)^{\mathfrak{d}(d)}$$

give the desired result (the initial factor $z^{\mathfrak{d}(0)}$ appears only if $0 \in T$). In 1876, Weierstrass extended this product construction to transcendental entire functions: for a prescribed sequence $d_\nu \in \mathbb{C}^\times$ with $\lim d_\nu = \infty$, he constructs products of the form

$$z^m \prod_{\nu \geq 1} \left[\left(1 - \frac{z}{d_\nu}\right) \exp\left(\frac{z}{d_\nu} + \frac{1}{2}\left(\frac{z}{d_\nu}\right)^2 + \cdots + \frac{1}{k_\nu}\left(\frac{z}{d_\nu}\right)^{k_\nu} \right) \right]$$

and forces their normal convergence in $\mathbb{C}$ by an appropriate choice of natural numbers k_ν. The novelty of this construction is the use of nonvanishing *convergence-producing factors* (for historical details, see 1.6).

We study Weierstrass's construction in detail in Section 1 and discuss its applications in Section 2.

§1. The Weierstrass Product Theorem for $\mathbb{C}$

The goal of this section is the proof of the Weierstrass product theorem for the plane. To formulate it properly, we make use of the concept of divisors. In Subsection 1, with a view toward later generalizations, we define divisors for arbitrary regions D in $\mathbb{C}$. Theorem 2 describes the simple principle by which *Weierstrass products* are used to obtain holomorphic functions with prescribed divisors. In order to apply this theorem, we introduce the *Weierstrass factors* $E_n(z)$ in Subsection 3. They are used in Subsection 4 to construct the classical Weierstrass products for the case $D = \mathbb{C}$. Subsection 5 contains elementary but important consequences of the product theorem.

1. Divisors and principal divisors. A map $\mathfrak{d} : D \to \mathbb{Z}$ whose *support* $S := \{z \in D : \mathfrak{d} \neq 0\}$ is locally finite in D is called a *divisor on* D. Every function h meromorphic in D whose zero set $Z(h)$ and pole set $P(h)$ are discrete in D determines, by $z \mapsto o_z(h)$, a divisor (h) on D with support $Z(h) \cup P(h)$; such divisors are called *principal divisors on* D. The problem posed in the introduction to this chapter is now contained in the following problem:

Prove that every divisor is a principal divisor.

We begin by making a few general observations. Divisors $\mathfrak{d}$, $\widehat{\mathfrak{d}}$ (as maps into $\mathbb{Z}$) can be *added* in a natural way; the sum $\mathfrak{d} + \widehat{\mathfrak{d}}$ is again a divisor (why?). It follows easily:

The set $\mathrm{Div}(D)$ *of all the divisors on* D *is an abelian group, with addition as group operation.*

A divisor $\mathfrak{d}$ is called *positive* and written $\mathfrak{d} \geq 0$ if $\mathfrak{d}(z) \geq 0$ for all $z \in D$; for obvious reasons, positive divisors are also called distributions of zeros. Holomorphic functions f have positive divisors (f). The set $\mathcal{M}(D)^\times$ of all functions meromorphic in D that have discrete zero sets is a *multiplicative abelian* group; more precisely, $\mathcal{M}(D)^\times$ is the *group of units* of the ring

$\mathcal{M}(D)$. If $D = G$ is a domain, then $\mathcal{M}(G)$ is a field; thus $\mathcal{M}(G)^\times = \mathcal{M}(G)\backslash\{0\}$ (cf. I.10.3.3).

The following is immediate.

The map $\mathcal{M}(D)^\times \to \mathrm{Div}(D)$, $h \mapsto (h)$, *is a group homomorphism. Moreover,*

1) $f \in \mathcal{M}(D)^\times$ *is holomorphic in* $D \Leftrightarrow (f) \geq 0$;
2) $f \in \mathcal{M}(D)^\times$ *is a unit in* $\mathcal{O}(D) \Leftrightarrow (f) = 0$.

Every divisor $\mathfrak{d}$ is the difference of *two positive divisors:*

$\mathfrak{d} = \mathfrak{d}^+ - \mathfrak{d}^-$, where $\mathfrak{d}^+(z) := \max(0, \mathfrak{d}(z))$, $\mathfrak{d}^-(z) := \max(0, -\mathfrak{d}(z))$, $z \in D$.

It follows immediately from this that

$\mathfrak{d}$ *is a principal divisor on* D *if* $\mathfrak{d}^+$ *and* $\mathfrak{d}^-$ *are principal divisors on* D.

Proof. Let $\mathfrak{d}^+ = (f)$, $\mathfrak{d}^- = (g)$, with $f, g \in \mathcal{O}(D)$. Then, for $h := f/g \in \mathcal{M}(D)^\times$, we have $(h) = (f/g) = (f) - (g) = \mathfrak{d}^+ - \mathfrak{d}^- = \mathfrak{d}$. □

The problem stated above is thus reduced to the following:

For every positive divisor $\mathfrak{d}$ *on* D, *construct a function* $f \in \mathcal{O}(D)$ *with* $(f) = \mathfrak{d}$.

Such functions can be constructed with the aid of special products, which we now introduce.

2. Weierstrass products. Let $\mathfrak{d} \neq 0$ be a *positive* divisor on D. The support $T \neq \emptyset$ of $\mathfrak{d}$ is *at most countable* (since T is locally finite in D). From the points of $T\backslash\{0\}$ we form, *in some fashion*, a finite or infinite sequence $d_1, d_2, \ldots$ such that every point $d \in T\backslash\{0\}$ appears exactly $\mathfrak{d}(d)$ *times* in this sequence. We call $d_1, d_2, \ldots$ *a sequence corresponding to* $\mathfrak{d}$. A product

$$(*) \qquad f = z^{\mathfrak{d}(0)} \prod_{\nu \geq 1} f_\nu, \quad f_\nu \in \mathcal{O}(D),$$

is called a *Weierstrass product for the divisor* $\mathfrak{d} \geq 0$ *in* D if the following conditions hold.

1) f_ν *has no zeros in* $D\backslash\{d_\nu\}$ *and* $o_{d_\nu}(f_\nu) = 1$, $\nu \geq 1$.

2) *The product* $\prod_{\nu \geq 1} f_\nu$ *converges normally in* D.

This terminology will turn out to be especially convenient; the next result is immediate.

Proposition. *If* f *is a Weierstrass product for* $\mathfrak{d} \geq 0$, *then* $(f) = \mathfrak{d}$; *that is, the zero set of* $f \in \mathcal{O}(D)$ *is the support* T *of* $\mathfrak{d}$, *and every point* $d \in T$ *is a zero of* f *of order* $\mathfrak{d}(d)$.

Proof. By 2), $f \in \mathcal{O}(D)$. Every point $d \in T$, $d \neq 0$, occurs exactly $\mathfrak{d}(d)$ times in the sequence d_ν; hence 1) and Theorem 1.2.2 (applied to the connected components of D) imply that $o_z(f) = \mathfrak{d}(z)$ for all $z \in D$. Therefore $(f) = \mathfrak{d}$. □

The next statement follows immediately from the definition.

If $z^{\mathfrak{d}(0)} \prod f_\nu$ and $z^{\tilde{\mathfrak{d}}(0)} \prod \tilde{f}_\nu$ are Weierstrass products for $\mathfrak{d} \geq 0$ and $\tilde{\mathfrak{d}} \geq 0$, respectively, then $z^{\mathfrak{d}(0)+\tilde{\mathfrak{d}}(0)} \prod g_\nu$ is a Weierstrass product for $\mathfrak{d} + \tilde{\mathfrak{d}}$, where $g_{2\nu-1} := f_\nu$ and $g_{2\nu} := \tilde{f}_\nu$.

We will construct Weierstrass products for every positive divisor $\mathfrak{d}$. (This involves more than finding functions $f \in \mathcal{O}(D)$ with $(f) = \mathfrak{d}$.) In the construction, the "only" thing that matters is choosing the factors $f_\nu \in \mathcal{O}(D)$ in such a way that 1) and 2) hold. When $D = \mathbb{C}$, such factors can be specified explicitly.

3. Weierstrass factors. The entire functions

$$E_0(z) := 1 - z, \quad E_n(z) := (1-z)\exp\left(z + \frac{z^2}{2} + \frac{z^3}{3} + \cdots + \frac{z^n}{n}\right), \quad n \geq 1,$$

are called *Weierstrass factors.* We observe immediately that

(1) $E_n'(z) = -z^n \exp\left(z + \dfrac{z^2}{2} + \cdots + \dfrac{z^n}{n}\right)$ for $n \geq 1$;

(2) $E_n(z) = 1 + \sum\limits_{\nu > n} a_\nu z^\nu$, where $\sum\limits_{\nu>n} |a_\nu| = 1$, for $n \geq 0$.

Proof. Let $t_n(z) := z + z^2/2 + \cdots + z^n/n$; then $(1-z)t_n'(z) = 1 - z^n$.

ad (1): Write $E_n'(z) = -\exp t_n(z) + (1-z)t_n'(z)\exp t_n(z) = -z^n \exp t_n(z)$.

ad (2): Let $\sum a_\nu z^\nu$ be the Taylor series for E_n about 0. The case $n = 0$ is trivial. For $n \geq 1$, we have $\sum \nu a_\nu z^{\nu-1} = -z^n \exp t_n(z)$ by (1). Since the function on the right-hand side has an nth-order zero at 0 and all the Taylor coefficients of $\exp t_n(z)$ about 0 are positive, we see that

$$a_1 = \cdots = a_n = 0 \text{ and } a_\nu \leq 0; \quad \text{thus } |a_\nu| = -a_\nu \text{ for } \nu > n.$$

(2) follows because $a_0 = E_n(0) = 1$ and $0 = E_n(1) = 1 + \sum_{\nu>n} a_\nu$. □

From (2), we immediately obtain

(3) $|E_n(z) - 1| \leq |z|^{n+1}$, $n = 0, 1, 2, \ldots$, *for all* $z \in \mathbb{C}$ *with* $|z| \leq 1$.

A second proof of (3), *using only* (1). Since $|e^w| \leq e^{|w|}$, $w \in \mathbb{C}$, it follows immediately that

$$|E_n'(tz)| \leq -|z|^n E_n'(t) \quad \text{for all } (t, z) \in [0, \infty) \times \overline{\mathbb{E}}.$$

Since $f(z) - f(0) = z\int_0^1 f'(tz)dt$ for all $f \in \mathcal{O}(\mathbb{C})$ and all $z \in \mathbb{C}$,

$$|E_n(z) - 1| \leq |z| \int_0^1 |E_n'(tz)|dt \leq -|z|^{n+1} \int_0^1 E_n'(t)dt, \quad z \in \overline{\mathbb{E}}.$$

The integral on the right-hand side is equal to -1. □

In the next subsection, Weierstrass products will be formed from Weierstrass factors; the estimate (3) will be crucial for the proof of convergence.

Historical note. The sequence E_n appears in [W₁] (p. 94). From the equation

$$1 - z = \exp(\log(1-z)) = \exp\left(-\sum_{\nu\geq 1} \frac{z^\nu}{\nu}\right), \quad z \in \mathbb{E},$$

he obtains the formula $E_n(z) = \exp\left(-\sum_{\nu>n} z^\nu/\nu\right)$, $z \in \mathbb{E}$, which plays the role of the estimate (3) in his reasoning. — The first proof of (3) given above is attributed to L. Fejér; cf. [Hi], vol. 1, p. 227, as well as [F], vol. 2, pp. 849–850. But the argument appears as early as 1903, in a paper of L. Orlando; cf. [O].

4. The Weierstrass product theorem. In this subsection, $\mathfrak{d} \neq 0$ denotes a *positive divisor on* $\mathbb{C}$ and $(d_\nu)_{\nu\geq 1}$ a sequence corresponding to $\mathfrak{d}$.

Lemma. *If* $(k_\nu)_{\nu\geq 1}$ *is any sequence of natural numbers such that*

$$\text{(1)} \qquad \sum_1^\infty |r/d_\nu|^{k_\nu+1} < \infty \quad \textit{for every real } r > 0,$$

then $z^{\mathfrak{d}(0)}\prod_{\nu\geq 1} E_{k_\nu}(z/d_\nu)$ *is a Weierstrass product for* $\mathfrak{d}$.

Proof. We may assume that $\mathfrak{d}$ is not finite. By 3(3),

$$|E_{k_\nu}(z/d_\nu) - 1| \leq |r/d_\nu|^{k_\nu+1} \text{ for all } z \in B_r(0) \text{ and all } \nu \text{ with } |d_\nu| \geq r.$$

Since $\lim |d_\nu| = \infty$, for every $r > 0$ there exists an $n(r)$ such that $|d_\nu| \geq r$ for $\nu > n(r)$. Hence

$$\sum_{\nu>n(r)} |E_{k_\nu}(z/d_\nu) - 1|_{B_r(0)} \leq \sum_{\nu>n(r)} |r/d_\nu|^{k_\nu+1} < \infty \quad \text{for every } r > 0,$$

proving the normal convergence of the product. Since the factor $E_{k_\nu}(z/d_\nu) \in \mathcal{O}(\mathbb{C})$ has no zeros in $\mathbb{C}\backslash\{d_\nu\}$ and has a first-order zero at d_ν, we have a Weierstrass product for $\mathfrak{d}$. □

Product theorem. *For every divisor* $\mathfrak{d} \geq 0$ *on* $\mathbb{C}$, *there exist Weierstrass products, e.g.*

$$z^{\mathfrak{d}(0)} \prod_{\nu \geq 1} E_{\nu-1}(z/d_\nu)$$
$$= z^{\mathfrak{d}(0)} \prod_{\nu \geq 1} \left[\left(1 - \frac{z}{d_\nu}\right) \exp\left(\frac{z}{d_\nu} + \frac{1}{2}\left(\frac{z}{d_\nu}\right)^2 + \cdots + \frac{1}{\nu-1}\left(\frac{z}{d_\nu}\right)^{\nu-1} \right) \right].$$

Proof. Given $r > 0$, choose $m \in \mathbb{N}$ such that $|d_\nu| > 2r$ for $\nu > m$. It follows that $\sum_{\nu>m} |r/d_\nu|^\nu < \sum_{\nu>m} 2^{-\nu} < \infty$. Thus (1) holds for $k_\nu := \nu - 1$. □

The choice $k_\nu := \nu - 1$ is *not optimal.* It suffices, for example, just to require that $k_\nu > \alpha \log \nu$ with $\alpha > 1$: since $|d_\nu| > e \cdot r$ for all but finitely many ν, we have $|r/d_\nu|^{k_\nu+1} < \nu^{-\alpha}$, so that (1) holds.

5. Consequences. The product theorem 4 has important corollaries.

Existence theorem. *Every divisor on $\mathbb{C}$ is a principal divisor.*

Factorization theorem. *Every entire function $f \neq 0$ can be written in the form*

$$f(z) = e^{g(z)} z^m \prod_{\nu \geq 1} \left[\left(1 - \frac{z}{d_\nu}\right) \exp\left(\frac{z}{d_\nu} + \frac{1}{2}\left(\frac{z}{d_\nu}\right)^2 + \cdots + \frac{1}{k_\nu}\left(\frac{z}{d_\nu}\right)^{k_\nu} \right) \right],$$

where $g \in \mathcal{O}(\mathbb{C})$ and and $z^m \prod_{\nu \geq 1} \cdots$ is a (possibly empty) Weierstrass product for the divisor (f).

Only the factorization theorem needs justification. By the product theorem, there exists a Weierstrass product $\widehat{f}$ for the divisor (f). Then $f/\widehat{f}$ is a function without zeros, and thus of the form $\exp g$ with $g \in \mathcal{O}(\mathbb{C})$ (cf. I.9.3.2). □

The next result is a simple consequence of the existence theorem.

Theorem *(Quotient representation of meromorphic functions). For every function h meromorphic in $\mathbb{C}$, there exist two entire functions f and g, without common zeros in $\mathbb{C}$, such that $h = f/g$.*

Proof. Let $h \neq 0$. Positive divisors on $\mathbb{C}$ with *disjoint* supports are defined by $\mathfrak{d}^+(z) := \max\{0, o_z(h)\}$ and $\mathfrak{d}^-(z) := \max\{0, -o_z(h)\}$; they satisfy $(h) = \mathfrak{d}^+ - \mathfrak{d}^-$. Let $g \in \mathcal{O}(\mathbb{C})$ be chosen with $(g) = \mathfrak{d}^-$. Then $g \neq 0$. For $f := gh$, it follows that $(f) = (g) + (h) = \mathfrak{d}^+ \geq 0$, whence f is holomorphic in $\mathbb{C}$. By construction, $Z(f) \cap Z(g)$ is empty. □

In particular, we have proved the following:

The field $\mathcal{M}(\mathbb{C})$ of functions meromorphic in $\mathbb{C}$ is the quotient field of the integral domain $\mathcal{O}(\mathbb{C})$ of functions holomorphic in $\mathbb{C}$.

The theorem contains more than this last statement: for an arbitrary quotient f/g, the numerator and denominator may have infinitely many common zeros; without the existence theorem, it is not clear that these zeros *all* cancel out.

We conclude by noting a

Root criterion. *The following statements about an entire function $f \neq 0$ and a natural number $n \geq 1$ are equivalent:*

i) *There exists a holomorphic* n*th root of f; that is, there exists a $g \in \mathcal{O}(\mathbb{C})$ with $g^n = f$.*
ii) *Every natural number $o_z(f)$, $z \in \mathbb{C}$, is divisible by n.*

Proof. Only the implication ii) $\Rightarrow$ i) must be proved. By hypothesis, there exists a positive divisor $\mathfrak{d}$ on $\mathbb{C}$ with $n\mathfrak{d} = (f)$. Let $\widehat{g} \in \mathcal{O}(\mathbb{C})$ be chosen such that $(\widehat{g}) = \mathfrak{d}$. Then $\widehat{u} := f/\widehat{g}^n$ is holomorphic and nonvanishing in $\mathbb{C}$; hence there exists $u \in \mathcal{O}(\mathbb{C})$ with $\widehat{u} = u^n$ (existence theorem for holomorphic roots; cf. I.9.3.3). The function $g := u\widehat{g}$ is an nth root of f. □

The existence theorem allows us to prescribe the location and order of the poles of meromorphic functions. We will see in Chapter 6 that, in doing so, we can also arbitrarily prescribe *all principal parts*. But the following is immediate from the product theorem 4, by logarithmic differentiation of Weierstrass products.

(1) *Let* $0, d_1, d_2, \ldots$ *be a sequence of pairwise distinct points in $\mathbb{C}$ that have no accumulation point in $\mathbb{C}$. Then the function*

$$\frac{1}{z} + \sum_{\nu \geq 1} \left(\frac{1}{z - d_\nu} + \frac{1}{d_\nu} + \frac{z}{d_\nu^2} + \cdots + \frac{z^{\nu-2}}{d_\nu^{\nu-1}} \right)$$

is meromorphic in $\mathbb{C}$ and holomorphic in $\mathbb{C}\backslash\{0, d_1, d_2, \ldots\}$; it has principal part $(z - d_\nu)^{-1}$ at d_ν, $\nu \geq 1$.

6. On the history of the product theorem. Weierstrass developed his theory in 1876 ([W$_1$], pp. 77–124). His main objective was to establish the "general expression" for all functions meromorphic in $\mathbb{C}$ except at finitely many points. "[Dazu] hatte ich jedoch ... zuvor eine in der Theorie der transcendenten ganzen Functionen bestehende ... Lücke auszufüllen, was mir erst nach manchen vergeblichen Versuchen vor nicht langer Zeit in befriedigender Weise gelungen ist." (To do this, however, I ... first needed to fill in a gap ... in the theory of transcendental entire functions, which, after a number of futile attempts, I succeeded only recently in doing in a satisfactory way.) ([W$_1$], p. 85) The gap he mentioned was closed by the product theorem ([W$_1$], pp. 92–97). What was new and, for his contemporaries, sensational in Weierstrass's construction was the application of *convergence-producing factors* that have no influence on the behavior of the zeros. Incidentally, according to Weierstrass ([W$_1$], p. 91), the idea of forcing convergence by adjoining exponential factors came to him by way of the product formula

$$1/\Gamma(z) = z\prod_{\nu\geq 1}\left\{\left(1+\frac{z}{\nu}\right)\left(\frac{\nu+1}{\nu}\right)^{-z}\right\} = z\prod_{\nu\geq 1}\left\{\left(1+\frac{z}{\nu}\right)e^{-z\log\left(\frac{\nu+1}{\nu}\right)}\right\},$$

which he attributes to Gauss rather than Euler; cf. 2.2.2. In 1898 H. Poincaré, in his obituary for Weierstrass, assessed the discovery of the factors $E_n(z)$ as follows ([P$_2$], p. 8): "La principale contribution de Weierstraß aux progrès de la théorie des fonctions est la découverte des facteurs primaires." (Weierstraß's major contribution to the development of function theory is the discovery of primary factors.) Special cases of the product theorem had already appeared in the literature before 1876, for example in the work of E. Betti (cf. 2.1).

The awareness that there exist entire functions with "arbitrarily" prescribed zeros revolutionized the thinking of function theorists. Suddenly one could "construct" holomorphic functions that were not even hinted at in the classical arsenal. Of course, this freedom does not contradict the *solidarity of value behavior* of holomorphic functions required by the identity theorem: the "analytic cement" turns out to be pliable enough to globally bind locally prescribed data in an analytic way.

From his product theorem, Weierstrass immediately deduced the theorem on quotient representation of meromorphic functions ([W$_1$], p. 102). He attracted attention by this alone. No less a figure than H. Poincaré seized this observation of the "célèbre géomètre de Berlin" and carried it over to meromorphic functions of two variables [P$_1$]. With his theorem on the representability of every function meromorphic in $\mathbb{C}^2$ as the quotient $f(w,z)/g(w,z)$ of two entire functions in $\mathbb{C}^2$ (locally relatively prime everywhere), Poincaré initiated a theory that, through the work of P. Cousin, H. Cartan, K. Oka, J-P. Serre, and H. Grauert, is still alive today; see the glimpses in 4.2.5, 5.2.6, and 6.2.5.

§2. Discussion of the Product Theorem

When we apply the product lemma 1.4, we will choose the numbers k_ν as small as possible, in accordance with the idea that *the smaller k_ν is, the simpler the factor $E_{k_\nu}(z/d_\nu)$*. Situations in which all the k_ν can be chosen to be equal are especially nice; they lead to the concept of the *canonical* product (Subsection 1). In Subsection 2 we show that not only the Euler products of Chapter 1.4 but also the sine product and the product $H(z)$, so important for the theory of the gamma function, are canonical Weierstrass products.

In Subsections 3 and 4 we discuss the σ-product and the $\wp$-function. We prove that $\sigma(z;\omega_1,\omega_2)$ and $\wp(z;\omega_1,\omega_2)$ are holomorphic and meromorphic, respectively, in *all three variables.* Since the time of Eisenstein and Weierstrass, these functions have been central to the theory of *elliptic* functions. Subsection 5 contains an amusing observation of Hurwitz.

1. Canonical products. Let $\mathfrak{d}$ again denote a positive divisor on $\mathbb{C}$ and $d_1, d_2, \ldots$ a corresponding sequence. We first make a few observations.

(1) *If $f(z) = \prod(1 - z/d_\nu)e^{p_\nu(z)}$ converges normally in $\mathbb{C}$ and every function p_ν is a polynomial of degree $\leq k$, then $\sum |1/d_\nu|^{k+1}$ converges.*

Proof. Differentiating $f'(z)/f(z) = \sum [1/(z-d_\nu) + p'_\nu(z)]$ k times yields the series $\sum (-1)^k k!/(z-d_\nu)^{k+1}$, which converges absolutely at $0 \in \mathbb{C}$. □

We now ask when, for a given $\mathfrak{d}$, there exist Weierstrass products of the particularly simple form $z^{\mathfrak{d}(0)} \prod_{\nu \geq 1} E_k(z/d_\nu)$ with fixed $k \in \mathbb{N}$.

(2) *$z^{\mathfrak{d}(0)} \prod_{\nu\geq 1} E_k(z/d_\nu)$ is a Weierstrass product for the divisor $\mathfrak{d}$ if and only if $\sum |1/d_\nu|^{k+1} < \infty$.*

Proof. If the product in question is a Weierstrass product for $\mathfrak{d}$, then $\sum |1/d_\nu|^{k+1} < \infty$ by (1), since $E_k(z/d) = (1 - z/d)e^{p(z)}$ with a polynomial of degree k. Conversely, if $\sum |1/d_\nu|^{k+1} < \infty$, then the product is a Weierstrass product for $\mathfrak{d}$ by Lemma 1.4. □

If there exist Weierstrass products for $\mathfrak{d}$ as in (2), we can choose k to be *minimal;* in this case $z^{\mathfrak{d}(0)} \prod_{\nu\geq 1} E_k(z/d_\nu)$ is called *the canonical Weierstrass product for $\mathfrak{d}$.*

The following is clear by (2).

Proposition. *$z^{\mathfrak{d}(0)} \prod_{\nu\geq 1} E_k(z/d_\nu)$ is the canonical product for $\mathfrak{d}$ if and only if*

$$\sum |1/d_\nu|^k = \infty \quad \textit{and} \quad \sum |1/d_\nu|^{k+1} < \infty.$$

Examples of canonical products are given in the next two subsections. Such products depend only on the divisor $\mathfrak{d}$; the incidental choice of the sequence d_ν — in contrast to the general situation — plays no role. If the sequence d_ν grows too slowly, there is no canonical product: there is none, for example, if $\log(1+\nu)$ is a subsequence of the sequence d_ν. (Prove this.) It is thus easy to see that the function $1 - \exp(\exp z)$ has *no* canonical product. — We also note, without proof:

(3) *If $m > 0$ is such that $|d_\mu - d_\nu| \geq m$ for all $\mu \neq \nu$, then $\sum |1/d_\nu|^\alpha < \infty$ for $\alpha > 2$. In this case, there exists a canonical product for $\mathfrak{d}$ with $k \leq 2$.*

Historical note. E. Betti, in 1859–60, proved (3) in order to write elliptic functions as quotients of theta series; cf. the article by P. Ullrich ([U], p. 166).

2. Three classical canonical products. 1) The product

$$\prod_{\nu\geq 1}(1+q^\nu z) = \prod_{\nu\geq 1} E_0(-q^\nu z), \quad \text{where} \quad 0<|q|<1,$$

discussed in 1.4.3, is the *canonical* product for the divisor on $\mathbb{C}$ given by

$$\mathfrak{d}(-q^{-\nu}) := 1 \text{ for } \nu = 1,2,\ldots; \quad \mathfrak{d} := 0 \text{ otherwise.}$$

(Proposition 1 holds with $k := 0$.)

2) The function

$$H(z) = e^{-\gamma z}/\Gamma(z) = z\prod_{\nu\geq 1}(1+\frac{z}{\nu})e^{-z/\nu} = z\prod_{\nu\geq 1}E_1(-\frac{z}{\nu}),$$

considered in 2.1.1, is the *canonical* product for the divisor on $\mathbb{C}$ defined by

$$\mathfrak{d}(-\nu) := 1 \text{ for } \nu\in\mathbb{N}, \quad \mathfrak{d}(z) := 0 \text{ otherwise.}$$

(Proposition 1 holds with $k := 1$ but not with $k := 0$.)

3) The sine product

$$\begin{aligned} z\prod_{\nu\geq 1}(1-\frac{z^2}{\nu^2}) &= z\prod_{\nu\geq 1}[(1-\frac{z}{\nu})e^{z/\nu}(1+\frac{z}{\nu})e^{-z/\nu}] \\ &= z\prod_{\nu\geq 1}E_1(\frac{z}{\nu})E_1(-\frac{z}{\nu}) \end{aligned}$$

is the *canonical* product for the divisor on $\mathbb{C}$ defined by

$$\mathfrak{d}(\nu) := 1 \text{ for } \nu\in\mathbb{Z}; \quad \mathfrak{d}(z) = 0 \text{ otherwise.}$$

(Proposition 1 holds with $k := 1$ but not with $k := 0$; a corresponding sequence d_ν is $1, -1, 2, -2, \ldots$.)

In lectures and textbooks, these examples are sometimes given as examples of applications of the Weierstrass product theorem. *This is misleading. These products were known long before Weierstrass.* Of course, his theorem shows that the same construction principle underlies them all.

Exercises. Determine the canonical product for $\mathfrak{d}\geq 0$ in $\mathbb{C}$ corresponding to each of the following sequences:

a) $d_\nu := (-1)^\nu\sqrt[3]{\nu}$, $\nu\geq 1$;
b) $d_\nu := \mu i^\nu$, where $\mu\in\mathbb{N}$ with $4\mu-3\leq\nu\leq 4\mu$ for $\nu\geq 1$.

3. The σ-function. If $\omega_1, \omega_2 \in \mathbb{C}$ are *linearly independent over* $\mathbb{R}$, the set

$$\Omega := \mathbb{Z}\omega_1 + \mathbb{Z}\omega_2 = \{\omega = m\omega_1 + n\omega_2 : m, n \in \mathbb{Z}\}$$

is called a *lattice in* $\mathbb{C}$. Ω is locally finite in $\mathbb{C}$ and

$$\delta : \mathbb{C} \to \mathbb{N}, \quad z \mapsto \delta(z) := \begin{cases} 1 & \text{if } z \in \Omega, \\ 0 & \text{if } z \notin \Omega, \end{cases}$$

is a positive divisor on $\mathbb{C}$ with support Ω.

Proposition. *The entire function*

$$(1) \quad \sigma(z) := \sigma(z, \Omega) := z \prod_{0 \neq \omega \in \Omega} \left(1 - \frac{z}{\omega}\right) e^{\frac{z}{\omega} + \frac{1}{2}\left(\frac{z}{\omega}\right)^2} = z \prod_{0 \neq \omega \in \Omega} E_2\left(\frac{z}{\omega}\right)$$

is the canonical Weierstrass product for the lattice divisor δ.

The proposition is contained in Betti's result 1(3). We give a direct proof, which even yields the normal convergence of the σ-product (1) in all three variables z, ω_1, and ω_2. The set $U := \{(u, v) \in \mathbb{C}^2 : u/v \in \mathbb{H}\}$ is a domain in $\mathbb{C}^2$. For every point $(\omega_1, \omega_2) \in U$, the set $\Omega(\omega_1, \omega_2) := \mathbb{Z}\omega_1 + \mathbb{Z}\omega_2$ is a lattice in $\mathbb{C}$; conversely, *every* lattice $\Omega \subset \mathbb{C}$ has a basis in U. The following lemma is now crucial.

Convergence lemma. *Let $K \subset U$ be compact and let $\alpha > 2$. Then there exists a bound $M > 0$ such that*

$$\sum_{0 \neq \omega \in \Omega(\omega_1, \omega_2)} |\omega|^{-\alpha} \leq M \text{ for all } (\omega_1, \omega_2) \in K; \qquad \sum_{0 \neq \omega \in \Omega(\omega_1, \omega_2)} |\omega|^{-2} = \infty.$$

Proof. The function

$$q : (\mathbb{R}^2 \backslash \{(0,0)\}) \times U \to \mathbb{R}, \quad (x, y, \omega_1, \omega_2) \mapsto |x\omega_1 + y\omega_2| / \sqrt{x^2 + y^2}$$

is homogeneous in x, y; hence $q(\mathbb{R}^2 \backslash \{(0,0)\} \times U) = q(S^1 \times U)$. Since q is continuous, it has a maximum T and a minimum t on the compact set $S^1 \times K$. The $\mathbb{R}$-linear independence of ω_1, ω_2 implies that q is always positive; hence $t > 0$. Since

$$t\sqrt{m^2 + n^2} \leq |m\omega_1 + n\omega_2| \leq T\sqrt{m^2 + n^2}$$

for all $(\omega_1, \omega_2) \in K$ and all $(m, n) \in \mathbb{Z}^2$, the convergence of $\sum |\omega|^{-\alpha}$ is equivalent to the convergence of

$$\sum_{0 \neq (m,n) \in \mathbb{Z}^2} (m^2 + n^2)^{-\beta} = 4 \sum_{m=1}^{\infty} \frac{1}{m^\alpha} + 4 \sum_{m,n=1}^{\infty} \frac{1}{(m^2 + n^2)^\beta}, \text{ where } \beta := \tfrac{1}{2}\alpha.$$

Since $m^2 + n^2 \geq 2mn > mn > 0$ for all $m, n \geq 1$, it follows for $\alpha > 2$ that

$$\sum_{m,n=1}^{\infty} \frac{1}{(m^2+n^2)^\beta} < \sum_{m,n=1}^{\infty} \frac{1}{m^\beta n^\beta} = \left(\sum_{n=1}^{\infty} \frac{1}{n^\beta}\right)\left(\sum_{m=1}^{\infty} \frac{1}{m^\beta}\right) < \infty.$$ [1]

Divergence follows for $\beta := 1$, since the inequality $m^2 + n^2 \leq 2n^2$ for $1 \leq m \leq n$ implies that

$$\sum_{m,n=1}^{\infty} \frac{1}{m^2+n^2} > \sum_{n=1}^{\infty}\sum_{m=1}^{n} \frac{1}{m^2+n^2} \geq \frac{1}{2}\sum_{n=1}^{\infty}\sum_{m=1}^{n} \frac{1}{n^2} = \frac{1}{2}\sum_{n=1}^{\infty} \frac{1}{n} = \infty. \quad \square$$

Since $|E_2(z/\omega) - 1| < |z/\omega|^3$ for $|z| < |\omega|$, the lemma immediately yields not only the proposition but also:

(2) *In* $\mathbb{C} \times U$, *the* σ*-product* $\sigma(z; \omega_1, \omega_2) := \sigma(z, \Omega(z_1, z_2))$ *converges normally to a function holomorphic in* z, ω_1, *and* ω_2.

Historical remark. The trick of trivializing the proof by means of the inequality $m^2 + n^2 > mn$ is due to Weierstrass; he "dictated it to Herr F. Mertens in 1863" ([W$_2$], Foreword and p. 117). The *arithmetic-geometric inequality* $n_1^\beta + \cdots + n_d^\beta \geq d(n_1 \cdot \ldots \cdot n_d)^{\beta/d}$ even gives

$$\sum_{(n_1,\ldots,n_d)\neq 0} \frac{1}{(n_1^\beta + n_2^\beta + \cdots + n_d^\beta)^\alpha} < \infty$$

if $d \in \mathbb{N}\backslash\{0\}$, $\alpha > 0$, $\beta > 0$, $\alpha\beta > d$. Such series (with $\beta = 2$) were considered by Eisenstein in 1847 (*Werke*, pp. 361–363).

A *variant of the proof* was given in 1958 by H. Kneser ([Kn], pp. 201–202). He replaces q by the function $|x\omega_1 + y\omega_2| / \max(|x|, |y|)$. As above, there exist numbers $S \geq s > 0$ such that $s \leq |m\omega_1 + n\omega_2| / \max(|m|, |n|) \leq S$. The convergence of $\sum |\omega|^{-\alpha}$ is now equivalent to that of

$$\sum_{0 \neq (m,n) \in \mathbb{Z}^2} [\max(|m|, |n|)]^{-\alpha} = 4\sum_{m=1}^{\infty} \frac{1}{m^\alpha} + 4\sum_{m,n=1}^{\infty} [\max(m,n)]^{-\alpha}.$$

But the series on the right-hand side can be written as follows (!):

$$\sum_{n=1}^{\infty}\left(nn^{-\alpha} + \sum_{m=n+1}^{\infty} m^{-\alpha}\right) = \sum_{n=1}^{\infty} n^{1-\alpha} + \sum_{k=1}^{\infty}(k-1)k^{-\alpha} = \sum_{n=1}^{\infty}(2n^{1-\alpha} - n^{-\alpha});$$

this converges for $\alpha > 2$ and diverges for $\alpha = 2$.

---

[1] For $\gamma > 0$, noting that $\frac{\gamma}{n-1} \leq \left(\frac{n}{n-1}\right)^\gamma - 1$, we have

$$\sum_{2}^{\infty} \frac{\gamma}{n^{1+\gamma}} < \sum_{2}^{\gamma} \frac{\gamma}{(n-1)n^\gamma} \leq \sum_{2}^{\infty}\left(\frac{1}{(n-1)^\gamma} - \frac{1}{n^\gamma}\right) = 1.$$

4. The $\wp$-function. Since the product $\sigma(z;\omega_1,\omega_2) \in \mathcal{O}(\mathbb{C}\times U)$ converges normally by 3(2), it can be differentiated logarithmically with respect to z (Theorem 1.2.3):

$$(1)\qquad \begin{aligned}\zeta(z;\omega_1,\omega_2) &:= \frac{\sigma'(z;\omega_1,\omega_2)}{\sigma(z;\omega_1,\omega_2)}\\ &= \frac{1}{z} + \sum_{0\neq\omega\in\Omega(\omega_1,\omega_2)} \left(\frac{1}{z-\omega} + \frac{1}{\omega} + \frac{z}{\omega^2}\right) \in \mathcal{M}(\mathbb{C}\times U).\end{aligned}$$

This series (of meromorphic functions), which converges normally in $\mathbb{C}\times U$, is called the Eisenstein-Weierstrass ζ-function. Ordinary differentiation of (1) gives

$$(2)\qquad \begin{aligned}\wp(z;\omega_1,\omega_2) &:= -\zeta'(z;\omega_1,\omega_2)\\ &= \frac{1}{z^2} + \sum_{0\neq\omega\in\Omega(\omega_1,\omega_2)} \left(\frac{1}{(z-\omega)^2} - \frac{1}{\omega^2}\right) \in \mathcal{M}(\mathbb{C}\times U).\end{aligned}$$

This series also converges normally in $\mathbb{C}\times U$. Both the ζ-function and the $\wp$-function are holomorphic in $\mathbb{C}\backslash\Omega(\omega_1,\omega_2)$ for fixed ω_1, ω_2 and have poles of *first* and *second* order, respectively, at each lattice point. The $\wp$-function is *doubly periodic* (= *elliptic*), with $\Omega(\omega_1,\omega_2)$ as *period lattice.* In the theory of elliptic functions, it is fundamental that the $\wp$-function is *meromorphic in all three variables* z, ω_1, and ω_2; this is often not sufficiently emphasized in the literature.

In the case $\omega_2 := \infty$, the functions σ, ζ, and $\wp$ become *trigonometric* functions: Writing ω for $\omega_1 \in \mathbb{C}^\times$, we have

$$\sigma(z;\omega,\infty) := \frac{\omega}{\pi}e^{\frac{\pi^2}{6}\left(\frac{z}{\omega}\right)^2}\sin\pi\frac{z}{\omega}, \quad \zeta(z;\omega,\infty) := \frac{\pi^2}{3}\left(\frac{z}{\omega}\right)^2 + \frac{\pi}{\omega}\cot\pi\frac{z}{\omega}$$

$$\wp(z;\omega,\infty) := -\frac{1}{3}\left(\frac{\pi}{\omega}\right)^2 + \left(\frac{\pi}{\omega}\right)^2\left(\sin\left(\pi\frac{z}{\omega}\right)\right)^{-2}$$

$$\sigma(z;\infty,\infty) := z, \quad \zeta(z;\infty,\infty) := \frac{1}{z}, \quad \wp(z;\infty,\infty) := \frac{1}{z^2}.$$

Here we continue to use the notation $\zeta = \sigma'/\sigma$ and $\wp = -\zeta'$. With some effort, it can be shown that $\lim_{\omega_2\to\infty}\sigma(z;\omega_1,\omega_2) = \sigma(z;\omega_1,\infty)$, where the convergence is *compact;* the same holds for ζ and $\wp$. Thus the theory of elliptic functions contains the theory of trigonometric functions as a degenerate case.

5*. An observation of Hurwitz. *Every positive divisor $\mathfrak{d}$ on $\mathbb{C}$ is the divisor of an entire function $\sum a_\nu z^\nu$ whose coefficients all lie in the field $\mathbb{Q}(i)$ of rational complex numbers. In particular, if $\mathfrak{d}(\overline{z}) = \mathfrak{d}(z)$ for all $z\in\mathbb{C}$, then all the numbers a_ν can be chosen to lie in $\mathbb{Q}$.*

The following lemma is necessary for the proof.

Lemma. *Let f be holomorphic at $0 \in \mathbb{C}$. Then there exists an entire function g such that all the coefficients a_ν of the Taylor series of $f \exp g$ about 0 belong to $\mathbb{Q}(i)$. In particular, if all the coefficients of the Taylor series of f about 0 are real, then g can be chosen in such a way that all the a_ν lie in $\mathbb{Q}$.*

Proof. Let $f \neq 0$. Then $f(z) = z^s e^{h(z)}$, $s \in \mathbb{N}$, where $h(z) = b_0 + b_1 + \cdots + b_n z^n + \cdots$ is holomorphic in a neighborhood of 0. (Write $f(z) = z^s \widetilde{f}(z)$, where $\widetilde{f}$ is holomorphic and nonvanishing in a neighborhood of 0; then $\widetilde{f}$ can be put in the form e^h.) Since the field $\mathbb{Q}(i)$ is dense in $\mathbb{C}$, there exist numbers $q_1, q_2, \ldots \in \mathbb{Q}(i)$ such that $g(z) := -b_0 + \sum_{\nu \geq 1}(q_\nu - b_\nu)z^\nu$ is an entire function. We have

$$f(z)e^{g(z)} = z^s e^{q_1 z + q_2 z^2 + \cdots} = z^s \left[1 + \sum_{\nu \geq 1} \frac{1}{\nu!}(q_1 z + q_2 z^2 + \cdots)^\nu\right].$$

Expanding the right-hand side in powers of z gives Taylor coefficients a_ν that indeed lie in $\mathbb{Q}(i)$, since each a_ν is a polynomial with rational coefficients in finitely many of the $q_1, q_2, \ldots \in \mathbb{Q}(i)$. If the power series of f about 0 has only real coefficients, then all the b_ν with $\nu \geq 1$ are real. In this case, one can always choose $q_\nu \in \mathbb{Q}$ and hence $a_\nu \in \mathbb{Q}$. □

At this point, Hurwitz's observation is quickly proved. We choose $f \in \mathcal{O}(\mathbb{C})$ with $(f) = \mathfrak{d}$. Then $\mathfrak{d}$ is also the divisor of every function $q := f \exp g$, $g \in \mathcal{O}(\mathbb{C})$. By the lemma, g can be chosen in such a way that all the Taylor coefficients a_ν of q belong to $\mathbb{Q}(i)$.

If it always holds that $\mathfrak{d}(\overline{z}) = \mathfrak{d}(z)$, then $\mathfrak{d}$ is also the divisor of the entire function $\widetilde{q}$ whose Taylor coefficients are the numbers $\overline{a}_\nu$. Then $2\mathfrak{d}$ is the divisor of $q\widetilde{q}$; by the root criterion, there exists $\widehat{q} \in \mathcal{O}(\mathbb{C})$ with $\widehat{q}^2 = q\widetilde{q}$. Moreover, $(\widehat{q}) = \mathfrak{d}$. Since all the Taylor coefficients of $q\widetilde{q}$ are rational real numbers and the first nonzero coefficient is positive, all the Taylor coefficients of $\widehat{q}$ are rational real numbers. □

Hurwitz proved the preceding assertion in 1889. As an amusing corollary, he also noted the following:

Every (real or complex) number a (thus, for instance, e or π) is the root of an equation $0 = r_0 + r_1 z + r_2 z^2 + \cdots$ whose right-hand side is an entire function with rational coefficients (real or complex, respectively), which has no roots other than a.

Bibliography

[F] FEJÉR, L.: Über die Weierstrassche Primfunktion, *Ges. Arb.* 2, 849–850.

[Hi] HILLE, E.: *Analytic Function Theory*, 2 vols., Ginn and Company, 1959 and 1962.

[Hu] HURWITZ, A.: Über beständig convergierende Potenzreihen mit rationalen Zahlencoeffizienten und vorgeschriebenen Nullstellen, *Acta Math.* 14, 211–215 (1890–91); *Math. Werke* 1, 310–313.

[Kn] KNESER, H.: *Funktionentheorie*, Vandenhoeck & Ruprecht 1958.

[O] ORLANDO, L.: Sullo sviluppo della funzione $(1-z)e^{z+\frac{1}{2}z^2+\cdots+\frac{z^{p-1}}{p-1}}$, *Giornale Matem. Battaglini* 41, 377–378 (1903).

[P₁] POINCARÉ, H.: Sur les fonctions de deux variables, *Acta Math.* 2, 97–113 (1883); *Œuvres* 4, 147–161.

[P₂] POINCARÉ, H.: L'œuvre mathématique de Weierstraß, *Acta Math.* 22, 1–18 (1898); not in Poincaré's *Œuvres.*

[U] ULLRICH, P.: Weierstraß' Vorlesung zur "Einleitung in die Theorie der analytischen Funktionen," *Arch. Hist. Ex. Sci.* 40, 143–172 (1989).

[W₁] WEIERSTRASS, K.: Zur Theorie der eindeutigen analytischen Functionen, *Math. Werke* 2, 77–124.

[W₂] WEIERSTRASS, K.: Vorlesungen über die Theorie der elliptischen Funktionen, adapted by J. KNOBLAUCH, *Math. Werke* 5.

4*
Holomorphic Functions with Prescribed Zeros

We extend the results obtained in Chapter 3 for *entire* functions to functions holomorphic in *arbitrary* regions D in $\mathbb{C}$. Our goal is to prove that *every* divisor on D is a principal divisor (existence theorem 1.5). For this purpose we first construct, in Section 1, Weierstrass products for every *positive* divisor. As before, they are built up from Weierstrass factors E_n and converge normally in regions that contain $\mathbb{C}\backslash\partial D$ (product theorem 1.3). In Section 2 we develop, among other things, the theory of the greatest common divisor for integral domains $\mathcal{O}(G)$.

Blaschke products are a special class of Weierstrass products in $\mathbb{E}$; they are studied in Section 3 and serve in the construction of *bounded* functions in $\mathcal{O}(\mathbb{E})$ for prescribed positive divisors. In an appendix to Section 3 we prove Jensen's formula.

§1. The Product Theorem for Arbitrary Regions

A convergence lemma is proved in Subsection 1. In Subsection 2, Weierstrass products are constructed for some special divisors; the factors $E_n(z/d)$ are now replaced by factors of the form

$$E_n\left(\frac{d-c}{z-c}\right) = \left(\frac{z-d}{z-c}\right)\cdot\exp\left[\frac{d-c}{z-c}+\frac{1}{2}\left(\frac{d-c}{z-c}\right)^2+\cdots+\frac{1}{n}\left(\frac{d-c}{z-c}\right)^n\right],$$

$c \neq d$, which also vanish to first order at the point d. The general product theorem is derived in Subsection 3.

1. Convergence lemma. Let $\mathfrak{d}$ be a positive divisor on D with support T. From the points of the *countable* set T, we somehow construct a sequence (d_ν) in which *every* point $d \in T$ occurs exactly $\mathfrak{d}(d)$ times. (In contrast to our earlier discussion — in 3.1.2 — the origin, if it lies in T, is not excluded from the sequence.) The following is a substitute for Lemma 3.1.4.

Lemma. *Let $(c_\nu)_{\nu\geq 1}$ be a sequence in $\mathbb{C}\backslash D$ and $(k_\nu)_{\nu\geq 1}$ a sequence of natural numbers such that*

$$(1) \qquad \sum_{\nu=1}^{\infty} |r(d_\nu - c_\nu)|^{k_\nu+1} < \infty \quad \textit{for all } r > 0.$$

Then the product

$$\prod_{\nu\geq 1} E_{k_\nu}\left(\frac{d_\nu - c_\nu}{z - c_\nu}\right) = \prod_{\nu\geq 1}\left(\frac{z-d_\nu}{z-c_\nu}\right)\cdot \exp\left[\left(\frac{d_\nu - c_\nu}{z - c_\nu}\right) + \frac{1}{2}\left(\frac{d_\nu - c_\nu}{z - c_\nu}\right)^2 + \cdots + \frac{1}{k_\nu}\left(\frac{d_\nu - c_\nu}{z - c_\nu}\right)^{k_\nu}\right]$$

converges normally in $\mathbb{C}\backslash\overline{\{c_1, c_2, \ldots\}} \supset D$; it is a Weierstrass product in D for the divisor $\mathfrak{d}$.

Proof. We set $S := \overline{\{c_1, c_2, \ldots\}}$. For $f_\nu(z) := E_{k_\nu}[(d_\nu - c_\nu)/(z - c_\nu)]$, we have

$$(*) \qquad f_\nu \in \mathcal{O}(\mathbb{C}\backslash S), \quad f_\nu(z) \neq 0 \text{ if } z \neq d_\nu, \quad \text{and} \quad o_{d_\nu}(f_\nu) = 1.$$

Let K be a compact set in the region $\mathbb{C}\backslash S$. For all $z \in K$, $|z - c_\nu| \geq d(K, c_\nu) \geq d(K, S) > 0$; hence $|(d_\nu - c_\nu)/(z - c_\nu)|_K \leq r|d_\nu - c_\nu|$, where $r := d(K,S)^{-1}$. Since $\lim |d_\nu - c_\nu| = 0$ by (1), there exists $n(K) \in \mathbb{N}$ such that $r|d_\nu - c_\nu| < 1$ for $\nu > n(K)$. Since $|E_n(w) - 1| \leq |w|^{n+1}$ for $w \in \mathbb{E}$ by 3.1.3(3), it follows that

$$\sum_{\nu > n(K)} |f_\nu - 1|_K \leq \sum_{\nu > n(K)} |r(d_\nu - c_\nu)|^{k_\nu+1} < \infty.$$

This proves the normal convergence of $\prod f_\nu$ in $\mathbb{C}\backslash S$. By $(*)$, this product is a Weierstrass product for $\mathfrak{d}$ in D. □

Corollary to the lemma. *If $\sum |d_\nu - c_\nu|^{k+1} < \infty$ for some $k \in \mathbb{N}$, the product $\prod_{\nu\geq 1} E_k[(d_\nu - c_\nu)/(z - c_\nu)]$ is a Weierstrass product for $\mathfrak{d}$ in D.*

Proof. (1) holds with $k_\nu := k$. □

2. The product theorem for special divisors. In general, the region of convergence of the product constructed in Lemma 1 is larger than D. As the zero set of the product, T is closed in this larger region. We make a general observation, leaving the proof to the reader:

(1) *If T is a discrete set in $\mathbb{C}$, then the set $T' := \overline{T}\backslash T$ of all the accumulation points*[1] *of T in $\mathbb{C}$ is closed in $\mathbb{C}$. The region $\mathbb{C}\backslash T'$ is the largest subset of $\mathbb{C}$ in which T is closed.*

By (1), every positive divisor $\mathfrak{d}$ on D with support T can be viewed as a positive divisor on $\mathbb{C}\backslash T' \supset D$ with the same support (set $\mathfrak{d}(z) := 0$ for $z \in (\mathbb{C}\backslash T')\backslash D$). Clearly $T' \supset \partial D$. The next theorem now follows quickly from Lemma 1.

Product theorem. *Let $\mathfrak{d}$ be a positive divisor on D with corresponding sequence $(d_\nu)_{\nu\geq 1}$. Let a sequence $(c_\nu)_{\nu\geq 1}$ in T' be given such that $\lim |d_\nu - c_\nu| = 0$. Then the product $\prod E_{\nu-1}[(d_\nu - c_\nu)/(z - c_\nu)]$ is a Weierstrass product for $\mathfrak{d}$ in $\mathbb{C}\backslash T'$.*

Proof. Since $\lim |d_\nu - c_\nu| = 0$, it follows that $\sum |r(d_\nu - c_\nu)|^\nu < \infty$ for every $r > 0$. Hence 1(1) is satisfied with $k_\nu := \nu - 1$. Now we have $\{c_1, c_2, \ldots\} \subset T'$ (in fact, the two sets are equal!). Thus the claim follows from Lemma 1. □

Remark. In $\mathbb{C}^\times$, every divisor $\mathfrak{d}$ with $\lim d_\nu = 0$ has the "satellite sequence" $c_\nu := 0$. For such divisors on $\mathbb{C}^\times$, the product theorem holds with $\prod E_{\nu-1}(d_\nu/z)$. If we set $w := z^{-1}$, this is the Weierstrass product $\prod E_{\nu-1}(w/d_\nu^{-1})$ for the divisor $\mathfrak{d}'$ on $\mathbb{C}$ with the sequence $(d_\nu^{-1})_{\nu\geq 1}$. The product theorem 3.1.4 is thus contained in the product theorem above. □

"Satellite sequences" $(c_\nu)_{\nu\geq 1}$ with $c_\nu \in T'$ or just $c_\nu \in \mathbb{C}\backslash D$ do not exist in general; for example, they do not exist for divisors on $D := \mathbb{H}$ with support $T := \{i, 2i, 3i, \ldots\}$. However, the following does hold.

(2) *If T' is nonempty and every set $T(\varepsilon) := \{z \in T : d(T', z) \geq \varepsilon\}$, $\varepsilon > 0$, is finite, then there exists a sequence $(c_\nu)_{\nu\geq 1}$ in T' with $\lim |d_\nu - c_\nu| = 0$.*

Proof. Since T' is closed in $\mathbb{C}$, for every d_ν there exists $c_\nu \in T'$ such that $|d_\nu - c_\nu| = d(T', d_\nu)$. If $d_\nu - c_\nu$ did not converge to zero, there would exist $\varepsilon_0 > 0$ such that $|d_\nu - c_\nu| \geq \varepsilon_0$ for infinitely many ν. But then the set $T(\varepsilon_0)$ would be infinite. □

If T is *bounded and infinite*, then T' is nonempty and every set $T(\varepsilon)$, $\varepsilon > 0$, is finite. (Otherwise some set $T(\varepsilon_0)$, $\varepsilon_0 > 0$, would have an accumulation point $d^* \in T'$, which cannot occur since $|d^* - w| \geq d(T', w) \geq \varepsilon_0$ for all $w \in T(\varepsilon_0)$.) Hence:

[1] Following G. Cantor, we call T' the *derived set* of T in $\mathbb{C}$.

(3) *For every positive divisor* $\mathfrak{d}$ *on* D *with bounded infinite support, there exists a sequence* $(c_\nu)_{\nu\geq 1}$ *in* T' *with* $\lim |d_\nu - c_\nu| = 0$.

In particular, it is thus clear that on *bounded* regions every divisor is a principal divisor (special case of the existence theorem 5).

3. The general product theorem. *Let* D *be an arbitrary region in* $\mathbb{C}$. *Then, for every positive divisor* $\mathfrak{d}$ *on* D *with support* T, *there exist Weierstrass products in* $\mathbb{C}\backslash T'$.

The idea of the proof is to write the divisor $\mathfrak{d}$ as a sum of two divisors for which there exist Weierstrass products in $\mathbb{C}\backslash T'$. To do this, we need a lemma from set-theoretic topology, which will also be used in 6.2.2 in solving the analogous problem for principal part distributions.

Lemma. *Let* A *be a discrete set in* $\mathbb{C}$ *such that* $A' = \overline{A}\backslash A \neq \emptyset$. *Let*

$$A_1 := \{z \in A : |z|d(A', z) \geq 1\}, \quad A_2 := \{z \in A : |z|d(A', z) < 1\}.$$

Then A_1 *is closed in* $\mathbb{C}$. *Every set* $A_2(\varepsilon) := \{z \in A_2 : d(A', z) \geq \varepsilon\}$, $\varepsilon > 0$, *is finite.*

Proof. 1) If A_1 had an accumulation point $a \in \mathbb{C}$, it would follow that $a \in A'$ and there would exist a sequence $a_n \in A_1$ with $\lim a_n = a$. Since $d(A', a_n) \leq |a - a_n|$, the sequence $|a_n|d(A', a_n)$ would converge to zero, contradicting the definition of A_1. Thus $\overline{A_1} = A_1$.

2) $|z| < \varepsilon^{-1}$ for every $z \in A_2(\varepsilon)$. If there were an ε_0 with $A_2(\varepsilon_0)$ *infinite*, then $A_2(\varepsilon_0)$ would have an accumulation point $a \in A'$; but this is impossible since $|a - z| \geq d(A', z) \geq \varepsilon_0$ for all $z \in A_2(\varepsilon_0)$. □

Proof of the general product theorem. We take $\mathfrak{d}$ to be a positive divisor on $\mathbb{C}\backslash T'$. We may assume that $T' \neq \emptyset$. Let the sets T_1, T_2 be defined as in the lemma (with $A := T$). Then $T_1' = \emptyset$ and $T_2' = T'$. Since T_1 and T_2 are locally finite in $\mathbb{C}$ and $\mathbb{C}\backslash T'$, respectively, setting

$$\mathfrak{d}_j(z) := \mathfrak{d}(z) \text{ for } z \in T_j, \quad \mathfrak{d}_j(z) := 0 \text{ otherwise}, \quad j = 1, 2,$$

gives positive divisors $\mathfrak{d}_1$ on $\mathbb{C}$ with support T_1 and $\mathfrak{d}_2$ on $\mathbb{C}\backslash T'$ with support T_2. Moreover, $\mathfrak{d} = \mathfrak{d}_1 + \mathfrak{d}_2$ in $\mathbb{C}\backslash T'$ since $T_1 \cap T_2 = \emptyset$. By the product theorem 3.1.4 there exists a Weierstrass product for $\mathfrak{d}_1$ in $\mathbb{C}$. Since all the sets $T_2(\varepsilon)$ are finite, 2(2) and the product theorem 2 imply that there exists a Weierstrass product for $\mathfrak{d}_2$ in $\mathbb{C}\backslash T'$. Hence, by 3.1.2(1), there also exists a Weierstrass product for $\mathfrak{d} = \mathfrak{d}_1 + \mathfrak{d}_2$ in $\mathbb{C}\backslash T'$. □

4. Second proof of the general product theorem. Using a biholomorphic map v, we will first transport the divisor $\mathfrak{d}$ to a divisor $\mathfrak{d} \circ v^{-1}$ on another region in such a way that a Weierstrass product $\widehat{f}$ exists there for

$\mathfrak{d} \circ v^{-1}$; we will then transport this product back to a Weierstrass product $\widehat{f} \circ v$ for $\mathfrak{d}$. We assume that T is infinite, take $\mathfrak{d}$ to be a divisor on $\mathbb{C}\backslash T'$, fix $a \in \mathbb{C}\backslash\overline{T}$, and map $\mathbb{C}\backslash\{a\}$ biholomorphically onto $\mathbb{C}^\times$ by means of $v(z) := (z-a)^{-1}$. Then $0 \notin v(T)$ and $v(T)' = v(T')$. A positive divisor $\widehat{\mathfrak{d}}$ in $\mathbb{C}\backslash v(T)'$ with support $v(T)$ is defined by

$$\widehat{\mathfrak{d}}(w) := \mathfrak{d}(v^{-1}(w)), \quad w \in \mathbb{C}\backslash v(T'), \quad \widehat{\mathfrak{d}}(0) := 0.$$

If $(d_\nu)_{\nu \geq 1}$ is a sequence for $\mathfrak{d}$, then $(\widehat{d}_\nu)_{\nu\geq 1}$, with $\widehat{d}_\nu := v(d_\nu)$, is a sequence for $\widehat{d}$ (transporting the divisor by v). Since $v(T)$ is infinite and *bounded* (because $a \notin \overline{T}$), 2(3) and the product theorem 2 imply that

$$\prod \widehat{f}_\nu, \text{ where } \widehat{f}_\nu(w) := E_{\nu-1}[(\widehat{d}_\nu - c_\nu)/(w - c_\nu)] \text{ and } c_\nu \in v(T'),$$

is a Weierstrass product for $\widehat{\mathfrak{d}}$ in $\mathbb{C}\backslash v(T')$. We now set $f_\nu(z) := \widehat{f}_\nu(v(z))$ for $z \in \mathbb{C}\backslash(T' \cup \{a\})$ and set $f_\nu(a) := 1$. Then f_ν is holomorphic in $\mathbb{C}\backslash T'$ since $\lim_{z\to a} f_\nu(z) = \lim_{w\to\infty} \widehat{f}_\nu(w) = E_{\nu-1}(0) = 1$. The normal convergence of $\prod \widehat{f}_\nu$ in $\mathbb{C}\backslash v(T')$ implies the normal convergence of $\prod f_\nu$ in $\mathbb{C}\backslash(T' \cup \{a\})$. Since a is isolated in $\mathbb{C}\backslash T'$, the product converges normally throughout $\mathbb{C}\backslash T'$ (inward extension of convergence; cf. I.8.5.4). Since f_ν vanishes only at $d_\nu = v^{-1}(\widehat{d}_\nu)$, and vanishes there to first order, $\prod f_\nu$ is a Weierstrass product for $\mathfrak{d}$ in $\mathbb{C}\backslash T'$.

5. Consequences. The product theorem 3 has important consequences for arbitrary regions — as we saw in 3.1.5 for $\mathbb{C}$; the proofs are similar to those of 3.1.5.

Existence theorem. *On every region $D \subset \mathbb{C}$, every divisor is a principal divisor.*

Factorization theorem. *Every function $f \neq 0$ that is holomorphic in an arbitrary domain G can be written in the form*

$$f = u \prod_{\nu\geq 1} f_\nu,$$

where u is a unit in the ring $\mathcal{O}(G)$ and $\prod_{\nu\geq 1} f_\nu$ is a (possibly empty) Weierstrass product for the divisor (f) in G.

In general, the unit u is no longer an exponential function (although it is for (homologically) simply connected domains; cf. I.9.3.2).

Proposition. *(Quotient representation of meromorphic functions). For every function h meromorphic in G there exist two functions f and g, holomorphic in G and without common zeros there, such that $h = f/g$. In particular, the field $\mathcal{M}(G)$ is the quotient field of the integral domain $\mathcal{O}(G)$.*

Root criterion. *The following statements about a function $f \in \mathcal{O}(G)\backslash\{0\}$ and a natural number $n \geq 1$ are equivalent:*

i) *There exist a unit $u \in \mathcal{O}(G)$ and a function $g \in \mathcal{O}(G)$ such that $f = ug^n$.*

ii) *Every number $o_z(f)$, $z \in G$, is divisible by n.*

In general, the unit u is *no longer* an nth power. For (homologically) simply connected domains, one can always choose $u = 1$; cf. I.9.3.3.

In the older literature, the existence theorem was often expressed as follows:

Theorem. *Let T be an arbitrary discrete set in $\mathbb{C}$ and let an integer $n_d \neq 0$ be assigned to every point $d \in T$. Then in the region $\mathbb{C}\backslash T'$, where $T' := \overline{T}\backslash T$, there exists a meromorphic function h that is holomorphic and nonvanishing in $(\mathbb{C}\backslash T')\backslash T$ and for which*

$$o_d(h) = n_d \quad \textit{for all } d \in T.$$

$\mathbb{C}\backslash T'$ is the largest subregion of $\mathbb{C}$ in which there exists such a function.

Proof. By 2(1), $\mathbb{C}\backslash T'$ is the largest region in $\mathbb{C}$ in which T is closed. There exists a divisor $\mathfrak{d}$ on $\mathbb{C}\backslash T'$ with support T such that $\mathfrak{d}(d) = n_d$, $d \in T$. The existence theorem yields an $h \in \mathcal{M}(\mathbb{C}\backslash T')$ with $(h) = \mathfrak{d}$. □

§2. Applications and Examples

We first use the product theorem 1.3 to prove that in *every* integral domain $\mathcal{O}(G)$ there exists a *greatest common divisor* for every nonempty set. We then deal explicitly with a few Weierstrass products in $\mathbb{E}$ and $\mathbb{C}\backslash\partial\mathbb{E}$, including a product due to E. Picard, which is constructed with the aid of the group $SL(2, \mathbb{Z})$.

1. Divisibility in the ring $\mathcal{O}(G)$. Greatest common divisors. The basic arithmetic concepts are defined in the usual way: $f \in \mathcal{O}(G)$ is called *a divisor of* $g \in \mathcal{O}(G)$ if $g = f \cdot h$ with $h \in \mathcal{O}(G)$. Divisors of the identity are called *units.* A nonunit $v \neq 0$ is called a *prime* of $\mathcal{O}(G)$ if v divides a (finite) product only when it divides one of the factors. The functions $z - c$, $c \in G$, are — up to unit factors — precisely the primes of $\mathcal{O}(G)$. Functions $\neq 0$ in $\mathcal{O}(G)$ with infinitely many zeros in G cannot be written as the products of finitely many primes. Since such functions exist in every domain G by Theorem 1.3, we see:

No ring $\mathcal{O}(G)$ is factorial.

Despite this, all rings $\mathcal{O}(G)$ have a straightforward divisibility theory. The reason is that *assertions about divisibility* for elements f, $g \neq 0$ are equivalent to *assertions about order* for their divisors (f), (g). Writing $\mathfrak{d} \le \widehat{\mathfrak{d}}$ if $\widehat{\mathfrak{d}} - \mathfrak{d}$ is *positive*, we have the simple

Divisibility criterion. *Let* $f, g \in \mathcal{O}(G)\backslash\{0\}$. *Then*

$$f \text{ divides } g \Leftrightarrow (f) \le (g).$$

Proof. f divides g if and only if $h := g/f \in \mathcal{O}(G)$. But this occurs if and only if $o_z(h) = o_z(g) - o_z(f) \ge 0$ for all $z \in G$, i.e. if and only if $(f) \le (g)$. □

If S is a nonempty set in $\mathcal{O}(G)$, then $f \in \mathcal{O}(G)$ is called a *common divisor* of S if f divides every element g of S; a common divisor f of S is called a *greatest common divisor of* S if every common divisor of S is a divisor of f. Greatest common divisors — when they exist — are uniquely determined only up to unit factors; despite this, one speaks simply of *the* greatest common divisor f of S and writes $f = \gcd(S)$. A set $S \ne \emptyset$ is called *relatively prime* if $1 = \gcd(S)$.

$S \ne \emptyset$ is relatively prime if and only if the functions in S have no common zeros in G, i.e. if $\bigcap_{g \in S} Z(g) = \emptyset$.

The proof of the next result is a simple verification.

If $f = \gcd(S)$ *and* $g = \gcd(T)$, *then* $\gcd(S \cup T) = \gcd\{f, g\}$.

If $\mathfrak{D} \ne \emptyset$ is a set of positive divisors $\mathfrak{d}$ on G, then the map $G \to \mathbb{Z}$, $z \mapsto \min\{\mathfrak{d}(z) : \mathfrak{d} \in \mathfrak{D}\}$ is a divisor $\min\{\mathfrak{d} : \mathfrak{d} \in \mathfrak{D}\} \ge \mathfrak{o}$. The divisibility criterion implies the following:

Every function $f \in \mathcal{O}(G)$ *with* $(f) = \min\{(g) : g \in S, g \ne 0\}$ *is a gcd of* $S \ne \{0\}$.

The next statement is now an immediate consequence of Theorem 1.3.

Existence of the gcd. *In the ring* $\mathcal{O}(G)$, *every set* $S \ne \emptyset$ *has a* gcd.

Proof. Given $S \ne \{0\}$, choose $f \in \mathcal{O}(G)$ such that $(f) = \min\{(g) : g \in S,\ g \ne 0\}$. □

It may seem surprising that the product theorem 1.3 is needed to prove the existence of the gcd (even if S has only two elements!). But it should not be forgotten that there are integral domains with identity in which a gcd does not always exist: in the ring $\mathbb{Z}[\sqrt{-5}]$, for example, the two elements 6 and $2(1+\sqrt{-5})$ have no gcd.

In principal ideal rings such as $\mathbb{Z}$, $\mathbb{Z}[i]$, and $\mathbb{C}[z]$, every set S has a gcd; in fact, it is always a finite linear combination of elements of S. This assertion also holds for the ring $\mathcal{O}(G)$ if S is *finite*, as we will see in 6.3.3 by means of Mittag-Leffler's theorem.

Exercises. Define the concept of the least common multiple as in number theory and, if it exists, denote it by lcm. Prove:

1) Every set $S \neq \emptyset$ has a least common multiple. If $(f) = \mathrm{lcm}(S) \neq 0$, then $(f) = \max\{(g) : g \in S\}$.

2) If f and g are respectively a gcd and lcm of two functions $u, v \in \mathcal{O}(G)\backslash\{0\}$, then the products $f \cdot g$ and $u \cdot v$ differ only by a unit factor.

2. Examples of Weierstrass products. 1) *Let $\mathfrak{d} \geq 0$ be a divisor on $\mathbb{E}$ that satisfies $\mathfrak{d}(0) \neq 0$ and has sequence $(d_\nu)_{\nu\geq 1}$, and suppose that $\sum_{\nu\geq 1}(1-|d_\nu|) < \infty$. Then*

$$(*) \qquad \prod_{\nu\geq 1} E_0\left(\frac{d_\nu - \overline{d}_\nu^{\,-1}}{z - \overline{d}_\nu^{\,-1}}\right) = \prod_{\nu\geq 1} \overline{d}_\nu \frac{z-d_\nu}{\overline{d}_\nu z - 1} \in \mathcal{O}(\mathbb{C}\backslash\overline{\{\overline{d}_1^{\,-1}, \overline{d}_2^{\,-1}, \ldots\}})$$

is a Weierstrass product for $\mathfrak{d}$.

Proof. Let $c_\nu := 1/\overline{d}_\nu$; then $\sum |c_\nu - d_\nu| < \infty$ since

$$(+) \qquad |d_\nu - c_\nu| = |\overline{d}_\nu|^{-1}(1-|d_\nu|^2) \leq 2m^{-1}(1-|d_\nu|),$$

where $m := \min\{|d_\nu| : \nu \geq 1\}$. The assertion follows from Lemma 1.1 with $k_\nu := 0$. □

The products $(*)$ are *bounded* in $\mathbb{E}$; up to a normalization, they are Blaschke products (cf. 3.3). Because of their importance, we also give a *direct proof of convergence*. It follows from (+) that

$$\left|\overline{d}\frac{z-d}{\overline{d}z-1} - 1\right| = \frac{1-|d|^2}{|\overline{d}|\,|z-\overline{d}^{\,-1}|} \leq \frac{2}{m}\cdot\frac{1-|d|}{|z-\overline{d}^{\,-1}|}, \quad z \neq \overline{d}^{\,-1}, \ d \in \{d_1, d_2, \ldots\}.$$

Now for every compact set K in $\mathbb{C}\backslash\overline{\{\overline{d}_1^{\,-1}, \overline{d}_2^{\,-1}, \ldots\}}$ there exists $t > 0$ such that $|z - \overline{d}_\nu^{\,-1}| \geq t$ for all $z \in K$ and all $\nu \geq 1$. Thus

$$\sum_{\nu\geq 1}\left|\overline{d}_\nu\frac{z-d_\nu}{\overline{d}_\nu z - 1} - 1\right|_K \leq \frac{2t^{-1}}{m}\sum_{\nu\geq 1}|1-d_\nu| < \infty,$$

which implies the normal convergence of $(*)$ in $\mathbb{C}\backslash\overline{\{\overline{d}_1^{\,-1}, \overline{d}_2^{\,-1}, \ldots\}}$.

2) Let $r_\nu > 0$, $r_\nu \neq 1$, be a sequence of pairwise distinct real numbers with $\lim r_\nu = 0$. The set

$$T := \{d_{\nu p} := (1-r_\nu)c_{\nu p}, \ 0 \leq p < \nu, \ \nu = 1, 2, \ldots\},$$

where $c_{\nu p} := \exp(2p\pi/\nu) \in \partial\mathbb{E}$, is locally finite in $\mathbb{C}\backslash\partial\mathbb{E}$. Since $d_{\nu p} - c_{\nu p} = -r_\nu c_{\nu p}$ tends to 0, the product theorem 1.2 implies that

$$\prod_{\nu=1}^{\infty}\prod_{p=0}^{\nu-1} E_{\nu-1}\left(\frac{r_\nu c_{\nu p}}{c_{\nu p} - z}\right)$$

is a Weierstrass product in $\mathbb{C}\backslash\partial\mathbb{E}$, which vanishes to first order precisely at the points of S.

Since $\sum_{\nu=1}^{\infty}\sum_{p=0}^{\nu-1}|d_{\nu p}-c_{\nu p}|^{k+1}=\sum_{\nu=1}^{\infty}\nu r_{\nu}^{k+1}$ converges in the cases $r_\nu = 1/\nu$, $k=2$ and $r_\nu = 1/\nu^3$, $k=0$, the corollary to Lemma 1.1 gives the Weierstrass products

$$\prod_{\nu,p}\left(1+\frac{1}{\nu}\frac{c_{\nu p}}{z-c_{\nu p}}\right)\exp\left[\frac{c_{\nu p}}{\nu(c_{\nu p}-z)}+\frac{c_{\nu p}^2}{2\nu^2(c_{\nu p}-z)^2}\right] \quad \text{and}$$

$$\prod_{\nu,p}\left(1+\frac{1}{\nu^3}\frac{c_{\nu p}}{z-c_{\nu p}}\right), \quad \text{respectively.}$$

3. On the history of the general product theorem. Weierstrass left it to others to extend his product theorem to regions in $\mathbb{C}$. As early as 1881 ([Pi], pp. 69–71), E. Picard considers the region $\mathbb{C}\backslash\partial\mathbb{E}$; he discusses, among other things, the product

$$\prod E_1\left(\frac{A-B}{z-B}\right)=\prod\frac{z-A}{z-B}\exp\frac{A-B}{z-B},$$

where

$$A:=\frac{\beta+\gamma-(\alpha-\delta)i}{\alpha+\delta+(\beta-\gamma)i}, \quad B:=\frac{\beta+\delta i}{\delta+\beta i},$$

and α, β, γ, and δ run through all numbers in $\mathbb{Z}$ satisfying $\alpha\delta-\beta\gamma=1$. Picard's product is probably the first example of a Weierstrass product in a region $\neq\mathbb{C}$ where convergence-producing factors of the Weierstrass type were consciously used. Picard said nothing about the convergence of his product, but in 1893 made the following comment (cf. *Traité d'analyse*, vol. 1, p. 149): "..., c'est ce que l'on reconnaît en considérant à la place de la série une intégrale triple convenable dont la valeur reste finie quand les limites deviennent infinies." (..., it is what one recognizes if, instead of the series, one considers an appropriate triple integral whose value remains finite when the limits become infinite.) A year later, Picard studied slit regions ([Pi], pp. 91–93). He introduced the products $\prod E_\nu((d_\nu-c_\nu)/(z-c_\nu))$ in his notes. They were also used in 1884 by Mittag-Leffler to prove the existence theorem for general regions ([ML], especially pp. 32–38). Picard's notes are not mentioned by Mittag-Leffler; in 1918 Landau ([L], p. 157) speaks of the "well-known Picard-Mittag-Leffler product construction."

In their work [BSt$_1$], carried out in 1948 but not published until 1950, H. Behnke and K. Stein extended the existence theorem 1.5 to arbitrary noncompact Riemann surfaces (loc. cit., Satz 2, p. 158).

4. Glimpses of several variables. With his product theorem, Weierstrass opened the door to a development that led to new insights in higher-dimensional function theory as well. The product theorem was generalized to the case of several complex variables as early as 1895 by P. Cousin, a student of Poincaré, in

[Co]. The formulation of the concept of divisors already presented difficulties at this point, since the zeros of holomorphic functions in $\mathbb{C}^n$, $n \geq 2$, are no longer isolated but form real $(2n-2)$-dimensional surfaces. Cousin and his successors could derive the analogous theorem only for $\mathbb{C}^n$ itself and *polydomains* in $\mathbb{C}^n$ (product domains $G_1 \times G_2 \times \cdots \times G_n$, where each G_ν is a domain in $\mathbb{C}$). Cousin thought he had proved his theorem for *all* polydomains. But the American mathematician T. H. Gronwall discovered in 1917 that Cousin's conclusions hold only for special polydomains; at least $(n-1)$ of the n domains $G_1, \ldots, G_n$ must be *simply connected* (cf. [Gro], p. 53). Thus there exist — and this was a sensation — *topological obstructions!* It was soon conjectured that Cousin's theorem was valid for many topologically nice domains of holomorphy;[2] for example, H. Behnke and K. Stein proved in 1937 that the theorem holds for all *star-shaped* domains of holomorphy ([BSt$_2$], p. 188). The Japanese mathematician K. Oka achieved a breakthrough in 1939; he was able to show that, in arbitrary domains of holomorphy $G \subset \mathbb{C}^n$, a positive divisor is the divisor of a function *holomorphic* in G if and only if it is the divisor of a function *continuous* in G ([O], pp. 33–34). This statement is the famous *Oka principle*, which K. Stein generalized in 1951 to his manifolds, interpreted homologically, and made precise [St]. It was J-P. Serre who, in 1953, gave the final solution of the Cousin problem ([S], pp. 263–264):

In a Stein manifold X, a divisor $\mathfrak{d}$ is the divisor of a function meromorphic in X if and only if its Chern cohomology class $c(\mathfrak{d}) \in H^2(X, \mathbb{Z})$ vanishes. In particular, on a Stein manifold X with $H^2(X, \mathbb{Z}) = 0$, every divisor is a principal divisor.

The significance of the second cohomology group with integer coefficients for the solvability of the Weierstrass-Cousin problem becomes evident here. In the thirties, the fundamental group $\pi_1(X)$ was still thought to have great importance in this context; but in [S] (p. 265), Serre exhibited a simply connected domain of holomorphy in $\mathbb{C}^3$ where not all divisors are principal divisors.

The methods that Serre and Cartan developed jointly to prove Serre's theorem revolutionized mathematics: the theory of coherent analytic sheaves and their cohomology theory began their triumphant advance. Serre ([S], p. 265) also put the finishing touches on Poincaré's old theorem (cf. 3.1.6):

In a Stein manifold, every meromorphic function is the quotient of two holomorphic functions (which need not be locally relatively prime).

We must be satisfied with this sketch; a comprehensive presentation can be found in [GR].

The Oka principle was substantially extended by H. Grauert in 1957; he showed, among other things, that holomorphic fiber bundles over Stein manifolds are holomorphically trivial if and only if they are topologically trivial ([Gra], p. 268 in particular). Weierstrass, Poincaré, and Cousin would certainly have been impressed to see how their theories culminated in the twentieth century in the Oka-Grauert principle: *Locally prescribed analytic data with globally continuous solutions always have globally holomorphic solutions as well.*

---

[2] For the concept of domains of holomorphy and Stein manifolds, see also 5.2.6.

§3. Bounded Functions on $\mathbb{E}$ and Their Divisors

By the product theorem 1.3, the zeros of functions holomorphic in D can be arbitrarily assigned in D as long as they have no accumulation point there. The situation changes when growth conditions are imposed on the functions. Thus, for divisors $\mathfrak{d} \neq 0$ on $\mathbb{C}$, there never exist *bounded* functions f with $(f) = \mathfrak{d}$.

A Weierstrass product $\prod f_\nu$ is certainly bounded in D if $|f_\nu|_D \leq 1$ for all ν. Products with such nice factors are rare. In what follows, we study the case $D = \mathbb{E}$. For the functions

$$g_d(z) = \frac{z-d}{\bar{d}z - 1}, \quad d \in \mathbb{E},$$

which we recognize as automorphisms of $\mathbb{E}$, $|g_d|_{\mathbb{E}} = 1$. We will see that a divisor $\mathfrak{d} \geq 0$ on $\mathbb{E}$ with $\mathfrak{d}(0) = 0$ is a divisor of a bounded function on $\mathbb{E}$ if and only if $\mathfrak{d}$ has a Weierstrass product of the form $\prod(|d_\nu|/d_\nu)g_{d_\nu}(z)$, and that such products exist if and only if

$$\sum (1 - |d_\nu|) < \infty \quad \text{(Blaschke condition).}$$

The necessity of this condition, which implies an identity theorem, follows quickly (in Subsection 2) from Jensen's inequality (Subsection 1). Its sufficiency is proved in Subsection 3.

1. Generalization of Schwarz's lemma. *Let $f \in \mathcal{O}(\mathbb{E})$, and let $d_1, \ldots, d_n \in \mathbb{E}$ be pairwise distinct zeros of f. Then*

$$(1) \qquad |f(z)| \leq \left|\frac{z-d_1}{\bar{d}_1 z - 1}\right| \cdot \ldots \cdot \left|\frac{z-d_n}{\bar{d}_n z - 1}\right| \cdot |f|_{\mathbb{E}} \quad \textit{for all } z \in \mathbb{E}.$$

Proof. Let $z \in \mathbb{E}$ be fixed, and let $m := \max\{|z|, |d_1|, \ldots, |d_n|\}$. Set $h := \prod_1^n g_{d_\nu}$; then $g := f/h \in \mathcal{O}(\mathbb{E})$, and the maximum principle implies that $|g(z)| \leq |f|_{\mathbb{E}}/\min_{|w|=r}\{|h(w)|\}$ for all $r \in (m, 1)$. Now $|h(w)| = 1$ for all $w \in \partial\mathbb{E}$ (since $\overline{w} g_d(w) = (1 - d\overline{w})/(\bar{d}w - 1)$ for $w \in \partial\mathbb{E}$). Hence

$$\lim_{r \to 1} \min_{|w|=r} \{|h(w)|\} = 1, \quad \text{and thus} \quad |g(z)| \leq |f|_{\mathbb{E}}. \qquad \square$$

Remark. If $|f|_{\mathbb{E}} \leq 1$ and $n = 1$, $d_1 = 0$, (1) is Schwarz's lemma; cf. I.9.2.1. As in that case, we now have a sharpened version of the result:

If equality holds in (1) *for a point $d \in \mathbb{E} \setminus \{d_1, \ldots, d_n\}$, then*

$$f(z) = \eta |f|_{\mathbb{E}} \prod_1^n g_{d_\nu}(z), \quad \text{with } \eta \in S^1.$$

For $z = 0$, (1) becomes "Jensen's inequality":

$$(2) \qquad |f(0)| \leq |d_1 d_2 \ldots d_n| \cdot |f|_{\mathbb{E}}.$$

Inequality (2) is a special case of Jensen's formula, which we derive in the appendix to this section.

Historical note. The proof above dates back to C. Carathéodory and L. Fejér [CF]. Inequality (2) appears in the work of J. L. W. V. Jensen ([J], 1898–99) and of J. Petersen ([P], 1899).

2. Necessity of the Blaschke condition. *Let $f \neq 0$ be holomorphic and bounded in $\mathbb{E}$, and let $d_1, d_2, \ldots$ be a sequence for the divisor (f). Then*

$$\sum (1 - |d_\nu|) < \infty.$$

Proof. We may assume that $f(0) \neq 0$. Then $\sum(1 - |d_\nu|) = \infty$ would imply that $\lim |d_1 d_2 \ldots d_n| = 0$ by 1.1.1(d), and hence $f(0) = 0$ by 1(2). □

As a corollary, we point out the surprising

Identity theorem for bounded functions on $\mathbb{E}$. *Let $A = \{a_1, a_2, \ldots\}$ be a countable set in $\mathbb{E}$ such that $\sum(1-|a_\nu|) = \infty$. Suppose that $f, g \in \mathcal{O}(\mathbb{E})$ are bounded in $\mathbb{E}$ and $f|A = g|A$. Then $f = g$.*

Proof. The function $h := f - g \in \mathcal{O}(\mathbb{E})$ is bounded in $\mathbb{E}$. If h were not the zero function, then $\sum(1-|a_\nu|)$ would be a subseries of the series $\sum(1-|d_\nu|)$ determined by the sequence (d_ν) for the divisor of h. Hence it would be convergent by the statement above. □

Bounded holomorphic functions in $\mathbb{E}$ therefore vanish identically as soon as their zeros move too slowly toward the boundary of $\mathbb{E}$ (this is made precise by $\sum(1 - |a_\nu|) = \infty$). Thus $f \in \mathcal{O}(\mathbb{E})$ must be the zero function if it is bounded and vanishes at all points $1 - 1/n$, $n \geq 1$.

3. Blaschke products. For each point $d \in \mathbb{E}$, we set

$$(1) \qquad \begin{aligned} b(z,d) &:= \frac{|d|}{d} \frac{z-d}{\bar{d}z - 1} = |d|^{-1} E_0\left(\frac{d - \bar{d}^{-1}}{z - \bar{d}^{-1}}\right) \quad \text{if } d \neq 0, \\ b(z,0) &:= z. \end{aligned}$$

Then $b(z,d)$ is holomorphic on $\overline{\mathbb{E}}$ and nonvanishing in $\mathbb{E}\backslash\{d\}$, the point d is a first-order zero, and $|b(z,d)|_{\mathbb{E}} = 1$.

Now let $\mathfrak{d} \geq 0$ be a divisor on $\mathbb{E}$ and let $(d_\nu)_{\nu \geq 1}$ be a corresponding sequence. The product

$$b(z) := \prod_{\nu \geq 1} b(z, d_\nu)$$

is called the *Blaschke product* for $\mathfrak{d}$ if it converges normally in $\mathbb{E}$ (and hence even in $\mathbb{C}\backslash\partial\mathbb{E}$). Blaschke products are thus a special class of Weierstrass products.

(2) *If b is a Blaschke product, then $b \in \mathcal{O}(\mathbb{E})$, $(b) = \mathfrak{d}$, and $|b|_{\mathbb{E}} \le 1$. Moreover, if $b(0) \neq 0$, then*

$$b(z) = b(0)^{-1} \prod_{\nu=1}^{\infty} E_0[(d_\nu - \overline{d}_\nu^{-1})/(z - \overline{d}_\nu^{-1})], \quad \textit{with} \quad b(0) := \prod_{\nu=1}^{\infty} |d_\nu|.$$

Example 1) in Subsection 2.2 contains the existence theorem for Blaschke products:

(3) *If $\sum_{\nu=1}^{\infty}(1 - |d_\nu|) < \infty$, the Blaschke product for $\mathfrak{d}$ exists.*

The direct proof — without recourse to Lemma 1.1 — goes as follows: For $d \in \mathbb{E}\backslash\{0\}$, we have $b(z,d)-1 = (1-|d|)(d+|d|z)/[d(\overline{d}z-1)]$. Since $|\overline{d}z-1| \ge 1-|z|$ for $d \in \mathbb{E}$ and $(1+|z|)/(1-|z|) \le 2(1-r)^{-1}$ if $|z| \le r < 1$, it follows that

$$|b(z,d) - 1|_{B_r(0)} \le \frac{2}{1-r}(1 - |d|) \quad \text{for all } r \in (0,1) \text{ and all } d \in \mathbb{E}.$$

It is thus clear that $\sum_1^\infty |b(z,d_\nu) - 1|_{B_r(0)} < \infty$. Hence the Blaschke product converges normally in $\mathbb{E}$.

The next result is now immediate from (3) and Theorem 2.

Proposition. *The following assertions about a divisor $\mathfrak{d} \ge 0$ on $\mathbb{E}$ are equivalent:*

i) *$\mathfrak{d}$ is the divisor of a function in $\mathcal{O}(\mathbb{E})$ that is bounded in $\mathbb{E}$.*
ii) $\sum_{z \in \mathbb{E}} \mathfrak{d}(z)(1 - |z|) < \infty$ *(Blaschke condition).*
iii) *The Blaschke product $\prod_{\nu=1}^{\infty} b(z, d_\nu)$ exists for $\mathfrak{d}$.*

Because of ii) there does not exist, for example, any bounded function $f \in \mathcal{O}(\mathbb{E})$ that vanishes to nth order at each point $1 - 1/n^2$, $n \in \mathbb{N}\backslash\{0\}$. The following is immediate.

For every bounded function $f \in \mathcal{O}(\mathbb{E})$, there exist a Blaschke product b and a function $g \in \mathcal{O}(\mathbb{E})$ such that $f = e^g \cdot b$.

Historical note. W. Blaschke introduced his products and proved the existence theorem in 1915 ([Bl], p. 199). Of course — as the title of his paper indicates — Blaschke was then mainly interested in Vitali's convergence theorem; we go into this in 7.1.4. Edmund Landau reviewed Blaschke's work in 1918 and simplified the proof by using Jensen's inequality; cf. [L].

4. Bounded functions on the right half-plane. Setting $t(z) := (z-1)/(z+1)$ gives a biholomorphic map of $\mathbb{T} := \{z \in \mathbb{C} : \operatorname{Re} z > 0\}$ onto $\mathbb{E}$; we have

$$(1) \quad 1 - |t(z)|^2 = \frac{4 \operatorname{Re} z}{|z+1|^2} = \frac{4}{|1+z^{-1}|^2} \operatorname{Re}(1/z) \quad \text{for all } z \in \mathbb{C}\backslash\{0,-1\}.$$

The results obtained for $\mathbb{E}$ carry over easily to $\mathbb{T}$.

a) *A positive divisor $\mathfrak{d}$ on $\mathbb{T}$ with corresponding sequence $d_1, d_2, \ldots$ is the divisor of a bounded holomorphic function on $\mathbb{T}$ if and only if*

$$\sum_{\nu=1}^{\infty} \frac{\operatorname{Re} d_\nu}{|1+d_\nu|^2} < \infty \quad (\textit{Blaschke condition for } \mathbb{T}).$$

b) *Let the function $f \in \mathcal{O}(\mathbb{T})$ be bounded in $\mathbb{T}$ and vanish at the pairwise distinct points $d_1, d_2, \ldots$, where $\delta := \inf\{|d_n|\} > 0$ and $\sum_{\nu \geq 1} \operatorname{Re}(1/d_\nu) = \infty$. Then f vanishes identically on $\mathbb{T}$.*

Proof. a) The map $\mathfrak{d} \circ t^{-1} : \mathbb{E} \to \mathbb{N}$ is a positive divisor on $\mathbb{E}$ with corresponding sequence $\widehat{d}_n := t(d_n)$. If $f \in \mathcal{O}(\mathbb{T})$ is bounded in $\mathbb{T}$, then $(f) = \mathfrak{d}$ if and only if $(f \circ t^{-1}) = \mathfrak{d} \circ t^{-1}$, where $f \circ t^{-1} \in \mathcal{O}(\mathbb{E})$ is bounded in $\mathbb{E}$. By Proposition 3, this occurs if and only if $\sum(1 - |\widehat{d}_\nu|) < \infty$. The assertion now follows from (1) because

$$\frac{1}{2}(1-|w|^2) \leq 1 - |w| \leq 1 - |w|^2 \quad \text{for all } w \in \mathbb{E}.$$

b) Since $|1+w^{-1}|^{-2} \geq (1+\delta^{-1})^{-2}$ for all w with $|w| \geq \delta$, it follows from (1) that

$$\sum_{\nu \geq 1} \frac{\operatorname{Re} d_\nu}{|1+d_\nu|^2} \geq \frac{1}{(1+\delta^{-1})^2} \sum_{\nu \geq 1} \operatorname{Re}(1/d_\nu) = \infty.$$

By a), f must then vanish identically. □

Statement b) will be used in 7.4.3, in the proof of Müntz's theorem.

Analogues of a) and b) are valid for the *upper* half-plane $\mathbb{H}$. The map $\mathbb{H} \xrightarrow{\sim} \mathbb{T}$, $z \mapsto -iz$, is biholomorphic and $\operatorname{Re}(-iz) = \operatorname{Im} z$. In this situation, we thus obtain the convergence condition $\sum(\operatorname{Im} d_\nu/|i + d_\nu|^2) < \infty$ in a) and the divergence condition $\sum \operatorname{Im}(1/d_\nu) = -\infty$ in b).

Exercise. Define "Blaschke products" for $\mathbb{T}$ and $\mathbb{H}$ and prove the analogue of 3(3) for these half-planes.

Appendix to Section 3: Jensen's Formula

Jensen's inequality 3.1(2) can be improved to an equality:

Jensen's formula. *Let $f \in \mathcal{O}(\mathbb{E})$, $f(0) \neq 0$. Let $0 < r < 1$ and let $d_1, d_2, \ldots, d_n$ be all the zeros of f in $B_r(0)$, where each zero appears according to its order. Then*

$$(\mathrm{J}) \qquad \log|f(0)| + \log \frac{r^n}{|d_1 d_2 \ldots d_n|} = \frac{1}{2\pi} \int_0^{2\pi} \log|f(re^{i\theta})| d\theta.$$

The integral on the right-hand side is *improper* if f has zeros on the boundary of $B_r(0)$. The second summand on the left-hand side is zero if f has no zeros in $B_r(0)$. Since $\log x$ is monotone for $x > 0$, (J) immediately leads to the inequality

$$r^n|f(0)| \leq |d_1 d_2 \dots d_n| \cdot |f|_{\partial B_r(0)}.$$

Jensen's inequality 3.1(2) follows by passing to the limit as $r \to 1$.

We reproduce the proof given by J. L. W. V. Jensen in 1898–99 ([J], p. 362 ff.); he also admitted poles of the function f. The formula can also be found in an 1899 paper of J. Petersen ([Pe], p. 87).

We write B for $B_r(0)$. Our starting point is the following special case of (J).

(1) *If $g \in \mathcal{O}(\mathbb{E})$ has no zeros in $\overline{B}$, then*

$$\log|g(0)| = \frac{1}{2\pi}\int_0^{2\pi} \log|g(re^{i\theta})|d\theta.$$

Proof. There exist a disc U with $\overline{B} \subset U \subset \mathbb{E}$ and a function $h \in \mathcal{O}(U)$ such that $g|U = g(0)\exp h$ with $h(0) = 0$.[3] Since $h(z)/z \in \mathcal{O}(U)$,

$$0 = \int_{\partial B} \frac{h(\zeta)}{\zeta} d\zeta = i\int_0^{2\pi} h(re^{i\theta})d\theta.$$

Since $\operatorname{Re} h(z) = \log|g(z)/g(0)|$ for $z \in U$, it follows that

$$0 = \operatorname{Re}\int_0^{2\pi} h(re^{i\theta})d\theta = \int_0^{2\pi} \log|g(re^{i\theta})|d\theta - 2\pi\log|g(0)|. \qquad \square$$

Statement (1) is Poisson's mean-value equation for the function $\log|g(z)|$, which is *harmonic* in a neighborhood of $\overline{B}$; cf. also I.7.2.5*.

In order to reduce (J) to (1), we need the following result.

(2) $$\int_0^{2\pi} \log|1 - e^{i\theta}|d\theta = 0.$$

Proof. Since $|1 - e^{2i\varphi}| = 2\sin\varphi$ for $\varphi \in [0, \pi]$, we have (with $\theta = 2\varphi$)

$$\frac{1}{2}\int_0^{2\pi} \log|1 - e^{i\theta}|d\theta = \int_0^{\pi} \log(2\sin\varphi)d\varphi = \pi\log 2 + \int_0^{\pi} \log\sin\varphi d\varphi.$$

[3] Choose U such that $g|U$ has no zeros. Since U is star-shaped, $g = \exp\widehat{h}$ with $\widehat{h} \in \mathcal{O}(U)$. Then set $h := \widehat{h} - \widehat{h}(0)$.

The integral on the right-hand side exists and, by 1.3.2(1), equals $-\pi \log 2$. □

Formula (2) is usually derived by function-theoretic methods. The direct calculation above supports Kronecker's sardonic maxim on the occasionally "gute Früchte bringenden Glauben an die Unwirksamkeit des Imaginären" (fruitful belief in the inefficacy of the imaginary); cf. also I.14.2.3.

It is now easy to prove (J). If $c_1, \dots, c_m$ are all the zeros of f on ∂B, the function

$$g(z) := f(z) \prod_{\nu=1}^{n} \frac{\overline{d}_\nu z - r^2}{r(z - d_\nu)} \prod_{\mu=1}^{m} \frac{c_\mu}{c_\mu - z} \in \mathcal{O}(\mathbb{E})$$

has no zeros in $\overline{B}$. Since $g(0) = f(0)r^n/d_1 d_2 \dots d_n$ and $\left|\frac{\overline{d}_\nu z - r^2}{r(z - d_\nu)}\right| = 1$ for $z \in \partial B$, it follows from (1), if we set $c_\mu = re^{i\theta_\mu}$, that

$$(*) \qquad \begin{aligned} &\log|f(0)| + \log \frac{r^n}{|d_1 d_2 \dots d_n|} \\ &= \frac{1}{2\pi} \int_0^{2\pi} \log \left| f(re^{i\theta}) \prod_{\mu=1}^{m} (1 - e^{i(\theta - \theta_\mu)})^{-1} \right| d\theta. \end{aligned}$$

Since the integrand on the right-hand side is the difference $\log|f(re^{i\theta})| - \sum_{\mu=1}^{m} \log|1 - e^{i(\theta - \theta_\mu)}|$, (J) follows from $(*)$ because of (2). □

In applications, r can often be chosen so that f has no zeros on $\partial B_r(0)$ (as in the derivation of 3.1(2), for example). Then the factors $c_\mu/(c_\mu - z)$ drop out, and (J) follows directly from (1) — without using (2).

Jensen's formula has important applications in the theory of entire functions and the theory of Hardy spaces; for lack of space we cannot investigate these further.

Bibliography

[BSt₁] BEHNKE, H. and K. STEIN: Elementarfunktionen auf Riemannschen Flächen ..., *Can. Journ. Math.* 2, 152–165 (1950).

[BSt₁] BEHNKE, H. and K. STEIN: Analytische Funktionen mehrerer Veränderlichen zu vorgegebenen Null- und Polstellenflächen, *Jber. DMV* 47, 177–192 (1937).

[Bl] BLASCHKE, W.: Eine Erweiterung des Satzes von Vitali über Folgen analytischer Funktionen, *Ber. Verh. Königl. Sächs. Ges. Wiss. Leipzig* 67, 194–200 (1915); *Ges. Werke* 6, 187–193.

[CF] CARATHÉODORY, C. and L. FEJÉR: Remarques sur le théorème de M. Jensen, *C.R. Acad. Sci. Paris* 145, 163–165 (1907); CARATHÉODORY's *Ges. Math. Schrift.* 3, 179–181; FEJÉR's *Ges. Arb.* 1, 300–302.

[Co] COUSIN, P.: Sur les fonctions de n variables complexes, *Acta Math.* 19, 1–62 (1895).

[GR] GRAUERT, H. and R. REMMERT: *Theory of Stein Spaces*, trans. A. HUCKLEBERRY, Springer, New York 1979.

[Gra] GRAUERT, H.: Analytische Faserungen über holomorph-vollständigen Räumen, *Math. Ann.* 135, 263–273 (1958).

[Gro] GRONWALL, T. H.: On the expressibility of a uniform function of several variables as the quotient of two functions of entire character, *Trans. Amer. Math. Soc.* 18, 50–64 (1917).

[J] JENSEN, J. L. W. V.: Sur un nouvel et important théorème de la théorie des fonctions, *Acta Math.* 22, 359–364 (1898–99).

[L] LANDAU, E.: Über die Blaschkesche Verallgemeinerung des Vitalischen Satzes, *Ber. Verh. Königl. Sächs. Ges. Wiss. Leipzig* 70, 156–159 (1918); *Coll. Works* 7, 138–141.

[ML] MITTAG-LEFFLER, G.: Sur la représentation analytique des fonctions monogènes uniformes d'une variable indépendante, *Acta Math.* 4, 1–79 (1884).

[O] OKA, K.: III. Deuxième problème de Cousin, *Journ. Sci. Hiroshima Univ.*, Ser. A, 9, 7–19 (1939); *Coll. Papers*, trans. R. NARASIMHAN, 24–35, Springer, 1984.

[Pe] PETERSEN, J.: Quelques remarques sur les fonctions entières, *Acta Math.* 23, 85–90 (1899).

[Pi] PICARD, E.: *Œuvres* 1.

[S] SERRE, J-P.: Quelques problèmes globaux relatifs aux variétés de Stein, *Coll. Fonct. Plus. Var. Bruxelles* 57–68 (1953); *Œuvres* 1, 259–270.

[St] STEIN, K.: Analytische Funktionen mehrerer komplexer Veränderlichen zu vorgegebenen Periodizitätsmoduln und das zweite Cousinsche Problem, *Math. Ann.* 123, 201–222 (1951).

5

Iss'sa's Theorem. Domains of Holomorphy

We begin by giving two interesting applications of the Weierstrass product theorem that have not yet made their way into the German textbook literature. In Section 1 we discuss Iss'sa's theorem, discovered only in 1965; in Section 2 we show — once directly and once with the aid of the product theorem — that *every* domain in $\mathbb{C}$ is a domain of holomorphy. In Section 3 we conclude by discussing simple examples of functions whose domains of holomorphy have the form $\{z \in \mathbb{C} : |q(z)| < R\}$, $q \in \mathbb{C}[z]$; Cassini domains, in particular, are of this form.

§1. Iss'sa's Theorem

Every nonconstant holomorphic map $h : \widehat{G} \to G$ between domains in $\mathbb{C}$ lifts every function f meromorphic in G to a function $f \circ h$ meromorphic in $\widehat{G}$. Thus h induces the $\mathbb{C}$-algebra homomorphism

$$\varphi : \mathcal{M}(G) \to \mathcal{M}(\widehat{G}), \quad f \mapsto f \circ h,$$

which maps $\mathcal{O}(G)$ into $\mathcal{O}(\widehat{G})$ (cf. also I.10.3.3). Iss'sa's theorem says that *every* $\mathbb{C}$-algebra homomorphism $\mathcal{M}(G) \to \mathcal{M}(\widehat{G})$ is induced by a holomorphic map $\widehat{G} \to G$. In preparation, we prove that all $\mathbb{C}$-algebra homomorphisms $\mathcal{O}(G) \to \mathcal{O}(\widehat{G})$ are induced by holomorphic maps $\widehat{G} \to G$. The proof of this theorem of Bers is elementary; it is based on the fact that every *character* $\chi : \mathcal{O}(G) \to \mathbb{C}$ is an evaluation. The proof of Iss'sa's general theorem, however, requires not only the Weierstrass product theorem but also tools from *valuation theory*; in the background is the theorem that every valuation on $\mathcal{M}(G)$ is equivalent to the order function o_c of a point $c \in G$ (Theorem 5). — G and $\widehat{G}$ always denote domains in $\mathbb{C}$.

1. Bers's theorem. Every $\mathbb{C}$-algebra homomorphism $\mathcal{O}(G) \to \mathbb{C}$ is called a *character of* $\mathcal{O}(G)$. Every *evaluation* $\chi_c : \mathcal{O}(G) \to \mathbb{C}$, $f \mapsto f(c)$, $c \in G$, is a character. We prove that these are *all* the characters of $\mathcal{O}(G)$.

(1) *For every character χ of $\mathcal{O}(G)$, $\chi = \chi_c$ with $c := \chi(id_G) \in G$.*

Proof. Set $e(z) := z - c$; then $\chi(e) = \chi(id_G) - c = 0$. It follows that $c \in G$, since otherwise e would be a unit in $\mathcal{O}(G)$ and we would have $1 = \chi(e \cdot e^{-1}) = \chi(e)\chi(e^{-1}) = 0$. Now let $f \in \mathcal{O}(G)$ be arbitrary. Then $f(z) = f(c) + e(z)f_1(z)$, with $f_1 \in \mathcal{O}(G)$. It follows that

$$\chi(f) = \chi(f(c)) + \chi(e)\chi(f_1) = f(c) = \chi_c(f); \quad \text{hence} \quad \chi = \chi_c. \qquad \square$$

The theorem now follows quickly from (1).

Bers's theorem. *For every $\mathbb{C}$-algebra homomorphism $\varphi : \mathcal{O}(G) \to \mathcal{O}(\widehat{G})$, there exists exactly one map $h : \widehat{G} \to G$ such that $\varphi(f) = f \circ h$ for all $f \in \mathcal{O}(G)$. In fact, $h = \varphi(id_G) \in \mathcal{O}(\widehat{G})$. — φ is bijective if and only if h is biholomorphic.*

Proof. Since h is to satisfy $\varphi(f) = f \circ h$ for all f, it must satisfy $\varphi(id_G) = id_G \circ h = h$. We show that the theorem does in fact hold for $h := \varphi(id_G)$. Since $\chi_a \circ \varphi$, $a \in \widehat{G}$, is always a character of $\mathcal{O}(G)$, it follows from (1) that

$$\chi_a \circ \varphi = \chi_c, \quad \text{with} \quad c = (\chi_a \circ \varphi)(id_G) = \chi_a(h) = h(a), \quad a \in \widehat{G}.$$

Hence $\varphi(f) = f \circ h$ for all $f \in \mathcal{O}(G)$, since we now have

$$\varphi(f)(a) = \chi_a(\varphi(f)) = (\chi_a \circ \varphi)(f) = \chi_{h(a)}(f) = f(h(a)) = (f \circ h)(a)$$

for all $a \in \widehat{G}$. The last statement of the theorem is immediate. $\square$

Bers's theorem contains some real surprises:

- if the function algebras $\mathcal{O}(G)$ and $\mathcal{O}(\widehat{G})$ are *algebraically isomorphic*, then the domains G and $\widehat{G}$ are *biholomorphically isomorphic*;
- every $\mathbb{C}$-algebra homomorphism $\varphi : \mathcal{O}(G) \to \mathcal{O}(\widehat{G})$ is *automatically continuous* (if a sequence in $\mathcal{O}(G)$ converges compactly in G to f, then the image sequence converges compactly in $\widehat{G}$ to $\varphi(f)$).

2. Iss'sa's theorem. *Let $\varphi : \mathcal{M}(G) \to \mathcal{M}(\widehat{G})$ be any $\mathbb{C}$-algebra homomorphism. Then there exists exactly one holomorphic map $h : \widehat{G} \to G$ such that $\varphi(f) = f \circ h$ for all $f \in \mathcal{M}(G)$.*

Because of Bers's theorem and because $\mathcal{M}(G)$ is the quotient field of $\mathcal{O}(G)$ (cf. 4.1.5), it suffices to prove the following lemma.

Lemma. *For every field homomorphism $\varphi : \mathcal{M}(G) \to \mathcal{M}(\widehat{G})$,*

$$\varphi(\mathcal{O}(G)) \subset \mathcal{O}(\widehat{G}).$$

The proof is carried out in the next subsection. We use methods from valuation theory (well known to algebraists, but less familiar to classical function theorists). We write $\mathcal{M}(G)^\times$ for the *multiplicative group* $\mathcal{M}(G)\backslash\{0\}$. A map $v : \mathcal{M}(G)^\times \to \mathbb{Z}$ is called a *valuation on* $\mathcal{M}(G)$ if for every f, $g \in \mathcal{M}(G)^\times$,

B1) $v(fg) = v(f) + v(g)$ *(product rule)*,

B2) $v(f+g) \geq \min\{v(f), v(g)\}$ if $f \neq -g$.

Our next result is immediate.

If v is a valuation on $\mathcal{M}(G)$, then $v(c) = 0$ for all $c \in \mathbb{C}^\times$.

Proof. For every $n \geq 1$ there exists $c_n \in \mathbb{C}^\times$ such that $(c_n)^n = c$. It follows from B1) that $v(c) = nv(c_n) \in n\mathbb{Z}$ for all $n \geq 1$; but this is possible only if $v(c) = 0$. □

Condition B2) can be sharpened:

B2') $v(f+g) = \min\{v(f), v(g)\}$ *if $f \neq -g$ and $v(f) \neq v(g)$.*

Proof. Let $v(f) \leq v(g)$. Since $v(-g) = v(g)$, it follows from B2) that

$$v(f) \geq \min\{v(f+g), v(g)\} \geq \min\{v(f), v(g)\} = v(f);$$

thus $\min\{v(f+g), v(g)\} = v(f)$. Hence $v(f+g) = v(f)$ if $v(f) < v(g)$. □

The valuations on $\mathcal{M}(G)$ that are important in function theory are the order functions o_c, $c \in G$, which assign to each function $f \in \mathcal{M}(G)^\times$ its order at the point c; see, for example, I.10.3.4. The following is immediate:

Holomorphy criterion. *A meromorphic function $f \in \mathcal{M}(G)^\times$ is holomorphic in G if and only if $o_c(f) \geq 0$ for all $c \in G$.*

3. Proof of the lemma. The core of the proof is contained in the next lemma.

Auxiliary lemma. *If v is a valuation on $\mathcal{M}(\mathbb{C})$, then $v(z) \geq 0$.*

Proof (cf. [Is], pp. 39–40). Suppose that $v(z) = -m$ with $m \geq 1$. Since $v(c) = 0$ for all $c \in \mathbb{C}^\times$, it follows from B2′) that

$$(1) \qquad v(z - c) = -m \quad \text{for all } c \in \mathbb{C}^\times.$$

Now let $d \in \mathbb{N}$, $d \geq 2$. By the existence theorem 3.1.5, there exists a function $q \in \mathcal{O}(\mathbb{C})$ which has no zeros in $\mathbb{C}\backslash\mathbb{N}$ and vanishes to order d^k at $k \in \mathbb{N}$. Set $q_n(z) := q(z) \Big/ \prod_0^{n-1}(z-\nu)^{d^\nu}$, $n \geq 1$. Then $q_n \in \mathcal{O}(\mathbb{C})$, and B1) and 1) imply that

$$(2) \qquad v(q_n) = v(q) + m \sum_0^{n-1} d^\nu = v(q) + \frac{m}{d-1}(d^n - 1).$$

By the construction of q_n, d^n divides every number $o_z(q_n)$, $z \in \mathbb{C}$; hence, by the root criterion 3.1.5, there exists $g_n \in \mathcal{O}(\mathbb{C})$ with $g_n^{d^n} = q_n$. Thus $d^n v(g_n) = v(q_n)$ and, by (2),

$$(3) \qquad v(q) + \frac{m}{d-1}(d^n - 1) \in d^n\mathbb{Z} \quad \text{for all } n \geq 1.$$

Therefore $(d-1)v(q) - m \in d^n\mathbb{Z}$ for all $n \geq 1$, which is possible only for $v(q) = m/(d-1)$. Since $d \geq 2$ was arbitrary, this gives the contradiction $m = 0$. □

The next statement is immediate from the auxiliary lemma:

($*$) *If v is any valuation on $\mathcal{M}(G)$, then $v(f) \geq 0$ for all $f \in \mathcal{O}(G)\backslash\{0\}$.*

Proof of ($*$). A verification shows that for every $f \neq 0$ in $\mathcal{O}(G)$, the map $v_f : \mathcal{M}(\mathbb{C})^\times \to \mathbb{Z}$, $g \mapsto v(g \circ f)$, is a valuation on $\mathcal{M}(\mathbb{C})$. Since $v_f(z) = v(f)$, ($*$) follows from the auxiliary lemma. □

After these preliminaries, the *proof of the lemma* is easy: Since φ is injective as a homomorphism of fields, $\varphi(f) \neq 0$ for all $f \in \mathcal{M}(G)^\times$. Hence, for every $c \in \widehat{G}$, setting

$$v_c(f) := o_c(\varphi(f)), \quad f \in \mathcal{M}(G)^\times,$$

defines a valuation on $\mathcal{M}(G)$. By ($*$), $o_c(\varphi(f)) \geq 0$ for all $c \in \widehat{G}$ if $f \in \mathcal{O}(G)^\times$. The holomorphy criterion 2 then implies that $\varphi(f) \in \mathcal{O}(\widehat{G})$, and the assertion follows. □

4. Historical remarks on the theorems of Bers and Iss'sa. The American mathematician Lipman Bers discovered his theorem in 1946 and published it in 1948 [Ber]. Bers considers only isomorphisms; he works with the maximal principal ideals of the rings $\mathcal{O}(G)$ and $\mathcal{O}(\widehat{G})$. Incidentally, Bers proves more: he proceeds from *ring* isomorphisms $\varphi : \mathcal{O}(G) \to \mathcal{O}(\widehat{G})$ and shows ingeniously that φ induces either the identity or conjugation on $\mathbb{C}$; $h : \widehat{G} \to G$ is correspondingly biholomorphic or anti-biholomorphic.

Before Bers, C. Chevalley and S. Kakutani had already studied the difficult case of the algebra of bounded holomorphic functions (unpublished). A historical survey can be found in [BuSa] (p. 84).

Bers's theorem is also valid for holomorphic functions of several variables, if their domains of definition are assumed to be normal Stein spaces. But the proof becomes rather demanding; one must use cohomological methods and resort to the theory of coherent analytic sheaves (cf. [GR], Chapter V, §7).

Hej Iss'sa (the pseudonym of a well-known Japanese mathematician) extended Bers's theorem to fields of functions in 1965. He immediately handles the case of *complex spaces* (cf. 2.6); his result is the following ([I], Theorem II, p. 34):

Let G be a normal complex space and $\widehat{G}$ a reduced Stein space, and let $\varphi : \mathcal{M}(G) \to \mathcal{M}(\widehat{G})$ be any $\mathbb{C}$-algebra homomorphism. Then there exists exactly one holomorphic map $h : \widehat{G} \to G$ such that $\varphi(f) = f \circ h$ for all $f \in \mathcal{M}(G)$.

Once again, the hard part of the proof is to show that φ maps the ring $\mathcal{O}(G)$ into $\mathcal{O}(\widehat{G})$. Iss'sa's theorem should also be compared with the 1968 paper [Ke] of J. J. Kelleher.

5*. Determination of all the valuations on $\mathcal{M}(G)$. The algebraically inclined reader will ask whether there actually exist valuations on $\mathcal{M}(G)$ that are not order functions. Certainly the function mo_c is a valuation on $\mathcal{M}(G)$ for every point $c \in G$ and every $m \in \mathbb{N}$. We prove that there are no other valuations.

Proposition. *For every valuation $v \neq 0$ on $\mathcal{M}(G)$, there exists exactly one point $c \in G$ such that $v(z-c) \geq 1$. Moreover, $v(h) = mo_c(h)$ for all $h \in \mathcal{M}(G)^\times$, where $m := v(z-c)$.*

Proof (cf. [I], pp. 40–41). First, $v(e) = 0$ for every unit $e \in \mathcal{O}(G)$, since $0 = v(1) = v(e \cdot 1/e) = v(e) + v(1/e)$ and since both $v(e) \geq 0$ and $v(1/e) \geq 0$ hold by 3(*).

We now set $A := \{a \in G : v(z-a) > 0\}$ and claim that

(#) $v(f) = 0$ *for every* $f \in \mathcal{O}(G) \setminus \{0\}$ *such that* $Z(f) \cap A = \emptyset$.

If $Z(f)$ is finite, then $f(z) = e(z)\prod_{\nu=1}^{n}(z - c_\nu)$, where e is a unit in $\mathcal{O}(G)$. Since $v(z - c_\nu) \geq 0$ by 3(*) and since $c_\nu \notin A$, it follows from B1) that $v(f) = v(e) = 0$. On the other hand, if $Z(f) = \{c_1, c_2, \ldots\}$ is infinite, we use the existence theorem 4.1.5 to choose $h \in \mathcal{O}(G)$ such that

$$Z(h) = Z(f) \quad \text{and} \quad o_{c_\nu}(h) = o_{c_\nu}(f) \cdot (\nu! - 1), \ \nu = 1, 2, \ldots.$$

For $h_n := h \cdot f \Big/ \prod_{\nu=1}^{n-1}(z-c_\nu)^{o_{c_\nu}(f)\cdot\nu!} \in \mathcal{O}(G)$, we then have

$Z(h_n) = \{c_n, c_{n+1}, \ldots\}$, $v(h_n) = v(h)+v(f)$, and $o_{c_\nu}(h_n) = o_{c_\nu}(f)\cdot\nu!$ for $\nu \geq n$.

Therefore $n!$ divides every number $o_z(h_n)$, $z \in G$; thus, by the root criterion 4.1.5, there exists $g_n \in \mathcal{O}(G)$ such that $h_n/g_n^{n!}$ is a unit in $\mathcal{O}(G)$. It follows that $v(h_n) = n!v(g_n)$ and hence that $v(h)+v(f) = v(h_n) \in n!\mathbb{Z}$, $n = 1, 2, \ldots$. This implies that $v(h)+v(f) = 0$. Since $v(h) \geq 0$ and $v(f) \geq 0$ by 3(*), we obtain $v(f) = 0$. This proves (#).

It follows immediately from (#) that A is *nonempty*, since otherwise $v(f) = 0$ would hold for all $f \in \mathcal{O}(G)\backslash\{0\}$, and therefore — since, by 4.1.5, $\mathcal{M}(G)$ is the quotient field of $\mathcal{O}(G)$ — v would have to be the zero valuation.

Hence there exists some $c \in A$. There exist no other points $c' \in A$, $c' \neq c$, since the equation $r(z-c') - r(z-c) = 1$, where $r := (c-c')^{-1} \in \mathbb{C}^\times$, would lead to the contradiction

$$0 = v(1) \geq \min\{v(z-c'), v(z-c)\} > 0.$$

Thus $A = \{c\}$, proving the first statement of the proposition. Now let $m := v(z-c)$. Then, if $f \neq 0$ is in $\mathcal{O}(G)$ and $n := o_c(f)$, the function $g := f/(z-c)^n \in \mathcal{O}(G)$ has no zero in A. It follows from (#) that

$$v(g) = 0, \quad \text{i.e.} \quad v(f) = v((z-c)^n) = m o_c(f).$$

Taking quotients gives $v(h) = m o_c(h)$ for all $h \in \mathcal{M}(G)^\times$. □

§2. Domains of Holomorphy

> Es giebt analytische Functionen, die nur für einen Theil der Ebene existieren und für den übrigen Theil der Ebene gar keine Bedeutung haben. (There are analytic functions which exist only for part of the plane and have no meaning at all for the rest of the plane.)
>
> — Weierstrass, 1884

1. A domain G in $\mathbb{C}$ is called the *domain of holomorphy* of a function f holomorphic in G if, for every point $c \in G$, the disc about c in which the Taylor series of f converges lies in G. The following is immediate:

If G is the domain of holomorphy of f, then G is the "maximal domain of existence" of f; in other words, if $\widehat{G} \supset G$ is a domain in which there exists a function $\widehat{f} \in \mathcal{O}(\widehat{G})$ such that $\widehat{f}|G = f$, then $\widehat{G}$ coincides with G.

If a *disc* is the maximal domain of existence of f, then it is also the domain of holomorphy of f (prove this); the definition given in I.5.3.3 for discs is thus consistent with the definition above. *In general, however, domain of holomorphy means more than maximal domain of existence.* For example, the slit plane $\mathbb{C}^-$ is the maximal domain of existence of the functions $\sqrt{z}$ and $\log z \in \mathcal{O}(\mathbb{C}^-)$ but not their domain of holomorphy: the Taylor series for $\sqrt{z}$ and $\log z$ about $c \in \mathbb{C}^-$ have $B_{|c|}(c)$ as their disc of convergence, and $B_{|c|}(c) \not\subset \mathbb{C}^-$ if $\operatorname{Re} c < 0$. (The functions $\sqrt{z}$ and $\log z$ can be continued holomorphically "from above and below" to every point on the negative real axis, but all the boundary points of $\mathbb{C}^-$ are "singular" for $\sqrt{z}$ and $\log z$, in the sense that none has a neighborhood U with a function $h \in \mathcal{O}(U)$ that coincides in $U \cap \mathbb{C}^-$ with $\sqrt{z}$ or $\log z$; cf. I.5.3.3 and Subsection 3 of this section.)

2. The domains $\mathbb{C}$, $\mathbb{C}^\times$, and $\mathbb{E}$ are the domains of holomorphy of z, z^{-1}, and $\sum z^{2^\nu}$, respectively (for the last example, cf. I.5.3.3). The focal point of this section is the following general theorem.

Existence theorem. *For every domain in $\mathbb{C}$, there exists a function f holomorphic in G such that G is the domain of holomorphy of f.*

There are two ways to prove this. A function $f \in \mathcal{O}(G)$ is constructed that either tends to ∞ as the boundary of G is approached or has zero set $Z(f) \neq G$ accumulating at every boundary point. Difficulties arise if the boundary ∂G is tricky (if it has accumulations of spikes, for example, as in Figure 8.4, page 176). We must ensure that the boundary is approached "from all directions inside G." To define this kind of approach to the boundary, we introduce the concepts of *well-distributed boundary set* and *peripheral set*. The first proof then goes through with "Goursat series"; the second exploits the existence theorem 4.1.5.

In what follows, we use concepts and methods of proof from set-theoretic topology. We use the fact that every point of an open set D in $\mathbb{C}$ lies in a uniquely determined connected component of D (cf. for example I.0.6.4), which for brevity we call a *component* of D; every such component is a *nonempty maximal subdomain of D*.

Remark. The following weak form of the existence theorem is easy to obtain:

Every domain G is the maximal domain of existence of a function $f \in \mathcal{O}(G)$.

Proof. Choose a set A that is locally finite in G and accumulates at every boundary point of G. By the general product theorem, there exists an $f \in \mathcal{O}(G)$ with $Z(f) = A$. By the identity theorem, there is no holomorphic extension of f to a domain $\widehat{G} \supset G$. □

1. A construction of Goursat. We fix a sequence $a_1, a_2, \ldots$ in $\mathbb{C}^\times$ with $\sum |a_\nu| < \infty$ and a sequence $b_1, b_2, \ldots$ of pairwise distinct points in $\mathbb{C}$. We denote by A the closure in $\mathbb{C}$ of the set $\{b_1, b_2, \ldots\}$.

(1) *The series* $f(z) = \sum_{\nu=1}^{\infty} \frac{a_\nu}{z - b_\nu}$ *converges normally in* $\mathbb{C} \backslash A$.

Proof. If $K \subset \mathbb{C}\setminus A$ is compact, the distance d between K and A is positive. Since $|z-b_\nu| \geq d$ for $z \in K$, it follows that $\sum |a_\nu/(z-b_\nu)|_K \leq d^{-1}\sum|a_\nu| < \infty$. □

The function $f \in \mathcal{O}(\mathbb{C}\setminus A)$ defined by (1) becomes arbitrarily large as the points of A are approached radially. The next lemma makes this precise.

Lemma (Goursat). *Let B be a disc in $\mathbb{C}\setminus A$ such that an element b_n of the sequence lies on ∂B. Then $\lim_{w\to b_n} f(w) = \infty$ if w approaches b_n along the radius of B to b_n.*

Proof. If w lies on the radius of B to b_n, then (!)

$$(\circ) \qquad |w-b_n| < |w-b_\nu| \quad \text{for all } \nu \neq n.$$

Let $p > n$ be chosen such that $\sum_{\nu=p+1}^\infty |a_\nu| \leq \frac{1}{2}|a_n|$. We rewrite (1) in the form

$$f(z) := \frac{a_n}{z-b_n} + g(z) + \sum_{\nu=p+1}^\infty \frac{a_\nu}{z-b_\nu}, \quad \text{where } g(z) := \left(\sum_1^p \frac{a_\nu}{z-b_\nu}\right) - \frac{a_n}{z-b_n}.$$

By $(\circ)$, for all w on the radius of B to b_n,

$$|f(w)| \geq \frac{|a_n|}{|w-b_n|} - |g(w)| - \sum_{\nu=p+1}^\infty \frac{|a_\nu|}{|w-b_\nu|} \geq \frac{1}{2}\frac{|a_n|}{|w-b_n|} - |g(w)|.$$

Since $|g(w)|$ remains finite as w approaches b_n, the assertion follows. □

Remark. The statement of the lemma is not obvious once the point b_n is an accumulation point of other points b_k. The growth of the "pole terms" $a_n/(z-b_n)$ about b_n could then be offset by the infinitely many terms $a_k/(z-b_k)$ corresponding to the b_k that accumulate at b_n. This phenomenon actually does occur for other series. For instance, every summand of the series

$$g(z) = \sum_{\nu=0}^\infty 2^\nu z^{2^\nu - 1}/(1+z^{2^\nu}),$$

which converges normally in $\mathbb{E}$, has poles on $\partial\mathbb{E}$; different summands never have equal poles (so that nothing cancels here), the poles of all the summands are dense in $\partial\mathbb{E}$, yet in the limit these poles cancel completely. The limit function, far from having infinitely many singularities on $\partial\mathbb{E}$, is just $g(z) = 1/(1-z)$; this can be seen at once by logarithmic differentiation of the product $\prod_{\nu=0}^\infty(1+z^{2^\nu})$ (cf. Exercise 2 in 1.2.1).

Historical note. E. Goursat used series of the type $\sum a_\nu/(z-b_\nu)$ in 1887 to construct functions with natural boundaries [Gour]. A. Pringsheim studied such series intensively; cf. [Pr], pp. 982–990.

2. Well-distributed boundary sets. First proof of the existence theorem. If b is a boundary point of G, a disc $V \subset G$ is called a *visible disc for* b if $b \in \partial V$; b is called a *visible (from G) boundary point of G*. In general, domains have boundary points that are not visible. Thus, in squares, the vertices are not visible; in domains with spikes (Figure 8.4, p. 176), there exist boundary curves none of whose points are visible.

A set M of visible boundary points of G is called *well distributed* if the following holds:

(∗) *Let $c \in G$ and let B be a disc about c that intersects ∂G. Then in the component of $B \cap G$ containing c there is a visible disc V for some point $b \in M \cap B$.*

With the aid of this concept, we obtain a

First criterion for domains of holomorphy. *If $\{b_1, b_2, \ldots\}$ is a countable well-distributed boundary set, then G is the domain of holomorphy of every function*

$$f(z) = \sum_1^\infty a_\nu/(z - b_\nu), \quad z \in G, \quad \textit{where} \quad a_\nu \in \mathbb{C}^\times, \quad \sum_1^\infty |a_\nu| < \infty.$$

Proof. $f \in \mathcal{O}(G)$ by 1(1) since $\overline{\{b_1, b_2, \ldots\}} \subset \partial G$. Let $c \in G$ and let B be the disc of convergence of the Taylor series h of f about c. Suppose that $B \cap \partial G \neq \emptyset$. Then, by (∗), in the component W of $B \cap G$ containing c there is a visible disc V for some point $b_n \in B$. Since $h|W = f|W$, Lemma 1 implies that h tends to ∞ as b_n is approached along the radius of V to b_n. Then $b_n \notin B$. It follows that $B \subset G$. □

The following statement is not obvious:

(1) *Any $G \neq \mathbb{C}$ has a countable well-distributed boundary set M.*

Proof. Let R be countable and dense in G, e.g. $R = (\mathbb{Q} + i\mathbb{Q}) \cap G$. For each $\zeta \in R$, choose $b \in \partial G$ on the boundary of the *largest* disc $V \subset G$ about ζ. The set M of all these visible boundary points b is countable.

Now let B be a disc about $c \in G$ that intersects ∂G. If $\zeta \in R$ is chosen close enough to c, then the largest disc $V \subset G$ about ζ lies, together with ∂V, in B, and $c \in V$. By the construction of M, V is the visible disc for some point $b \in M$. Since $b \in \partial V \subset B$, and since $V \subset B \cap G$ lies in the component of $B \cap G$ containing c because $c \in V$, (1) is proved. □

The construction of the set M by way of $(\mathbb{Q} + i\mathbb{Q}) \cap G$ is motivated by the Poincaré-Volterra theorem, which says, among other things, that the Taylor series of f about all the rational complex points yield all possible holomorphic continuations of f. (These "function elements" are dense in the "analytic structure" of f.)

The existence theorem follows immediately from (1) and the criterion above.

Historical note. The proof of the existence theorem given here can be found in Pringsheim's 1932 book [Pr] (pp. 986–988); he credits an oral communication of F. Hartogs. Pringsheim works only with dense sets of visible boundary points. In 1938, J. Besse pointed out drawbacks to such a choice of boundary points (see the following exercises) and eliminated them ([Bes], pp. 303–305). H. Kneser, in his *Funktionentheorie*, 2nd ed. (pp. 158–159), discusses only the weak form of the existence theorem (see the introduction to this section); his approach is the same as Pringsheim's.

Exercises. Prove: a) Well-distributed boundary sets for G are *dense* in ∂G.

b) There exist domains for which *not* every dense set of visible boundary points is well distributed.

c) If G is *convex*, then every dense set of visible boundary points is well distributed.

3. Discussion of the concept of domains of holomorphy. A function $f \in \mathcal{O}(G)$ is called *holomorphically extendible (or continuable) to a boundary point p of G* if there exist a neighborhood U of p and a holomorphic function $g \in \mathcal{O}(U)$ such that f and g coincide on a component W of $U \cap G$, with $p \in \partial W$; otherwise p is called a *singular* point of f. In general, the neighborhood U is "large": Thus, for the boundary point 0 of the domain $G := \mathbb{H} \setminus \bigcup_{n=1}^{\infty} \{(-\infty, n] \times \{i/n\}\}$, there does not exist any disc $B \neq \mathbb{C}$ such that 0 lies in the boundary of a component of $B \cap G$; every function $f := g|G$ with $g \in \mathcal{O}(\mathbb{C})$ is of course holomorphically extendible to 0 (with $U := \mathbb{C}$).

If there exist discs $B \subset U$ about p such that $B \cap G$ is connected, then we can choose U to be such a disc. If G is a *convex* domain, then for every disc B the region $B \cap G$ is again convex and thus a domain; the next statement follows.

(1) *If G is convex and $f \in \mathcal{O}(G)$ can be extended holomorphically to $p \in \partial G$, then there exist a disc B about p and a function $g \in \mathcal{O}(U)$ such that $g|B \cap G = f|B \cap G$.*

By (1), the expression "singular point of f" just introduced agrees for discs with that introduced in I.5.3.3. We now make precise the notion that a domain of holomorphy is the *maximal* domain in which a function is holomorphic.

Theorem. *The following statements about a function $f \in \mathcal{O}(G)$ are equivalent:*

i) *The domain G is the domain of holomorphy of f.*
ii) *There do not exist a domain $\widehat{G} \not\subset G$ and a function $\widehat{f} \in \mathcal{O}(\widehat{G})$ such that the set $\{z \in G \cap \widehat{G} : f(z) = \widehat{f}(z)\}$ has interior points.*
iii) *Every boundary point of G is a singular point of f.*

Condition ii) sharpens the maximality property discussed in the introduction, where we required that $\widehat{G} \supset G$. To prove this theorem, we need a lemma.

Lemma. *Let G and $\widehat{G}$ be domains in $\mathbb{C}$, and let W be a component of $G \cap \widehat{G}$. Then $\widehat{G} \cap \partial W \subset \partial G$. If $\widehat{G} \not\subset G$, then $\widehat{G} \cap \partial W$ is nonempty (see Figure 5.1).*

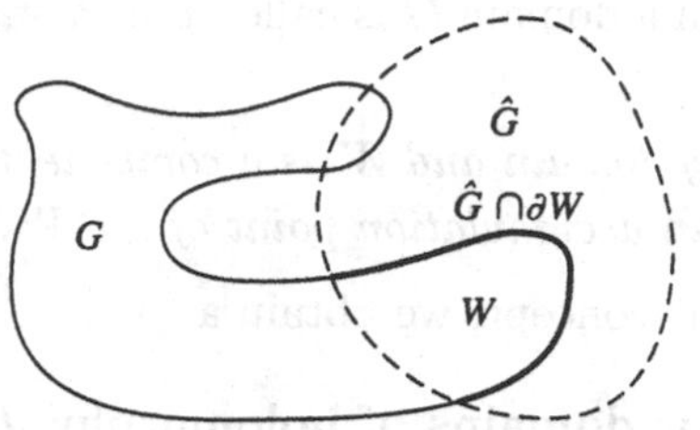

FIGURE 5.1.

Proof. 1) Let $q \in \widehat{G} \cap \partial W$. Since $\partial W \subset \overline{W} \subset \overline{G}$, it follows that $q \in \overline{G}$. If q were in G, then, since $q \in \widehat{G}$, it would follow that $q \in W$; but this contradicts $q \in \partial W$. Hence $q \in \overline{G} \backslash G = \partial G$.

2) Let $\widehat{G} \not\subset G$. Then $\widehat{G} \backslash W$ is nonempty, for otherwise the inclusion $W \subset \widehat{G}$ would imply that $W = \widehat{G}$; but this would give the contradiction $\widehat{G} \subset G$ since $W \subset G$. Moreover, $\widehat{G} \backslash W$ is not open in $\mathbb{C}$ since $\widehat{G} = W \cup (\widehat{G} \backslash W)$, W is open, and $\widehat{G}$ is connected. Let $p \in \widehat{G} \backslash W$ but not an interior point of $\widehat{G} \backslash W$. Then $U \cap W \neq \emptyset$ for every neighborhood U of p; that is, $p \in \partial W$. It follows that $p \in \widehat{G} \cap \partial W$. □

We now prove the equivalence of the statements of the theorem in the form "*non* i) $\Rightarrow$ *non* ii) $\Rightarrow$ *non* iii) $\Rightarrow$ *non* i)."

non i) $\Rightarrow$ *non* ii). There exists a $c \in G$ such that the disc of convergence of the Taylor series $\widehat{f}$ of f about c does *not* lie in G. Since $\widehat{f} \in \mathcal{O}(\widehat{G})$ and $f|W = \widehat{f}|W$ on the component W of $G \cap \widehat{G}$ containing c, *non* ii) follows.

non ii) $\Rightarrow$ *non* iii). Suppose that $\widehat{G} \not\subset G$, $\widehat{f} \in \mathcal{O}(\widehat{G})$, and W_1 is a component of $G \cap \widehat{G}$ such that $f|W_1 = \widehat{f}|W_1$. By the lemma, there exists a point $p \in \widehat{G} \cap \partial W_1 \subset \partial G$. We may assume that p is a visible boundary point of W_1 (by density; cf. Exercise 2.a)). We choose a disc $U \subset \widehat{G}$ about p and a visible circle $V \subset W_1$ for $p \in \partial W_1$. Then $U \cap V$ lies in a component W of $G \cap U$. That $p \in \partial V \cap \partial G$ implies that $p \in \partial W$. Set $g := \widehat{f}|U$; then $g|W = f|W$ since $V \cap U \subset W_1$. Thus p is not a singular boundary point of f.

non iii) $\Rightarrow$ *non* i). Suppose that $p \in \partial G$ is not singular for f, and let U, g, and W be chosen accordingly. Let r be the radius of convergence of the Taylor series of g about p. We choose $c \in W$ with $|c-p| < r/2$. The disc of convergence of the Taylor series of g about c then contains the point $p \in \partial G$. Since f and g have the same Taylor series about c, G is not the domain of holomorphy of f. □

Exercise. If G is convex and is the maximal domain of existence of $f \in \mathcal{O}(G)$ in the sense of the introduction, then G is the domain of holomorphy of f.

4. Peripheral sets. Second proof of the existence theorem. A set A that is locally finite in a domain G is called *peripheral in* G if the following holds:

($*$) *If* $\widehat{G} \subset \mathbb{C}$ *is any domain and* W *is a component of* $G \cap \widehat{G}$, *then every point of* $\widehat{G} \cap \partial W$ *is an accumulation point of* $A \cap W$.

With the aid of this concept, we obtain a

Second criterion for domains of holomorphy. *If the zero set* $Z(f)$ *of* $f \in \mathcal{O}(G)$ *is peripheral in* G, *then* G *is the domain of holomorphy of* f.

Proof. We show that statement iii) of Theorem 3 holds. For contradiction, suppose there exist a point $p \in \partial G$, a disc U about p, and a function $g \in \mathcal{O}(U)$ such that $f|W = g|W$ on a component W of $G \cap U$ with $p \in \partial W$. Since $Z(f)$ is peripheral in G, p is an accumulation point of $Z(f) \cap W$. Since $Z(f) \cap W = Z(g) \cap W$, the identity theorem implies that $g \equiv 0$. It follows that $f \equiv 0$, which is impossible since $Z(f)$, as a peripheral set, is discrete in G. □

It is not obvious that peripheral sets always exist.

(1) *If* $G \neq \mathbb{C}$, *then there exist peripheral sets* A *in* G.

Proof. Let the set $(\mathbb{Q}+i\mathbb{Q}) \cap G$ be arranged in a sequence $\zeta_1, \zeta_2, \ldots$. In the largest disc $B^\nu \subset G$ about ζ_ν, choose a point a_ν with $d(a_\nu, \partial G) < 1/\nu$. Let $A := \{a_1, a_2, \ldots\}$. Since every compact set $K \subset G$ has a positive distance to the boundary $d(K, \partial G)$, the set $A \cap K$ is always finite; in other words, A is locally finite in G.

Now let $\widehat{G}$ and W be as in ($*$), and let $p \in \widehat{G} \cap \partial W$. Then, for every $\varepsilon > 0$ with $B_\varepsilon(p) \subset \widehat{G}$, there exists a rational point $\zeta_k \in B_\varepsilon(p) \cap W$ with $|p - \zeta_k| < \frac{1}{2}\varepsilon$. The largest disc B^k about ζ_k that is contained in G now lies in $B_\varepsilon(p)$, since $p \in \partial G$ by Lemma 3. $B^k \cap W \neq \emptyset$; hence $B^k \subset W$ since W is a maximal subdomain of $G \cap \widehat{G}$. For the point $a_k \in B^k$ corresponding to ζ_k, it now follows that $a_k \in B_\varepsilon(p) \cap A \cap W$. Since $\varepsilon > 0$ is arbitrary, ($*$) holds. □

The existence theorem again follows from (1) and the criterion, since there exists $f \in \mathcal{O}(G)$ with $Z(f) = A$ by the existence theorem 4.1.5. □

Besides domains of holomorphy, one also considers domains of meromorphy. G is called the *domain of meromorphy* of a function h meromorphic in G if there exists no domain $\widehat{G} \not\subset G$ with a function $\widehat{h} \in \mathcal{M}(\widehat{G})$ such that $\widehat{h}$ and h agree on a component of $G \cap \widehat{G}$. Clearly $\mathbb{C}^\times$ is the domain of meromorphy of $\exp(1/z)$ (but not of $1/z$). The reader should convince himself that we have actually proved the following:

Every domain G in $\mathbb{C}$ is the domain of meromorphy of a function holomorphic in G.

Exercise. Prove: If G is convex, then every set that is locally finite in G and accumulates at every boundary point of G is peripheral.

5. On the history of the concept of domains of holomorphy. As early as 1842, Weierstrass was well aware that holomorphic functions could have "natural boundaries" ([W$_1$], p. 84). He pointed this out in his lectures from 1863 on. At the same time Kronecker knew that $\mathbb{E}$ is the domain of holomorphy of the theta series $1 + 2\sum q^{\nu^2}$; cf. 11.1.4. The first reference in print to the appearance of natural boundaries occurs in 1866 in a treatise of Weierstrass (*Monatsber. Akad. Wiss. Berlin*, p. 617; in Weierstrass's *Werke*, which are hardly a faithful reproduction of the original papers, this passage is omitted).

Weierstrass claimed in 1880 that all domains in $\mathbb{C}$ are domains of holomorphy; he says ([W$_2$], p. 223): "Es ist leicht, ... selbst für einen beliebig begrenzten Bereich ... die Existenz von [holomorphen] Functionen [anzugeben], die über diesen Bereich nicht [holomorph] fortgesetzt werden können." (It is easy, ... even for an arbitrary bounded region, to show the existence of [holomorphic] functions that cannot be continued [holomorphically] beyond this region.) He gave no precise statement, far less a proof, of this claim. A few years later, in 1885, Runge gave a proof using the approximation theorem, which he had devised especially for this purpose (see Chapters 12 and 13); in Runge's paper ([R], p. 229) is the assertion "dass der Gültigkeitsbereich einer eindeutigen analytischen Function ... keiner andern Beschränkung unterliegt als derjenigen, zusammenhängend zu sein" (that the region of validity of a single-valued analytic function ... is subject to no other constraint than that of being connected).

Mittag-Leffler stated in a footnote to Runge's work (loc. cit., p. 229) that Runge's result had already appeared in 1884 in his own work [ML]; the existence theorem, however, does not appear there explicitly. The proof by means of the product theorem is given in many textbooks; for example, in 1912 in that of Osgood, in 1932 in that of Pringsheim, in 1934 in that of Bieberbach, and in 1956 in that of Behnke-Sommer. (Cf. [O], pp. 481–482; [Pr], pp. 713–716; [Bi], p.295; and [BS], pp. 253–255). In 1938 J. Besse, a student of G. Pólya, noted that the problem of approach to the boundary is overlooked in the first three books; he gave his elegant solution of the problem in [Bes].

6. Glimpse of several variables. The concept of domain of holomorphy can also — almost verbatim as in the introduction — be introduced for holomorphic functions of several variables. Surprisingly, it turns out that not all domains in $\mathbb{C}^n$, $n \geq 2$, are domains of holomorphy: for example, punctured domains $G \backslash \{p\}$, where $p \in G$, are never domains of holomorphy, since holomorphic functions of $n \geq 2$ variables cannot have isolated singularities; Hurwitz mentioned this as early as 1897 ([Hu], p. 474 in particular). Soon thereafter, in 1903, F. Hartogs discovered the famous "Kugelsatz," which appeared in his dissertation: If G is a bounded domain in $\mathbb{C}^n$, $2 \leq n < \infty$, with connected boundary ∂G, then *every* function holomorphic in a neighborhood of ∂G can be extended holomorphically to all of G ([Har], p. 231 in particular).

The best-known examples of domains that are not domains of holomorphy are "notched" bidiscs in $\mathbb{C}^2$, which are produced by removing sets of the form $\{(w, z) \in D : |w| \geq r, |z| \leq s\}$, $0 < r, s < 1\}$ from the unit bidisc $D := \{(w, z) \in \mathbb{C}^2 : |w| < 1, |z| < 1\}$: every function $f \in \mathcal{O}(Z)$ can be extended to all of D.

In 1932, H. Cartan and P. Thullen recognized *holomorphic convexity* as a characteristic property of domains of holomorphy; cf. [CT]. In the period that followed, a theory developed that brought deep insights into the nature of singularities of holomorphic functions of several variables and is still active today. The reader can find details in the monograph [BT] of H. Behnke and P. Thullen and in the textbook [GF] of H. Grauert and K. Fritzsche.

In 1951 K. Stein, in his notable paper [St], discovered complex spaces that have properties similar to those of domains of holomorphy. A complex space X is called a *Stein space* if many holomorphic functions live on it; more precisely, one imposes the following conditions.

a) *For any two points p, $q \in X$, $p \neq q$, there exists $f \in \mathcal{O}(X)$ with $f(p) \neq f(q)$ (separation axiom).*

b) *For every locally finite infinite set A, there exists $f \in \mathcal{O}(X)$ with* $\sup\{|f(x)| : x \in A\} = \infty$ *(convexity axiom).*

A domain G in $\mathbb{C}^n$ is a domain of holomorphy if and only if it is a Stein space. It turns out that many theorems of complex analysis can be proved at once for Stein spaces (see also 4.2.5 and 6.2.5). Readers who would like to go more deeply into these matters are referred to [GR].

§3. Simple Examples of Domains of Holomorphy

For domains G with complicated boundary, one can seldom *explicitly* determine holomorphic functions that have G as domain of holomorphy. But for discs, Cassini domains, and, more generally, domains of the form $\{z \in \mathbb{C} : |q(z)| < R\}$, where q is a nonconstant entire function, there are simple constructions, as we will now see.

1. Examples for $\mathbb{E}$. The disc $\mathbb{E}$ is the domain of holomorphy of $\sum z^{2^\nu}$ (cf. I.5.3.3); more generally, the circle of convergence of Hadamard lacunary series is their natural boundary (see 11.2.3 and 11.1.4). We give examples of a different kind.

1) *Let $a \in \mathbb{C}$, $|a| > 1$; let $\omega \in \mathbb{R}\backslash\mathbb{Q}\pi$. Then the domain of holomorphy of the "Goursat series"*

$$f(z) := \sum_1^\infty \frac{a^{-\nu}}{z - e^{i\nu\omega}} \in \mathcal{O}(\mathbb{E})$$

is the disc $\mathbb{E}$. Moreover, $ae^{i\omega} f(e^{i\omega} z) = (z-1)^{-1} + f(z)$.

Proof. $\{e^{i\nu\omega},\ \nu \geq 1\}$ is a well-distributed boundary set since $\omega \notin \mathbb{Q}\pi$. Hence the claim follows by Theorem 2.2. □

2) *The domain of holomorphy of the power series*

$$f(z) = \sum_{\lambda=0}^\infty \frac{z^\lambda}{e^{1+i\lambda} - 1}$$

is the disc $\mathbb{E}$.

Proof. For $G = \mathbb{E}$, the "Goursat series" $\sum_1^\infty a_\nu/(z - b_\nu)$ has the following Taylor series in $\mathbb{E}$ about 0:

$$-\sum_1^\infty \frac{a_\nu}{b_\nu} \cdot \frac{1}{1 - z/b_\nu} = -\sum_{\nu=1}^\infty \frac{a_\nu}{b_\nu} \sum_{\lambda=0}^\infty \left(\frac{z}{b_\nu}\right)^\lambda = -\sum_{\lambda=0}^\infty \left(\sum_{\nu=1}^\infty \frac{a_\nu}{b_\nu^{\lambda+1}}\right) z^\lambda, \quad z \in \mathbb{E}.$$

Setting $b_\nu := e^{i\nu}$ and $a_\nu := -e^{-\nu} b_\nu$ gives 2). □

3) *The domain of holomorphy of the product*

$$f(z) = \prod_{\nu=0}^\infty (1 - z^{2^\nu}) \in \mathcal{O}(\mathbb{E})$$

is the disc $\mathbb{E}$.

(Sketch of) Proof. Near every 2^nth root of unity ζ, f assumes arbitrarily small values. This holds for $\zeta = 1$ since $f(t) = (1-t)(1-t^2)(1-t^4)\ldots < 1-t$ for all t, $0 < t < 1$; it holds in general since

$$f(z) = f(z^{2^n}) \prod_{\nu=0}^{n-1} (1 - z^{2^\nu}), \quad \text{whence} \quad |f(z)| < 2^n |f(z^{2^n})|.$$

But the 2^nth roots ζ are dense in $\partial\mathbb{E}$, and 3) follows. □

4) The domain of holomorphy of the products in Example 2 of 4.2.2 is $\mathbb{E}$ if r_n is always chosen to be less than 1.

2. Lifting theorem. *Let $\widetilde{G}$ be a domain, let $q \in \mathcal{O}(\mathbb{C})$ be nonconstant, and let G be a component of $q^{-1}(\widetilde{G})$. If $\widetilde{G}$ is the domain of holomorphy of $\widetilde{f}$, then G is the domain of holomorphy of $f := \widetilde{f} \circ q|G$.*

Proof. Suppose that f could be extended holomorphically to a point $p \in \partial G$. Then there would exist a disc U about p and a function $g \in \mathcal{O}(U)$ such that $g|W = f|W$ on a component W of $U \cap G$ with $p \in \partial W$.

Suppose first that $q'(p) \neq 0$. Then q is locally biholomorphic about p. We choose U so small that q maps U biholomorphically onto a domain $\widehat{G}$. Since $q(p) \in \partial\widetilde{G}$, we have $\widehat{G} \not\subset \widetilde{G}$. For $\widehat{f} := g \circ (q|U)^{-1} \in \mathcal{O}(\widehat{G})$, it follows that $\{z \in \widetilde{G} \cap \widehat{G} : \widetilde{f}(z) = \widehat{f}(z)\} \supset q(W)$. By Theorem 2.3, $\widetilde{G}$ is not the domain of holomorphy of $\widetilde{f}$.

Suppose now that $q'(p) = 0$. Since $U \cap \partial W \subset \partial G$ by Lemma 2.3, g is a holomorphic extension of f to all the boundary points $\widetilde{p}$ of G that lie in $U \cap \partial W$. Since p is an *isolated* zero of q', by what has already been proved there do not exist such points $\widetilde{p}$ arbitrarily near p. Hence p is an *isolated* boundary point of G and $\widetilde{p} := q(p)$ is an isolated boundary point of $\widetilde{G}$. But now f is bounded in a neighborhood of p; therefore $\widetilde{f}$ is bounded in a neighborhood of $\widetilde{p}$, which is impossible since $\widetilde{G}$ is the domain of holomorphy of $\widetilde{f}$. □

Applications of the lifting theorem are obvious. In the next subsection, we discuss a situation that plays an important role in the theory of overconvergence; cf. also 11.3.1–11.3.4.

3. Cassini regions and domains of holomorphy. Regions $D := \{z \in \mathbb{C} : |z - z_1|\,|z - z_2| < \text{const.}\}$, z_1, z_2 fixed, are called Cassini regions, after the Italian-French astronomer G. D. Cassini (1625–1712), who — in contrast to Kepler — chose Cassini curves (lemniscates) $|z - z_1|\,|z - z_2| < \text{const.}$ rather than ellipses as the path of the planets around the sun. A normal form is

$$(1) \qquad |z - a|\,|z + a| = R^2 \quad \text{with} \quad a, R \in \mathbb{R},\ a > 0,\ R > 0.$$

These Cassini curves have only ordinary points except in the case $a = R$, where 0 is a double point (left-hand part of Figure 5.2). The Cassini region D corresponding to (1) has two components if $a \geq R$ and is connected if $a < R$. We prove a more precise result.

If $a < R$, then the Cassini region D is a star-shaped domain with center 0.

Proof. In polar coordinates, (1) has the form $r^4 - 2a^2r^2\cos 2\varphi = R^4 - a^4$. With $g(t) := (t - a^2\cos 2\varphi)^2 + a^4\sin^2 2\varphi - R^4$, it follows that $D = \{re^{i\varphi} \in \mathbb{C} : g(r^2) < 0\}$. Since $R^4 > a^4$, g has exactly one positive and one negative zero; hence $g(\rho^2) < 0$ for all $\rho \in [0, r]$ whenever $g(r^2) < 0$. Thus if $re^{i\varphi}$ lies in D, so do all points $tre^{i\varphi}$, $0 \leq t \leq 1$. □

The next statement follows immediately from the lifting theorem 2.

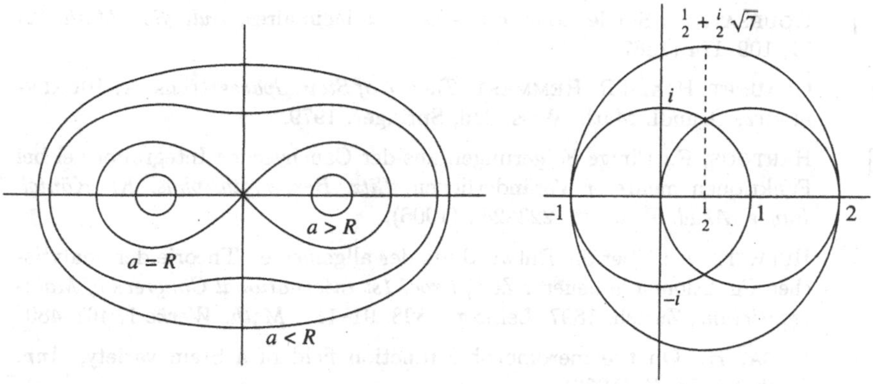

FIGURE 5.2.

For every $p \in \mathbb{N}\backslash\{0\}$, the series $\sum_0^\infty [\frac{1}{2}z(z-1)]^{2^\nu p}$ converges compactly in the Cassini domain $W := \{z \in \mathbb{C} : |z(z-1)| < 2\}$. The limit function has W as domain of holomorphy. Moreover, $W \supset (\mathbb{E}\backslash\{-1\}) \cup \overline{B_1(1)}\backslash\{2\})$.

The right-hand part of Figure 5.2 shows the domain W. (Writing $z + 1/2$ instead of z gives the normal form (1) for W, with $a = 1/2$, $R = \sqrt{2}$.) We will come across the Cassini domain W again, in the theory of overconvergence in 11.2.1.

Bibliography

[Ber] BERS, L.: On rings of analytic functions, *Bull. Amer. Math. Soc.* 54, 311–315 (1948).

[Bes] BESSE, J.: 303–305. Sur le domaine d'existence d'une fonction analytique, *Comm. Math. Helv.* 10, 302–305 (1938).

[Bi] BIEBERBACH, L.: *Lehrbuch der Funktionentheorie*, vol. 1, 4th ed., Teubner, Leipzig, 1934.

[BS] BEHNKE, H. and F. SOMMER: *Theorie der analytischen Funktionen einer komplexen Veränderlichen*, Springer, 2nd ed. 1962.

[BT] BEHNKE, H. and P. THULLEN: *Theorie der Funktionen mehrerer komplexer Veränderlichen*, 2nd ed., with appendices by W. BARTH, O. FORSTER, H. HOLMANN, W. KAUP, H. KERNER, H.-J. REIFFEN, G. SCHEJA, and K. SPALLEK, Erg. Math. Grenzgeb. 51, Springer, 1970.

[BuSa] BURCKEL, R. B. and S. SAEKI: Additive mappings on rings of holomorphic functions, *Proc. Amer. Math. Soc.* 89, 79–85 (1983).

[CT] CARTAN, H. and P. THULLEN: Zur Theorie der Singularitäten der Funktionen mehrerer komplexen Veränderlichen, *Math. Ann.* 106, 617-647 (1932); CARTAN's *Œuvres* 1, 376–406.

[GF] GRAUERT, H. and K. FRITZSCHE: *Several Complex Variables*, GTM 38, Springer, 1976.

[Gour] GOURSAT, E.: Sur les fonctions à espaces lacunaires, *Bull. Sci. Math.* (2) 11, 109–114 (1887).

[GR] GRAUERT, H. and R. REMMERT: *Theory of Stein Spaces*, trans. A.HUCKLEBERRY, Grundl. Math. Wiss. 236, Springer, 1979.

[Har] HARTOGS, F.: Einige Folgerungen aus der Cauchyschen Integralformel bei Funktionen mehrerer Veränderlichen, *Sitz. Ber. math.-phys. Kl. Königl. Bayer. Akad. Wiss.* 36, 223–242 (1906).

[Hu] HURWITZ, A.: Über die Entwicklung der allgemeinen Theorie der analytischen Funktionen in neuerer Zeit, *Proc. 1st International Congress of Mathematicians*, Zurich, 1897, Leipzig, 1898, 91–112; *Math. Werke* 1, 461–480.

[I] ISS'SA, H.: On the meromorphic function field of a Stein variety, *Ann. Math.* 83, 34–46 (1966).

[Ke] KELLEHER, J. J.: On isomorphisms of meromorphic function fields, *Can. Journ. Math.* 20, 1230–1241 (1968).

[ML] MITTAG-LEFFLER, G.: Sur la représentation analytique des fonctions monogènes d'une variable indépendante, *Acta Math.* 4, 1–79 (1884).

[O] OSGOOD, W. F.: *Lehrbuch der Funktionentheorie*, vol. 1, Teubner, 1907.

[Pr] PRINGSHEIM, A.: *Vorlesungen über Funktionenlehre. Zweite Abteilung: Eindeutige analytische Funktionen*, Teubner, Leipzig, 1932.

[R] RUNGE, C.: Zur Theorie der eindeutigen analytischen Functionen, *Acta Math.* 6, 229–244 (1885).

[St] STEIN, K.: Analytische Funktionen mehrerer komplexer Veränderlichen zu vorgegebenen Periodizitätsmoduln und das zweite Cousinsche Problem, *Math. Ann.* 123, 201–222 (1951).

[W_1] WEIERSTRASS, K.: Definition analytischer Functionen einer Veränderlichen vermittelst algebraischer Differentialgleichungen, *Math. Werke* 1, 75–84.

[W_2] WEIERSTRASS, K.: Zur Functionenlehre, *Math. Werke* 2, 201–223.

6

Functions with Prescribed Principal Parts

If h is meromorphic in the region D, its pole set $P(h)$ is *locally finite* in D. By the existence theorem 4.1.5, every set that is locally finite in D is the pole set of some function $h \in \mathcal{M}(D)$ (see also 3.1.5(1)). We now pose the following problem:

Let $T = \{d_1, d_2, \ldots\}$ be a set that is locally finite in D, and let every point $d_\nu \in T$ be somehow assigned a "finite principal part" $q_\nu(z) = \sum_{\mu=1}^{m_\nu} a_{\nu\mu}(z - d_\nu)^{-\mu} \neq 0$. Construct a function meromorphic in D that has T as its pole set and moreover has principal part q_ν at each point d_ν.

It is not at all clear that such functions exist. Of course, if T is finite, the "partial fraction series"

$$\sum q_\nu(z) = \sum_{\nu,\mu} a_{\nu\mu}(z - d_\nu)^{-\mu}$$

solves the problem. But if T is infinite, this series diverges in general. Mittag-Leffler, in the last century, forced convergence by subtracting a convergence-producing summand $g_\nu \in \mathcal{O}(D)$ from each principal part: the *Mittag-Leffler series* $\sum(q_\nu(z) - g_\nu(z))$ is then a meromorphic function in D with the desired poles and principal parts.

In this chapter we first treat Mittag-Leffler's theorem for the plane (Section 1). In Section 2 we discuss the case of arbitrary regions. In Section 3, as an application, we develop the basics of ideal theory in rings of holomorphic functions.

§1. Mittag-Leffler's Theorem for $\mathbb{C}$

The goal of this section is the proof of Mittag-Leffler's theorem for the plane. In order to formulate it conveniently, we use the concept of *principal part distributions*, which we discuss in Subsection 1, with a view toward later generalizations, for arbitrary regions in $\mathbb{C}$. The simple principle of using Mittag-Leffler series to find meromorphic functions with prescribed principal parts is described by Theorem 2. The classical Mittag-Leffler series for $D = \mathbb{C}$ are constructed in Subsection 3.

1. Principal part distributions. Every Laurent series $\sum_1^\infty b_\mu (z-d)^{-\mu} \in \mathcal{O}(\mathbb{C}\backslash\{d\})$ is called a *principal part at* $d \in \mathbb{C}$; a principal part is called *finite* if almost all the b_μ vanish.

(1) *$q \in \mathcal{O}(\mathbb{C}\backslash\{d\})$ is a principal part at d if and only if* $\lim_{z\to\infty} q(z) = 0$.

Proof. Every $q \in \mathcal{O}(\mathbb{C}\backslash\{d\})$ has a Laurent representation (cf. I.12.1.2–3):

$$q = q^+ + q^-, \quad \text{with} \quad q^+ \in \mathcal{O}(\mathbb{C}), \quad q^-(z) = \sum_{\mu=1}^{\infty} b_\mu (z-d)^{-\mu} \in \mathcal{O}(\mathbb{C}\backslash\{d\}),$$

and $\lim_{z\to\infty} q^-(z) = 0$. By Liouville's theorem, $\lim_{z\to\infty} q(z) = 0$ if and only if $q^+ \equiv 0$. □

A map φ which assigns to every point $d \in D$ a (finite) principal part q_d at d is called a *distribution of (finite) principal parts on D* if its support $T := \{z \in D : \varphi(z) \neq 0\}$ is locally finite in D. For brevity, we also call φ a *principal part distribution* on D.

Every function h that is holomorphic in D except for isolated singularities has a well-defined principal part $h^- \in \mathcal{O}(\mathbb{C}\backslash\{d\})$ at each such singularity; cf., for instance, I.12.1.3. Thus every such function determines a principal part distribution $PD(h)$ on D whose support is the set of nonremovable singularities of h in D. $PD(h)$ is a distribution of finite principal parts on D if and only if h is *meromorphic* in D; then the pole set $P(h)$ of h is the support of $PD(h)$.

Principal part distributions on D can be added and subtracted in a natural way, and thus (like divisors) form an *additive abelian group*.

There is a simple connection between divisors and principal part distributions:

(2) *If $\mathfrak{d}(d)$ is the divisor of $f \in \mathcal{O}(D)$, then $\varphi(d) := \mathfrak{d}(d)/(z-d)$, $d \in D$, is the principal part distribution of the logarithmic derivative $f'/f \in \mathcal{M}(D)$.*

This is clear: from $f(z) = (z-d)^n g(z)$, with $g(d) \neq 0$, it follows immediately that $f'(z)/f(z) = n/(z-d) + v(z)$, where v is holomorphic at d. □

The problem posed in the introduction to this chapter is now contained in the following problem:

For every principal part distribution φ in D with support T, construct a function $h \in \mathcal{O}(D\backslash T)$ with $PD(h) = \varphi$.

The key to the construction of such functions is provided by special series, which we now introduce.

2. Mittag-Leffler series. The support T of a principal part distribution φ, like the support of a divisor, is always at most countable. We arrange the points of T in a sequence $d_1, d_2, \ldots$, in which, however — unlike the case of divisors — each point of T occurs *exactly once.* We stipulate once and for all that $d_1 = 0$ if the origin belongs to T. The principal part distribution φ is uniquely described by the sequence (d_ν, q_ν), where $q_\nu := \varphi(d_\nu)$.

A series $h = \sum_1^\infty (q_\nu - g_\nu)$ is called a *Mittag-Leffler series for the principal part distribution (d_ν, q_ν) on D* if

1) *g_ν is holomorphic in D;*
2) *the series h converges normally in $D\backslash\{d_1, d_2, \ldots\}$.*

This terminology will turn out to be especially convenient, as we now see.

Proposition. *If h is a Mittag-Leffler series for (d_ν, q_ν), then*

$$h \in \mathcal{O}(D\backslash\{d_1, d_2, \ldots\}) \quad \textit{and} \quad PD(h) = (d_\nu, q_\nu).$$

Proof. By 2), h is holomorphic in $D\backslash\{d_1, d_2, \ldots\}$. Since all the summands $q_\nu - g_\nu$, $\nu \neq n$, are holomorphic in a neighborhood $U \subset D$ of d_n, the series $\sum_{\nu\neq n}(q_\nu - g_\nu)$ converges compactly in U, inclusive of d_n, to a function $\widehat{h}_n \in \mathcal{O}(U)$ (inward extension of convergence; cf. I.8.5.4). Since $h - q_n = \widehat{h}_n - g_n$ in $U\backslash\{d_n\}$ and $\widehat{h}_n$ and g_n are holomorphic at d_n, it follows that q_n is the principal part of h at d_n, $n \geq 1$. This proves that $PD(h) = (d_\nu, q_\nu)$. □

We also note:

If $\sum(q_\nu - g_\nu)$ is a Mittag-Leffler series for a distribution of *finite* principal parts, then every summand $q_\nu - g_\nu$ is *meromorphic in D* and the series is a *normally convergent series of meromorphic functions in D* (in the sense of I.11.1.1).

The terms g_ν in $\sum(q_\nu - g_\nu)$ force the convergence of the series, without disturbing the singular behavior of the series about d_ν that is prescribed by q_ν. The functions $g_1, g_2, \ldots$ are called *convergence-producing summands of the Mittag-Leffler series.*

We will construct Mittag-Leffler series for *every* principal part distribution φ. (This involves more than just finding functions h with $PD(h) = \varphi$;

compare the theorems on partial fraction decomposition in 4 and 2.3.) For this purpose, we need a way to determine convergence-producing summands. This is relatively simple in the case $D = \mathbb{C}$.

3. Mittag-Leffler's theorem. In this subsection $(d_\nu, q_\nu)_{\nu\geq 1}$ denotes a principal part distribution on $\mathbb{C}$. Every function $q_\nu \in \mathcal{O}(\mathbb{C}\backslash\{d_\nu\})$ has a Taylor series about 0, which converges in the disc of radius $|d_\nu|$, $\nu \geq 2$ (observe that $d_1 = 0$ is possible). We denote by $p_{\nu k}$ the kth Taylor polynomial for q_ν about 0 ($\deg p_{\nu k} \leq k$) and show that these polynomials can serve as convergence-producing summands.

Mittag-Leffler's theorem. *For every principal part distribution $(d_\nu, q_\nu)_{\nu\geq 1}$ in $\mathbb{C}$, there exist Mittag-Leffler series in $\mathbb{C}$ of the form*

$$q_1 + \sum_{\nu=2}^{\infty}(q_\nu - p_{\nu k_\nu}), \text{ where } p_{\nu k_\nu} := k_\nu\text{th Taylor polynomial of } q_\nu \text{ about } 0.$$

Proof. Since the sequence $(p_{\nu k})_{k\geq 1}$ converges compactly in $B_{|d_\nu|}(0)$ to q_ν, for every $\nu \geq 2$ there exists a $k_\nu \in \mathbb{N}$ such that $|q_\nu(z) - p_{\nu k_\nu}| \leq 2^{-\nu}$ for all z with $|z| \leq \frac{1}{2}|d_\nu|$. Since $\lim d_\nu = \infty$, every compact set K in $\mathbb{C}$ lies in almost all the discs $B_{\frac{1}{2}|d_\nu|}(0)$. Hence

$$\sum_{\nu\geq n} |q_\nu - p_{\nu k_\nu}|_K \leq \sum_{\nu\geq n} 2^{-\nu} < \infty \quad \text{for appropriate } n = n(K).$$

This proves the normal convergence of the series in $\mathbb{C}\backslash\{d_1, d_2, \ldots\}$. Since it always holds that $p_{\nu k_\nu} \in \mathcal{O}(\mathbb{C})$, the series in question is a Mittag-Leffler series in $\mathbb{C}$ for $(d_\nu, q_\nu)_{\nu\geq 1}$. □

The series

$$\frac{1}{z} + \sum_{\nu=2}^{\infty}\left(\frac{1}{z - d_\nu} + \frac{1}{d_\nu} + \frac{z}{d_\nu^2} + \cdots + \frac{z^{\nu-2}}{d_\nu^{\nu-1}}\right)$$

is a Mittag-Leffler series (with $q_\nu(z) = (z - d_\nu)^{-1}$ and $p_{\nu,\nu-2}(z)$); it appeared in 3.1.5(1).

4. Consequences. Mittag-Leffler's theorem has important corollaries.

Existence theorem. *Every principal part distribution on $\mathbb{C}$ with support T is the principal part distribution of a function holomorphic in $\mathbb{C}\backslash T$.*

Theorem on the partial fraction decomposition of meromorphic functions. *Every function h that is meromorphic in $\mathbb{C}$ can be represented by a series $\sum h_\nu$ that converges normally in $\mathbb{C}$, where each summand h_ν is rational and has at most one pole in $\mathbb{C}$.*

The existence theorem is clear; the second theorem is proved as follows: By Theorem 3, corresponding to the principal part distribution $PD(h)$ there is a Mittag-Leffler series $\widehat{h}$ in $\mathbb{C}$ whose convergence-producing summands are polynomials. Since all the principal parts of $\widehat{h}$ are finite, all the summands of this series are rational functions that have exactly one pole in $\mathbb{C}$. The difference $h - \widehat{h}$ is an entire function and therefore a normally convergent series of polynomials in $\mathbb{C}$ (Taylor series). □

The Weierstrass product theorem can be obtained from Mittag-Leffler's theorem. We sketch the proof for the case that $\mathfrak{d}$ is a positive divisor on $\mathbb{C}$ satisfying $\mathfrak{d}(0) = 0$ and prescribing only first-order zeros. If $d_1, d_2, \dots$ is a sequence corresponding to $\mathfrak{d}$, we consider the principal part distribution $(d_\nu, 1/(z-d_\nu))_{\nu\geq 1}$ on $\mathbb{C}$. Then $-(d_\nu)^{-1}\sum_{\kappa=0}^{k}(z/d_\nu)^\kappa$ is the kth Taylor polynomial; Mittag-Leffler series for this distribution look like

$$h(z) = \sum_{\nu=1}^{\infty} h_\nu(z), \quad \text{with} \quad h_\nu(z) := \frac{1}{z-d_\nu} + \frac{1}{d_\nu}\sum_{\kappa=0}^{k_\nu}\left(\frac{z}{d_\nu}\right)^\kappa.$$

Every $f \in \mathcal{O}(\mathbb{C})$ with $f'/f = h$ now has $\mathfrak{d}$ as divisor. Since

$$f_\nu'/f_\nu = h_\nu \text{ for } f_\nu(z) := \left(1 - \frac{z}{d_\nu}\right)\left[\exp\sum_{\kappa=0}^{k_\nu}\frac{1}{\kappa+1}\left(\frac{z}{d_\nu}\right)^{\kappa+1}\right], \quad \nu \in \mathbb{N},$$

Weierstrass's assertion that $f = \prod E_{k_\nu}(z/d_\nu)$ follows automatically. Of course, it must still be shown that this product converges. This can be done, for example, by integrating f_ν'/f_ν; for details, see [FL], pp. 176–177.

5. Canonical Mittag-Leffler series. Examples. In applying Theorem 3, we will choose the numbers k_ν as small as possible — as we did for Weierstrass products. If all the k_ν can be chosen to be equal, the Mittag-Leffler series $q_1 + \sum_2^\infty (a_\nu - p_{\nu k})$ with the smallest $k \geq 0$ is called the *canonical* series for the principal part distribution $(d_\nu, q_\nu)_{\nu\geq 1}$ in $\mathbb{C}$. We give four examples.

1) The *Eisenstein series* $\varepsilon_m(z) := \sum_{-\infty}^{\infty}(z+\nu)^{-m}$, $m \geq 2$, is the *canonical* series for the principal part distribution $(-\nu, 1/(z+\nu)^m)_{\nu\in\mathbb{Z}}$; convergence-producing summands are unnecessary here. We recall the explicit formulas (I.11.2.3)

$$\frac{\pi^2}{\sin^2 \pi z} = \sum_{-\infty}^{\infty}\frac{1}{(z+\nu)^2}, \qquad \pi^3\frac{\cot \pi z}{\sin^2 \pi z} = \sum_{-\infty}^{\infty}\frac{1}{(z+\nu)^3}.$$

2) The *cotangent series*

$$\pi \cot \pi z = \varepsilon_1(z) = \frac{1}{z} + \sum_{-\infty}^{\infty}{}'\left(\frac{1}{z+\nu} - \frac{1}{\nu}\right)$$

is the *canonical* series for the principal part distribution $(-\nu, 1/(z+\nu))_{\nu\in\mathbb{Z}}$; here $k = 0$.

3) The series

$$\frac{\Gamma'(z)}{\Gamma(z)} = -\gamma - \frac{1}{z} - \sum_{\nu=1}^{\infty}\left(\frac{1}{z+\nu} - \frac{1}{\nu}\right) \quad \text{(cf. 2.2.3)},$$

if we disregard the term $-\gamma$, is the *canonical* series for the principal part distribution $(-\nu, -1/(z+\nu))_{\nu\geq 0}$. Again $k = 0$.

4) The *Eisenstein-Weierstrass series*

$$\wp(z) := \frac{1}{z^2} + \sum_{\omega\neq 0}\left(\frac{1}{(z+\omega)^2} - \frac{1}{\omega^2}\right)$$

is the *canonical* series for the principal part distribution $(-\omega, 1/(z+\omega)^2)_{\omega\in\Omega}$, where Ω denotes a lattice in $\mathbb{C}$; cf. 3.2.4. Here also, $k = 0$.

These examples were well known before Mittag-Leffler proved his theorem. The theorem shows that the same construction principle underlies all four examples.

6. On the history of Mittag-Leffler's theorem for $\mathbb{C}$. In connection with the research of Weierstrass, which was published in 1876 ([Wei], pp. 77–124), Mittag-Leffler published his theorem in 1876–77 for the case that all the principal parts are finite, in Swedish, in the *Reports of the Royal Swedish Academy of Sciences, Stockholm* (cf. [ML], p. 20 and p. 21). In his 1880 note [Wei], pp. 189–199, Weierstrass simplified the proof considerably by introducing the Taylor polynomials as convergence-producing summands; in this work, Weierstrass also drew attention to the theorem proved by Mittag-Leffler in 1877 on the partial fraction decomposition (cf. pp. 194–195):

> *Es lässt sich also jede eindeutige analytische Function $f(x)$, für die im Endlichen keine wesentliche singuläre Stelle existirt, als eine Summe von rationalen Functionen der Veränderlichen x dergestalt ausdrücken, dass jede dieser Functionen im Endlichen höchstens eine Unendlichkeits-Stelle hat.* (Thus every single-valued analytic function $f(x)$ for which no essential singularity exists in the finite plane can be expressed as a sum of rational functions of the variable x in such a way that each of these functions is infinite at no more than one point in the finite plane.)

The derivation of the Weierstrass product theorem for $\mathbb{C}$ from Mittag-Leffler's theorem for $\mathbb{C}$ by integrating logarithmic derivatives was communicated to Mittag-Leffler by Hermite in a letter in 1880 ([Her], especially pp. 48–52). This method of proof found its way into textbooks at the beginning of this century. A.Pringsheim expressed his indignation with this approach in 1915 ([P], p. 388): "Wenn nun aber einige Lehrbücher sich so weit von der Weierstraßchen Methode entfernen, daß sie den fraglichen Satz als Folgerung (!) aus dem Mittag-Lefflerschen Satze durch logarithmische Integration herleiten (und zwar dieses Verfahren nicht etwa nur in Form einer gelegentlichen, ja sehr nahe liegenden Bemerkung, sondern als einzigen und maßgebenden Beweis mitteilen), so dürfte diese Art, die Dinge auf dem Kopf zu stellen, wohl von niemandem gebilligt werden, der in der Mathematik etwas anderes sieht, als eine regellose Anhäufung mathematischer Resultate." (But even if some textbooks now deviate so far from Weierstrass's methods that, using logarithmic integration, they derive the theorem in question as a consequence (!) of Mittag-Leffler's theorem (and in fact present this method not just in the form of an incidental, indeed quite obvious, remark, but as the only and standard proof), this topsy-turvy way of doing things should not be sanctioned by anyone who sees mathematics as something other than a disordered heap of mathematical results.)

Interesting details of the history of ideas of Mittag-Leffler's theorem can be found in Y. Domar's article [D].

§2. Mittag-Leffler's Theorem for Arbitrary Regions

As always, D denotes a region in $\mathbb{C}$. Our goal is to construct Mittag-Leffler series in D for every principal part distribution in D. We consider only principal part distributions (d_ν, q_ν) with infinite supports.

Mittag-Leffler series for special principal part distributions are given in Subsection 1. The general case is handled in Subsection 2.

1. Special principal part distributions. We begin by proving the following:

Let $q(z) \in \mathcal{O}(\mathbb{C}\backslash\{d\})$ be a principal part at $d \in \mathbb{C}$, and let $c \in \mathbb{C}\backslash\{d\}$. Then, in the annulus $\{z \in \mathbb{C} : |z-c| > |d-c|\}$ about c, q has a Laurent expansion of the form $\sum_{\mu=-1}^{\infty} a_\mu (z-c)^\mu$.

Proof. For all $\rho > |d-c|$, the coefficients a_μ of the Laurent series of q in the given annulus satisfy the Cauchy inequalities

$$\rho^\mu |a_\mu| \leq M(\rho) := \max\{|q(z)| : z \in \partial B_\rho(c)\}, \quad \mu \in \mathbb{Z}.$$

$\lim_{\rho\to\infty} M(\rho) = 0$ since $\lim_{z\to\infty} q(z) = 0$ (cf. 1.1); hence $a_\mu = 0$ for all $\mu \geq 0$. □

We call $g_k(z) := \sum_{\mu=-1}^{-k} a_\mu (z-c)^\mu \in \mathcal{O}(\mathbb{C}\backslash\{c\})$ the *kth Laurent term of q about c*. We will see that in special situations such Laurent terms can serve as convergence-producing summands for Mittag-Leffler series.

Let $(d_\nu, q_\nu)_{\nu\ge1}$ be a principal part distribution on D with support T. As in 4.1.2, let $T' := \overline{T}\backslash T$ denote the set of all the accumulation points of T in $\mathbb{C}$; this is closed in $\mathbb{C}$. Then $(d_\nu, q_\nu)_{\nu\ge1}$ can be interpreted in a natural way as a principal part distribution on the region $\mathbb{C}\backslash T' \supset D$.

Proposition. *Let a sequence $(c_\nu)_{\nu\ge1}$ in T' be given with $\lim |d_\nu - c_\nu| = 0$. Let $g_{\nu k}$ denote the kth Laurent term of q_ν about c_ν. Then there exist (many) sequences $(k_\nu)_{\nu\ge1}$ of natural numbers such that $\sum_{\nu=1}^{\infty}(q_\nu - g_{\nu k_\nu})$ is a Mittag-Leffler series for $(q_\nu, d_\nu)_{\nu\ge1}$ in $\mathbb{C}\backslash T'$.*

Proof. Since the sequence $(g_{\nu k})_{k\ge0}$ converges uniformly to q in $\{z \in \mathbb{C} : |z - c_\nu| \ge 2|d_\nu - c_\nu|\}$, for every $\nu \ge 1$ there exists a $k_\nu \in \mathbb{N}$ such that

$$|q_\nu(z) - g_{\nu k_\nu}(z)| \le 2^{-\nu} \text{ for all } z \in \mathbb{C} \text{ with } |z - c_\nu| \ge 2|d_\nu - c_\nu|.$$

Now let K be a compact set in $\mathbb{C}\backslash\overline{T}$. Since $d(K, \overline{T}) > 0$ and $\lim |d_\nu - c_\nu| = 0$, there exists an $n(K)$ such that, for all $\nu \ge n(K)$,

$$K \subset \{z \in \mathbb{C} : |z - c_\nu| \ge 2|d_\nu - c_\nu|\}$$

and hence

$$\sum_{\nu\ge n(K)} |q_\nu - g_{\nu k_\nu}|_K \le \sum 2^{-\nu} \le \infty.$$

This proves the normal convergence of the series in $\mathbb{C}\backslash\overline{T}$. Since $g_{\nu k_\nu}$ is holomorphic in $\mathbb{C}\backslash\{c_\nu\} \supset \mathbb{C}\backslash T'$, $\sum(q_\nu - g_{\nu k_\nu})$ is a Mittag-Leffler series for $(d_\nu, q_\nu)_{\nu\ge1}$ in $\mathbb{C}\backslash T'$. □

2. Mittag-Leffler's general theorem. *Let D be any region in $\mathbb{C}$. Then for every principal part distribution φ on D with support T there exist Mittag-Leffler series in $\mathbb{C}\backslash T'$.*

Proof (similar to that of the general product theorem 4.1.3). We view φ as a principal part distribution on $\mathbb{C}\backslash T'$ and assume that $T' \ne \emptyset$. Let the sets T_1 and T_2 be defined as in Lemma 4.1.3 (with $A := T$). Then $T_1' = \emptyset$, $T_2' = T$, and T_1 and T_2 are locally finite in $\mathbb{C}$ and $\mathbb{C}\backslash T'$, respectively; hence

$$\varphi_j(z) := \varphi(z) \text{ for } z \in T_j, \quad \varphi_j(z) := 0 \text{ otherwise}, \quad j = 1, 2,$$

defines principal part distributions φ_1 on $\mathbb{C}$ with support T_1 and φ_2 on $\mathbb{C}\backslash T'$ with support T_2. Since $T_1 \cap T_2 = \emptyset$, we have $\varphi = \varphi_1 + \varphi_2$ in $\mathbb{C}\backslash T'$. By Theorem 1.3, there exists a Mittag-Leffler series $\sum(q_{1\nu} - g_{1\nu})$ for φ_1 in $\mathbb{C}$. Since all the sets $T_2(\varepsilon)$, $\varepsilon > 0$, are finite, 4.1.2(2) and Proposition 1 imply that there exists a Mittag-Leffler series $\sum(q_{2\nu} - g_{2\nu})$ for φ_2 in $\mathbb{C}\backslash T'$. A Mittag-Leffler series for φ in $\mathbb{C}\backslash T'$ can now be formed (in various ways) by rearranging the terms of the series $\sum(q_{1\nu} - g_{1\nu}) + \sum(q_{2\nu} - g_{2\nu})$, which converges normally in $\mathbb{C}\backslash\overline{T}$. □

Remark. A proof can also be modeled on the second proof in 4.1.4. One again works with the map $v(z) = (z-a)^{-1}$ and begins by proving:

(*) *If q is a principal part at $d \neq a$, then $\widehat{q} := q \circ v^{-1} - q(a)$ is a principal part at $v(d)$.*

Thus $(\widehat{d_\nu}, \widehat{q_\nu})_{\nu \geq 1}$, with $\widehat{d_\nu} := v(d_\nu)$ and $\widehat{q_\nu} := q_\nu \circ v^{-1} - q_\nu(a)$, is the principal part distribution transported by v to $\mathbb{C}\backslash v(T')$. By (*), $q_\nu - g_\nu = (\widehat{q_\nu} - \widehat{g_\nu}) \circ v$; it follows that $\sum(q_\nu - g_\nu)$ is a Mittag-Leffler series in $\mathbb{C}\backslash T'$, as desired. Details are left to the reader.

A third short proof, using Runge theory, is given in 13.1.1.

3. Consequences. We first note the following:

Existence theorem. *Every principal part distribution on an arbitrary region $D \subset \mathbb{C}$ with support T is the principal part distribution of a function holomorphic in $D\backslash T$.*

Every distribution of finite principal parts on D is the principal part distribution of a function meromorphic in D.

Theorem on the partial fraction decomposition of meromorphic functions. *Every function meromorphic in $D \subset \mathbb{C}$ can be represented by a partial fraction series; that is, by a series $\sum h_\nu$ of meromorphic functions on D that converges normally in D, where each function h_ν has at most one pole in D.*

By combining the general product theorem 4.1.3 with the general Mittag-Leffler theorem, we obtain

Mittag-Leffler's osculation theorem. *Let T be locally finite in D and let every point $d \in T$ be assigned a series $v_d(z) = \sum_{-\infty}^{n_d} a_{d\nu}(z-d)^\nu$ that converges normally in $\mathbb{C}\backslash T$, where $n_d \in \mathbb{N}$. Then there exists a function h, holomorphic in $D\backslash T$, whose Laurent expansion about d has the function v_d as "section"; that is, $o_d(h - v_d) > n_d$ for all $d \in T$.*

Proof. Let $f \in \mathcal{O}(D)$ be chosen such that $n_d < o_d(f) < \infty$ for all $d \in T$. We consider the principal part distribution $(d, q_d)_{d \in T}$ on D with support T, where q_d is the principal part of v_d/f at d. Let $g \in \mathcal{O}(D\backslash T)$ be a solution of this distribution. Then $h := f \cdot g$ is a function with the desired properties.

To begin with, it is clear that h is holomorphic in $D\backslash T$. The functions

$$p_d := g - q_d \quad \text{and} \quad r_d := (v_d/f) - q_d$$

are holomorphic in a neighborhood of $d \in T$. Since the equation

$$h = f \cdot (p_d + q_d) = f \cdot (v_d/f + p_d - r_d) = v_d + f \cdot (p_d - r_d)$$

obviously holds in a neighborhood of $d \in T$ and since $p_d - r_d$ is holomorphic in a neighborhood of d, it follows that $o_d(h - v_d) > n_d$ for all $d \in T$. □

A special case of the osculation theorem is the

Interpolation theorem for holomorphic functions. *Let T be locally finite in D; let a polynomial $p_d(z) = \sum_0^{n_d} a_{d\nu}(z-d)^\nu$ be assigned to every point $d \in T$. Then there exists a function f, holomorphic in D, whose Taylor series about d begins with the polynomial p_d, $d \in T$.*

For the special case $D = \mathbb{C}$, this theorem means that there always exist entire functions that have arbitrarily prescribed values on a sequence d_1, $d_2, \ldots$ without accumulation points in $\mathbb{C}$. This statement generalizes the well-known Lagrange interpolation theorem, which says that given n distinct points $d_1, \ldots, d_n$ and n arbitrary numbers $w_1, \ldots, w_n$, there always exists (exactly) one polynomial $p(z)$ of degree $\leq n-1$ with $p(d_\nu) = w_\nu$, $1 \leq \nu \leq n$, namely

$$p(z) = \sum_{\nu=1}^{n} w_\nu \prod_{\mu \neq \nu} (z - d_\mu)/(d_\nu - d_\mu).$$

(This is the Lagrange interpolation formula.)

Exercise. Let $(a_\nu)_{\nu \geq 0}$ be a sequence of pairwise distinct complex numbers with $a_0 = 0$ and $\lim a_\nu = \infty$. Assume that $f \in \mathcal{O}(\mathbb{C})$ has no zeros in $\mathbb{C} \backslash \{a_0, a_1, \ldots\}$ and that $o_{a_\nu}(f) = 1$ for all ν. Then for every sequence $(b_\nu)_{\nu \geq 0}$, $b_\nu \in \mathbb{C}$, there exists a sequence $(n_\nu)_{\nu \geq 0}$, with $n_\nu \in \mathbb{N}$, such that the series

$$\frac{b_0 f(z)}{z f'(0)} + \sum_{\nu=1}^{\infty} \frac{b_\nu f(z)}{f'(a_\nu)(z - a_\nu)} \left(\frac{z}{a_\nu}\right)^{n_\nu}$$

converges normally in $\mathbb{C}$ to a function $F \in \mathcal{O}(\mathbb{C})$. Moreover, $F(a_\nu) = b_\nu$ for $\nu \geq 0$.

4. On the history of Mittag-Leffler's general theorem. Theorem 2 was stated in 1884 by Mittag-Leffler ([ML], p. 8). With Weierstrass's encouragement, he had worked on these questions since 1876 and had published several papers about them in Swedish and French. Y. Domar writes ([D], p. 10): "The extensive paper [ML] is the final summing-up of Mittag-Leffler's theory .... The paper is rather circumstantial, with much repetition in the argumentation, and when reading it one is annoyed that Mittag-Leffler is so parsimonious with credits to other researchers in the field, Schering, Schwarz, Picard, Guichard, yes, even Weierstrass. But as a whole, the exposition is impressive, showing Mittag-Leffler's mastering of the subject."

In [ML] the osculation theorem is also treated for the case of finite principal parts (cf. p. 43 and pp. 53–54); our elegant proof can be found in the 1930 paper of H. Cartan ([Ca₁], pp. 114–131). The paper [ML], in which ideas of G. Cantor already appear, contributed to Mittag-Leffler's reputation in the mathematical world (see his short biography on p. 323). The results immediately exerted a great influence; C. Runge wrote his groundbreaking paper on approximation theory under the impact of Mittag-Leffler's work (cf. 13.1.4 and Runge's short biography on p. 325).

H. Behnke and K. Stein showed in 1948, with methods of the function theory of several variables, that Theorem 2 and hence its corollaries in 3 hold verbatim if one admits arbitrary noncompact Riemann surfaces instead of regions in $\mathbb{C}$ (cf. [BSt$_2$], Satz 1, p. 156); the osculation theorem can be found at the end of this paper as Hilfsatz C.

5. Glimpses of several variables. In his paper [Co], already mentioned in 4.2.4, Cousin extended Mittag-Leffler's theorem to polydomains in $\mathbb{C}^n$. As in the case of the product theorem, difficulties arise even in formulating the problem: the concept of the principal part distribution must be understood in a different way, since the poles of meromorphic functions are no longer isolated, but — like their zeros — form real $(2n-1)$-dimensional surfaces. Furthermore, the various surfaces can intersect each other; this occurs, for instance, for the function w/z.

Surprisingly, it turns out that despite these complications the situation is nicer than when the Weierstrass product theorem is extended to higher dimensions; cf. 4.2.4. *No topological obstructions appear!* H. Cartan first observed, in 1934, that a domain in $\mathbb{C}^2$ for which Mittag-Leffler's theorem holds must be a domain of holomorphy ([Ca$_1$], p. 472); a proof of this was given in 1937 by H. Behnke and K. Stein ([BSt$_1$], pp. 183–184). In the same year, K. Oka [O$_1$] succeeded in proving that Mittag-Leffler's theorem holds for *all* domains of holomorphy in $\mathbb{C}^n$. In 1953, H. Cartan and J-P. Serre finally proved — again using sheaf theory and cohomological methods — that in any Stein space there always exist meromorphic functions corresponding to prescribed principal part distributions ([Ca$_2$], p. 679). The reader can find precise forms of these statements in [GR$_1$], especially pp. 140–142; the monograph [BT] contains further historical details.

§3*. Ideal Theory in Rings of Holomorphic Functions

> Der Weierstraßsche Produktsatz lehrt uns, daß in den Bereichen der z-ebene alle [endlich erzeugten] Ideale Hauptideale sind. (The Weierstraß product theorem teaches us that in the regions of the z-plane all [finitely generated] ideals are principal ideals.)
>
> — H. Behnke, 1940

Recall that a subset $\mathfrak{a} \neq \emptyset$ of a commutative ring R with unit element 1 is called an *ideal in R* if $ra + sb \in \mathfrak{a}$ for all $a, b \in \mathfrak{a}$ and all $r, s \in R$. If $M \neq \emptyset$ is any subset of R, then the set of all finite linear combinations $\sum r_\nu f_\nu$, $f_\nu \in M$, is an ideal in R with *generating set M*. Ideals $\mathfrak{a}$ that have a finite generating set $\{f_1, \ldots, f_n\}$ are called *finitely generated*; the suggestive notation $\mathfrak{a} = Rf_1 + \cdots + Rf_n$ is used in this case. Ideals of the form Rf are called *principal ideals.* A ring R is called *Noetherian* if every ideal in R is finitely generated, and a *principal ideal ring* if every ideal is a principal ideal.

One of the goals of this section is to show that in the ring $\mathcal{O}(G)$ of all functions holomorphic in a domain $G \subset \mathbb{C}$, every finitely generated ideal is a principal ideal (Subsection 3). Our tools are a lemma of Wedderburn (Subsection 2) and Theorem 4.2.1, on the existence of the gcd; thus Mittag-Leffler series and Weierstrass products form the basis for the ideal theory of $\mathcal{O}(G)$.

1. Ideals in $\mathcal{O}(G)$ that are not finitely generated. Let A be an *infinite* locally finite set in G. The set

$$\mathfrak{a} := \{f \in \mathcal{O}(G) : f \text{ vanishes } \textit{almost} \text{ everywhere on } A\}$$

is an ideal in $\mathcal{O}(G)$. If $f_1, \ldots, f_n$ are arbitrary functions in $\mathfrak{a}$, the set of their common zeros again consists of almost all the points of A. By the existence theorem 4.1.5, for every point $a \in A$ there exists an $f \in \mathfrak{a}$ with $f(a) \neq 0$; hence the ideal $\mathfrak{a}$ is not finitely generated. We have proved the following result.

No ring $\mathcal{O}(G)$ is Noetherian; in particular, $\mathcal{O}(G)$ is never a principal ideal ring.

Exercise. Let $G := \mathbb{C}$ and let $\mathfrak{a}$ denote the ideal in $\mathcal{O}(G)$ generated by the functions

$$\sin \pi z \prod_{\nu=-n}^{n} (z-\nu)^{-1} \in \mathcal{O}(\mathbb{C}), \quad n \in \mathbb{N}.$$

Is $\mathfrak{a}$ finitely generated?

Because of what we have just proved, the ideal theory of the rings $\mathcal{O}(G)$ is necessarily more complicated than the ideal theory of $\mathbb{Z}$, $\mathbb{Z}[i]$, or the polynomial rings $\mathbb{C}[X_1, \ldots, X_n]$ in finitely many indeterminates. Nonetheless, we will see that $\mathcal{O}(G)$ has an interesting ideal-theoretic structure. Our starting point is

2. Wedderburn's lemma (representation of 1). *Let u, $v \in \mathcal{O}(G)$ be relatively prime. Then they satisfy an equation*

$$au + bv = 1 \quad \textit{with functions } a, b \in \mathcal{O}(G).$$

Proof. We may assume that $uv \neq 0$. Since $1 = \gcd\{u, v\}$ implies that $Z(u) \cap Z(v) = \emptyset$, the pole set of $1/uv$ is the *disjoint* union of the pole sets of $1/u$ and $1/v$. By rearranging a (normally convergent) partial fraction series for $1/uv$ (using Theorem 1.4), we thus obtain

$$1/uv = a_1 + b_1, \quad a_1, b_1 \in \mathcal{M}(G),$$

where a_1 has poles (of order $-o_c(v)$) only at the points c of $Z(v)$ and b_1 has poles (of order $-o_c(u)$) only at the points c of $Z(u)$. Then $a := va_1$ and $b := ub_1$ are holomorphic in G, and it follows that $au + bv = 1$. □

Historical note. The lemma just proved was published in 1915 by J. H. M. Wedderburn; we have reproduced his elegant proof, which is almost unknown in the literature ([Wed], p. 329). The trick of splitting the pole set of a meromorphic function h into two disjoint sets P_1 and P_2, and writing a Mittag-Leffler series for h as the sum $h_1 + h_2$ of two such series, with $P(h_1) = P_1$ and $P(h_2) = P_2$, had already been used in a different context by A. Hurwitz in 1897 ([Hu], p. 457).

Wedderburn's lemma reappears implicitly in a 1940 paper of O. Helmer for $G = \mathbb{C}$ ([Hel], pp. 351–352); he considers entire functions with coefficients in a prescribed fixed subfield of $\mathbb{C}$. Helmer is not familar with Wedderburn's work. □

We give a second proof of Wedderburn's lemma, using Mittag-Leffler's osculation theorem, which even gives a sharpened version:

If u, $v \in \mathcal{O}(G)$ are relatively prime, then there exist functions a, $b \in \mathcal{O}(G)$ such that

$$au + bv = 1, \quad a \text{ has no zeros in } G.$$

Proof. If $v \equiv 0$, set $a := 1/u$, $b := 0$. Suppose $v \not\equiv 0$. It suffices to show that there exist functions λ, $h \in \mathcal{O}(G)$ with $u - \lambda v = e^h$, since $a := e^{-h}$, $b := -\lambda e^{-h}$ will then give the desired result. Since $Z(u) \cap Z(v) = \emptyset$, for every $c \in Z(v)$ there exist a disc $U_c \subset G$ about c and a function $f_c \in \mathcal{O}(U_c)$ such that $u|U_c = e^{f_c}$. Since $Z(v)$ is locally finite in G, by the osculation theorem 2.3 there exists $h \in \mathcal{O}(G)$ such that $o_c(h - f_c) > o_c(v)$, $c \in Z(v)$. Noting that $o_c(e^q - 1) = o_c(q)$ whenever q vanishes at c, we see that

$$o_c(u - e^h) = o_c(e^{f_c} - e^h) = o_c(e^{f_c - h} - 1) = o_c(f_c - h), \quad c \in Z(v).$$

Hence $\lambda := (u - e^h)/v \in \mathcal{O}(G)$. □

The preceding proof was sketched in 1978 by L. A. Rubel, who, however, uses Wedderburn's lemma [Rub]. In general, it is impossible to choose *both* the functions a and b to be nonvanishing; for instance, when $G = \mathbb{C}$ and $u = 1$, $v = z$ ([Rub], p. 505).

Exercise. Prove that for every $g \in \mathcal{M}(G)$ there exists an $f \in \mathcal{O}(G)$ such that $f(z) \neq g(z)$ for all $z \in G$.

Hint. Start with a representation $g = f_1/f_2$ with relatively prime $f_1, f_2 \in \mathcal{O}(G)$.

3. Linear representation of the gcd. Principal ideal theorem. In 4.2.1 we saw that every nonempty set in $\mathcal{O}(G)$ has a gcd. Wedderburn's lemma makes it possible, in important cases, to represent the gcd additively.

Proposition. *If $f \in \mathcal{O}(G)$ is a gcd of the finitely many functions $f_1, \ldots, f_n \in \mathcal{O}(G)$, then there exist functions $a_1, \ldots, a_n \in \mathcal{O}(G)$ such that*

$$f = a_1 f_1 + a_2 f_2 + \cdots + a_n f_n.$$

Proof (by induction on n). Let $f \neq 0$. The case $n = 1$ is clear. Let $n > 1$ and let $\widehat{f} := \gcd\{f_2, \ldots, f_n\}$. By the induction hypothesis, $\widehat{f} = \widehat{a}_2 f_2 + \cdots + \widehat{a}_n f_n$, with $\widehat{a}_2, \ldots, \widehat{a}_n \in \mathcal{O}(G)$. Since $f = \gcd\{f_1, \widehat{f}\}$ by 4.2.1, it follows that $u := f_1/f$, $v := \widehat{f}/f \in \mathcal{O}(G)$ are relatively prime. Thus by Wedderburn there exist $a, b \in \mathcal{O}(G)$ such that $1 = au + bv$. Hence $f = a_1 f_1 + \cdots + a_n f_n$, with $a_1 := a$, $a_\nu := b\widehat{a}_\nu$ for $\nu \geq 2$. □

One important consequence of the proposition and the existence of the gcd is the

Principal ideal theorem. *Every finitely generated ideal $\mathfrak{a}$ in $\mathcal{O}(G)$ is a principal ideal: If $\mathfrak{a}$ is generated by $f_1, \ldots, f_n$, then $\mathfrak{a} = \mathcal{O}(G)f$, where $f = \gcd\{f_1, \ldots, f_n\}$.*

Proof. By the proposition, $\mathcal{O}(G)f \subset \mathfrak{a}$. Since f divides all the functions $f_1, \ldots, f_n$, we have $f_1, \ldots, f_n \in \mathcal{O}(G)f$; hence $\mathfrak{a} \subset \mathcal{O}(G)f$. □

The proposition and the principal ideal theorem, together with their proofs, are valid for any integral domain R with gcd in which the statement of Wedderburn's lemma is true. — Integral domains in which any finite collection $\{f_1, \ldots, f_n\}$ always has a gcd are sometimes called *pseudo-Bézout domains*; they are called *Bézout domains* if this gcd is moreover a linear combination of the $f_1, \ldots, f_n$ (cf. [Bo], pp. 550–551, Exercises 20 and 21). $\mathcal{O}(G)$ is thus a Bézout domain.

Exercise. Let A denote the set $\{\sin 2^{-n} z : n \in \mathbb{N}\}$ in $\mathcal{O}(\mathbb{C})$. Is the ideal generated by A in $\mathcal{O}(\mathbb{C})$ a principal ideal?

4. Nonvanishing ideals. A point $c \in G$ is called a *zero* of an ideal $\mathfrak{a}$ in $\mathcal{O}(G)$ if $f(c) = 0$ for all $f \in \mathfrak{a}$, i.e. if $\mathfrak{a} \subset \mathcal{O}(G) \cdot (z - c)$. We call $\mathfrak{a}$ *nonvanishing* if $\mathfrak{a}$ has no zeros in G. An ideal $\mathfrak{a}$ in $\mathcal{O}(G)$ is called *closed* if $\mathfrak{a}$ contains the limit function of every sequence $f_n \in \mathfrak{a}$ that converges compactly in G.

Proposition. *If $\mathfrak{a}$ is a closed nonvanishing ideal in $\mathcal{O}(G)$, then $\mathfrak{a} = \mathcal{O}(G)$.*

For the proof, we need a

Reduction rule. *Let $\mathfrak{a}$ be an ideal in $\mathcal{O}(G)$ for which the point $c \in G$ is not a zero. Let $f, g \in \mathcal{O}(G)$ be such that $fg \in \mathfrak{a}$ and, if f vanishes anywhere in G, it does so only at c. Then $g \in \mathfrak{a}$.*

Proof. Choose $h \in \mathfrak{a}$ with $h(c) \neq 0$. Let $n := o_c(f)$. If $n \geq 1$, then

$$\frac{f}{z-c} \cdot g = -\frac{1}{h(c)}\left[\frac{h(z)-h(c)}{z-c} \cdot fg - \frac{fg}{z-c}h\right] \in \mathfrak{a}.$$

Applying this n times gives $[f/(z-c)^n] \cdot g \in \mathfrak{a}$. Since $f/(z-c)^n$ is invertible in $\mathcal{O}(G)$, it follows that $g \in \mathfrak{a}$. □

The proof of the proposition now goes as follows: Let $f \in \mathfrak{a}$, $f \neq 0$. Let $\prod f_\nu$ be a factorization of f, where $f_\nu \in \mathcal{O}(G)$ has exactly one zero c_ν in G. Then the sequence $\widehat{f}_n := \prod_{\nu \geq n} f_\nu \in \mathcal{O}(G)$ converges compactly in G to 1 (cf. 1.2.2). We have $\widehat{f}_n = f_n \widehat{f}_{n+1}$. Since $\widehat{f}_0 = f \in \mathfrak{a}$ and f_n has no zeros in $G \backslash \{c_n\}$, it follows (inductively) by the reduction rule that $\widehat{f}_n \in \mathfrak{a}$ for all $n \geq 0$. Since $\mathfrak{a}$ is closed, it follows that $1 \in \mathfrak{a}$; hence $\mathfrak{a} = \mathcal{O}(G)$. □

The hypothesis that $\mathfrak{a}$ is closed is essential for the validity of the proposition: the ideals given in Subsection 1 are nonvanishing but not finitely generated and hence also not closed.

Exercise. Let $\mathfrak{a} \neq \mathcal{O}(G)$ be a nonvanishing ideal in $\mathcal{O}(G)$. Prove that every function $f \in \mathfrak{a}$ has infinitely many zeros in G.

5. Main theorem of the ideal theory of $\mathcal{O}(G)$. *The following statements about an ideal $\mathfrak{a} \subset \mathcal{O}(G)$ are equivalent.*

i) *$\mathfrak{a}$ is finitely generated.*
ii) *$\mathfrak{a}$ is a principal ideal.*
iii) *$\mathfrak{a}$ is closed.*

Proof. i) $\Rightarrow$ ii): by the principal ideal theorem; ii) $\Rightarrow$ i): trivial.

ii) $\Rightarrow$ iii): Let $\mathfrak{a} = \mathcal{O}(G)f \neq 0$, and let $g_n = a_n f \in \mathfrak{a}$ be a sequence that converges compactly to $g \in \mathcal{O}(G)$. Then $a_n = g_n/f$ converges compactly in $G \backslash Z(f)$ and therefore, by the sharpened version of the Weierstrass convergence theorem, in G, to a function $a \in \mathcal{O}(G)$ (cf. I.8.5.4). It follows that $g = af \in \mathfrak{a}$.

iii) $\Rightarrow$ ii): By Theorem 4.2.1, $\mathfrak{a}$ has a greatest common divisor f in $\mathcal{O}(G)$. Then $\mathfrak{a}' := f^{-1}\mathfrak{a}$ is a nonvanishing ideal in $\mathcal{O}(G)$. Since $\mathfrak{a}'$ is closed if $\mathfrak{a}$ is, Proposition 4 implies that $\mathfrak{a}' = \mathcal{O}(G)$. Hence $\mathfrak{a} = \mathcal{O}(G)f$. □

Corollary. *The following statements about an ideal $\mathfrak{m} \subset \mathcal{O}(G)$ are equivalent.*

i) $\mathfrak{m}$ *is closed and a maximal ideal in* $\mathcal{O}(G)$.[1]
ii) *There exists a point* $c \in G$ *such that* $\mathfrak{m} = \{f \in \mathcal{O}(G) : f(c) = 0\}$.
iii) *There exists a character* $\mathcal{O}(G) \to \mathbb{C}$ *with kernel* $\mathfrak{m}$.

Details are left to the reader, who should also convince himself that there exist uncountably many maximal ideals in $\mathcal{O}(G)$ that are not closed. □

One might think that for *maximal ideals* $\mathfrak{m}$ in $\mathcal{O}(G)$ that are *not closed*, the residue class fields $\mathcal{O}(G)/\mathfrak{m}$ would be complicated. In 1951, however, M. Henriksen used transfinite methods to prove (for $G = \mathbb{C}$) that $\mathcal{O}(G)/\mathfrak{m}$, as a field, is always isomorphic to $\mathbb{C}$ ([Hen], p. 183); these isomorphisms are extremely pathological.

All the results obtained in this section remain true if we admit noncompact Riemann surfaces instead of domains in $\mathbb{C}$. The theorems of Weierstrass and Mittag-Leffler are at our disposal in this situation (cf. 4.2.4 and 2.4); hence so are Wedderburn's lemma and the existence of a gcd for arbitrary noncompact Riemann surfaces, and we can argue just as for domains.

6. On the history of the ideal theory of holomorphic functions. The ideal theory of the ring $\mathcal{O}(G)$ was not developed until relatively late in the twentieth century. Mathematicians of the nineteenth and early twentieth centuries had no interest in it. As early as 1871, R. Dedekind had completely mastered the ideal theory of the ring of algebraic integers (cf. his famous supplement to the 2nd edition of Dirichlet's *Vorlesungen über Zahlentheorie*, and also Dedekind's *Gesammelte Mathematische Werke*, vol. 3, pp. 396–407). But even a great algebraist like Wedderburn, who was certainly familiar with Dedekind's theory, and who in 1912, with his lemma, already held the key to ideal theory in arbitrary domains $G \subset \mathbb{C}$, said nothing about ideal theory. His goal — as the very title of his work [Wed] indicates — was to obtain normal forms for holomorphic matrices (cf. also [N], p. 139 ff). It was not until Emmy Noether — who, incidentally, is credited with the statement: "Es steht alles schon bei Dedekind" (It's all in Dedekind already) — that the ideal theory of rings that are not Dedekind was also considered.

The ideal theory of the ring $\mathcal{O}(\mathbb{C})$ was first considered in 1940 by O. Helmer. Helmer admits subfields of $\mathbb{C}$. He first proves Hurwitz's observation (cf. 3.2.5*), without referring to Hurwitz ([Hel], p. 346). Helmer's main result is that finitely generated ideals are principal ideals ([Hel], p. 351); to this end, he proves Wedderburn's lemma — compared to Wedderburn, in a rather complicated way. Among function theorists in several variables, the principal ideal theorem 3 was already folklore by around 1940; the epigraph

---

[1]An ideal $\mathfrak{m} \neq R$ of a ring R is called *maximal* if R is the only ideal in R which properly contains $\mathfrak{m}$. With the aid of Zorn's lemma, it can be shown that every ideal $\mathfrak{a} \neq R$ is contained in a maximal ideal.

of this section is in Behnke's report (*Fortschr. Math.* 66, p. 385 (1940)) on Cartan's work ([Ca_2], pp. 539–564); cf. also Subsection 7.

A paper by O. F. G. Schilling, in which our Proposition 4 for the case $G = \mathbb{C}$ occurs as Lemma 1, appeared in 1946 ([Sch], p. 949). Further papers then appeared in rapid succession; in these, arbitrary domains in $\mathbb{C}$ and finally arbitrary noncompact Riemann surfaces were admitted. The approach at that time was somewhat different: the principal ideal theorem was proved first, and everything else was derived from it. Of these numerous publications, we have included in our bibliography only the 1979 paper [A] of N. L. Alling, which gives a good overview of the status quo. All these papers have a strong algebraic flavor, and offer little to the reader primarily interested in function theory.

7. Glimpses of several variables. The ideal theory of holomorphic functions of several variables — in contrast to that of one variable — has long been a focal point of research, and has been an essential factor in shaping the higher-dimensional theory. The *local* theory was already developed in 1931. W. Rückert, a student of W. Krull, then proved in his paper [Rü], published only in 1933 and now considered a classic, that the ring of all convergent power series in n variables, $1 \leq n < \infty$, is *Noetherian*; in other words, that every ideal is finitely generated. (For $n = 1$ the ring is even a principal ideal domain, as we know by I.4.4.4.) The analytic tool is the so-called *Weierstrass division theorem*; as to the rest, Rückert argues algebraically; he says proudly (p. 260), "[Es] wird gezeigt, daß eine sachgemäße Behandlung nur formale Methoden, also keine functionentheoretische Hilfsmittel benötigt. Als solche Methoden erweisen sich die allgemeine Idealtheorie ...." ([It] is shown that an adequate treatment requires only formal methods, thus no function-theoretic techniques. General ideal theory turns out to yield such methods ....) No attention was paid at first to Rückert's work; contemporary function theorists had little taste for algebra.

The verbatim analogue of Wedderburn's lemma appears in 1931 in [Ca_1], p. 279, for the case $G = \mathbb{C}^2$, but ideals are not yet mentioned. The systematic development of *global* ideal theory does not begin until 1940, in the paper [Ca_2] (pp. 539–564); Cartan writes cautiously about his patching lemma for holomorphic matrices (p. 540): "Notre théorème semble susceptible de jouer un rôle important dans l'étude *globale* des *idéaux de fonctions holomorphes*." (Our theorem seems likely to play an important role in the *global* study of *ideals of holomorphic functions*.) He was right. But even though Cartan immediately proved that for domains of holomorphy $G \subset \mathbb{C}^n$, finitely many functions $f_1, \ldots, f_p \in \mathcal{O}(G)$ without common zeros in G always generate the ideal $\mathcal{O}(G)$ (p. 560), it was still a long way to general ideal theory in Stein spaces. Rückert's local theory first had to be refined. Moreover, since the common zeros of systems of holomorphic functions are not necessarily isolated, one began by facing apparently insurmountable difficulties in global problems. Oka, in his 1948 paper with the significant title "Sur quelques notions arithmétiques" (not published until 1950), struggled with "idéaux de domaines indéterminés" ([O_2], p. 84 and p. 107). It was Cartan who, in 1950, was first able to formulate the problem clearly and adapt it to a calculation; he made systematic use of the concept of coherent analytic sheaves (see, for example, [Ca_2], p. 626). The general theory of coherent analytic sheaves in

Stein spaces then yields trivially that in any Stein space X, given finitely many functions $f_1, \ldots, f_p \in \mathcal{O}(X)$ without common zeros in X, there always exist functions $a_1, \ldots, a_p \in \mathcal{O}(X)$ such that $1 = \sum a_j f_j$ (see, for example, [Ca$_2$], p. 681). A detailed presentation of ideal theory in Stein spaces, with complete proofs, can be found in [GR$_1$] and [GR$_2$].

Bibliography

[A] ALLING, N. L.: Global ideal theory of meromorphic function fields, *Trans. Amer. Math. Soc.* 256, 241–266 (1979).

[BSt$_1$] BEHNKE, H. and K. STEIN: Analytische Funktionen mehrerer Veränderlichen zu vorgegebenen Null- und Polstellenflächen, *Jber. DMV* 47, 177–192 (1937).

[BSt$_2$] BEHNKE, H. and K. STEIN: Elementarfunktionen auf Riemannschen Flächen ..., *Can. Journ. Math.* 2, 152–165 (1950).

[BT] BEHNKE, H. and P. THULLEN: *Theorie der Funktionen mehrerer komplexer Veränderlichen*, 2nd ed., with appendices by W. BARTH, O. FORSTER, H. HOLMANN, W. KAUP, H. KERNER, H.-J. REIFFEN, G. SCHEJA, and K. SPALLEK, Erg. Math. Grenzgeb. 51, Springer, 1970.

[Bo] BOURBAKI, N.: *Commutative Algebra*, Chapter 7, Divisors, Addison-Wesley, Reading, 1972.

[Ca$_1$] CARTAN, H.: *Œuvres* 1, Springer, 1979.

[Ca$_2$] CARTAN, H.: *Œuvres* 2, Springer, 1979.

[Co] COUSIN, P.: Sur les fonctions de n variables complexes, *Acta Math.* 19, 1–62 (1895).

[D] DOMAR, Y.: Mittag-Leffler's theorem, *Dept. Math. Uppsala University, Report* No. 1, 1981.

[FL] FISCHER, W. and I. LIEB: *Funktionentheorie*, Vieweg u. Sohn, Braunschweig, 1980.

[GR$_1$] GRAUERT, H. and R. REMMERT: *Theory of Stein Spaces*, trans. A. HUCKLEBERRY, Grundl. math. Wiss. 236, Springer, 1979.

[GR$_2$] GRAUERT, H. and R. REMMERT: *Coherent Analytic Sheaves*, Grundl. math. Wiss. 265, Springer, 1984.

[Hel] HELMER, O.: Divisibility properties of integral functions, *Duke Math. Journ.* 6, 345–356 (1940).

[Hen] HENRIKSEN, M.: On the ideal structure of the ring of entire functions, *Pac. Journ. Math.* 2, 179–184 (1952).

[Her] HERMITE, C.: Sur quelques points de la théorie des fonctions (Extrait d'une lettre de M. Hermite à M. Mittag-Leffler), *Journ. reine angew. Math.* 91, 54–78 (1881); *Œuvres* 4, 48–75.

[Hu] HURWITZ, A.: Sur l'intégrale finie d'une fonction entière, *Acta Math.* 20, 285–312 (1897); *Math. Werke* 1, 436–459.

[ML] MITTAG-LEFFLER, G.: Sur la représentation analytique des fonctions monogènes uniformes d'une variable indépendante, *Acta Math.* 4, 1–79 (1884).

[N] NARASIMHAN, R.: *Complex Analysis in One Variable*, Birkhäuser, 1985.

[O_1] OKA, K.: II. Domaines d'holomorphie, *Journ. Sci. Hiroshima Univ.*, Ser. A, 7, 115–130 (1937); *Coll. Pap.*, trans. R. NARASIMHAN, 11–23, Springer, 1984.

[O_2] OKA, K.: Sur quelques notions arithmétiques, *Bull. Soc. Math. France* 78, 1–27 (1950); *Coll. Pap.* 80–108, Springer 1984.

[P] PRINGSHEIM, A.: Über die Weierstrass'sche Produktdarstellung ganzer transzendenter Funktionen und über bedingt convergente unendliche Produkte, *Sitz. Ber. math.-phys. Kl. Königl. Bayer. Akad. Wiss.* 1915, 387–400.

[Rub] RUBEL, L. A.: Linear compositions of two entire functions, *Amer. Math. Monthly* 85, 505–506 (1978).

[Rü] RÜCKERT, W.: Zum Eliminationsproblem der Potenzreihenideale, *Math. Ann.* 107, 259–281 (1933).

[Sch] SCHILLING, O. F. G.: Ideal theory on open Riemann surfaces, *Bull. Amer. Math. Soc.* 52, 945–963 (1946).

[Wed] WEDDERBURN, J. H. M.: On matrices whose coefficients are functions of a single variable, *Trans. Amer. Math. Soc.* 16, 328–332 (1915).

[Wei] WEIERSTRASS, K.: *Math. Werke* 2.

Part B

Mapping Theory

7

The Theorems of Montel and Vitali

In infinitesimal calculus, the principle of selection of convergent sequences in bounded subsets M of $\mathbb{R}^n$ is crucial: Every sequence of points in M has a subsequence that converges in $\mathbb{R}^n$ (Bolzano-Weierstrass property). The extension of this accumulation principle to sets of functions is fundamental for many arguments in analysis. But caution is necessary: There are sequences of real-analytic functions from the interval $[0, 1]$ into a *fixed bounded* interval that have no convergent subsequences. A nontrivial example is the sequence $\sin 2n\pi x$; cf. 1.1.

It is of the greatest significance for function theory that, in Montel's theorem, we have a powerful accumulation principle at our disposal. We formulate, prove, and discuss this theorem in Sections 1 and 2. In Section 3 we treat Vitali's convergence-propagation theorem. *The theorems of Montel and Vitali are equivalent*: each can be derived from the other (cf. 1.4 and 3.2). Section 4 contains amusing applications of Vitali's theorem.

In analysis, sets of functions are usually called *families*;[1] we follow this practice.

---

[1] At the turn of the century, the concept of sets was still very narrow; the word "set" was reserved mainly for sets in $\mathbb{R}$ or $\mathbb{R}^n$. Mathematicians thought that sets of functions were more complicated and devised the term "family," which is still in use today.

§1. Montel's Theorem

> Une suite infinie de fonctions analytiques et bornées à l'intérieur d'un domaine simplement connexe, admet au moins une fonction limite à l'intérieur de ce domaine. (An infinite sequence of functions that are analytic and bounded in the interior of a simply connected domain admits at least one limit function in the interior of this domain.) — P. Montel, 1907

If the sequence $f_0, f_1, f_2, \ldots$ of functions defined in a region D of $\mathbb{C}$ is bounded at a point $a \in D$, then — since $\mathbb{C}$ has the Bolzano-Weierstrass property — it has a subsequence that converges at a. Since passage to subsequences does not destroy existing convergence, Cantor's diagonal process leads to the following insight:

(*) *Let $f_n : D \to \mathbb{C}$, $n \in \mathbb{N}$, be a sequence of functions that is bounded at every point of D. Then for every countable subset A of D there exists a subsequence g_n of the sequence f_n that converges pointwise in A.*

Proof. Let $a_0, a_1, a_2, \ldots$ be an enumeration of A. For every $l \in \mathbb{N}$, there exists a subsequence $f_{l0}, f_{l1}, f_{l2}, \ldots$ of the sequence $f_0, f_1, f_2, \ldots$ such that

a) the sequence $(f_{ln})_{n\geq 0}$ converges at a_l;
b) the sequence $(f_{ln})_{n\geq 0}$, $l \geq 1$, is a subsequence of $(f_{l-1,n})_{n\geq 0}$.

We argue inductively. Given the sequences $(f_{kn})_{n\geq 0}$, $k < l$, choose a subsequence $(f_{ln})_{n\geq 0}$ of the sequence $(f_{l-1,n})_{n\geq 0}$ which converges at a_l. Then a) and b) are satisfied for all sequences $(f_{kn})_{n\geq 0}$, $k \leq l$.

From the sequences $f_{l0}, f_{l1}, f_{l2}, \ldots$, we now construct the *diagonal sequence* $g_0, g_1, g_2, \ldots$, with $g_n := f_{nn}$, $n \in \mathbb{N}$. It converges at every point $a_m \in A$ since, by b), from the term g_m on it is a subsequence of the sequence $f_{m0}, f_{m1}, f_{m2}, \ldots$, which converges at a_m by a). □

The set A must be countable for (*) to hold. The subsequence g_n obtained above cannot be expected to converge pointwise *everywhere in D*. With suitable hypotheses on the sequence $f_0, f_1, f_2, \ldots$, however, this does occur. We can even obtain compact convergence, as will now be shown.

1. Montel's theorem for sequences. A family $\mathcal{F} \subset \mathcal{O}(D)$ is called *bounded* in a subset $A \subset D$ if there exists a real number $M > 0$ such that $|f|_A \leq M$ for all $f \in \mathcal{F}$ (equivalently, $\sup_{f\in\mathcal{F}} \sup_{z\in A} |f(z)| < \infty$).

The family $\mathcal{F}$ is called *locally bounded in D* if every point $z \in D$ has a neighborhood $U \subset D$ such that $\mathcal{F}$ is bounded in U: this occurs if and only if the family $\mathcal{F}$ is bounded on every compact set in D. In particular, a family $\mathcal{F} \subset \mathcal{O}(B)$ in a disc $B = B_r(c)$, $r > 0$, is locally bounded in B if and only if it is bounded in every disc $B_\rho(c)$, $\rho < r$.

Bounded families are locally bounded; the converse is not true, as is shown, for instance, by the family $\{nz^n \in \mathcal{O}(\mathbb{E}),\ n \in \mathbb{N}\}$. A sequence $f_0, f_1, f_2, \ldots$ of functions $f_n \in \mathcal{O}(D)$ is called (locally) bounded in D if the family $\{f_0, f_1, f_2, \ldots\}$ is (locally) bounded in D. The following theorem now holds.

Montel's theorem *(for sequences). Every sequence f_0, f_1, $f_2, \ldots$ of holomorphic functions in D that is locally bounded in D has a subsequence that converges compactly in D.*

Warning. The assertion of the theorem is *false for sequences of real-analytic functions:* the sequence $\sin nx$, $n \in \mathbb{N}$, which is bounded in $\mathbb{R}$, does not even have *pointwise convergent* subsequences. In fact:

The set $\{x \in \mathbb{R} : \lim_{k\to\infty} \sin n_k x$ exists$\}$ has Lebesgue measure zero for every sequence $n_1 < n_2 < \cdots$ in $\mathbb{N}$.

This statement is consistent with the Arzelà-Ascoli theorem since the sequence $\sin nx$ is *not locally equicontinuous*; cf. 2.2.

Montel's theorem will be proved in the next subsection; we use the following lemma.

Lemma. *Let $\mathcal{F} \subset \mathcal{O}(D)$ be a locally bounded family in D. Then for every point $c \in D$ and every $\varepsilon > 0$ there exists a disc $B \subset D$ about c such that*

$$|f(w) - f(z)| \le \varepsilon \quad \text{for all } f \in \mathcal{F} \text{ and all } w,\ z \in B.$$

Proof. We choose $r > 0$ so small that $B_{2r}(c) \subset D$. We set $\widetilde{B} := B_r(c)$ and $B' := B_{2r}(c)$. It follows from the Cauchy integral formula

$$\begin{aligned} f(w) - f(z) &= \frac{1}{2\pi i}\int_{\partial B'} f(\zeta)\left[\frac{1}{\zeta - w} - \frac{1}{\zeta - z}\right] d\zeta \\ &= \frac{w - z}{2\pi i}\int_{\partial B'} \frac{f(\zeta)}{(\zeta - w)(\zeta - z)} d\zeta \end{aligned}$$

and the standard estimate — since $|(\zeta - w)(\zeta - z)| \ge r^2$ for all w, $z \in \widetilde{B}$, $\zeta \in \partial B'$ — that

$$|f(w) - f(z)| \le |w - z|\ \frac{2}{r}\ |f|_{B'} \quad \text{for all } w,\ z \in \widetilde{B} \text{ and all } f \in \mathcal{F}.$$

Since $\mathcal{F}$ is locally bounded, $M := (2/r) \cdot \sup\{|f|_{B'} : f \in \mathcal{F}\} < \infty$; we may assume that $M > 0$. It now suffices to set $B := B_\delta(c)$ with $\delta := \min\{\varepsilon/(2M), r\}$. □

The lemma says that locally bounded families are *locally equicontinuous*; cf. 2.2.

2. Proof of Montel's theorem. We choose a *countable dense* set $A \subset D$, for instance the set of all rational complex numbers in D. By $(*)$ of the introduction, there exists a subsequence g_n of the sequence f_n that converges pointwise on A. We claim that the sequence g_n converges *compactly* in D. To prove this, we need only prove that it converges *continuously* in D,[2] i.e. that

$\lim g_n(z_n)$ *exists for every sequence* $z_n \in D$ *with* $z_n = z^* \in D$.

Let $\varepsilon > 0$ be given. By Lemma 1, there exists a disc $B \subset D$ about z^* such that $|g_n(w) - g_n(z)| \leq \varepsilon$ for all n if w, $z \in B$. Since A is dense in D, there exists a point $a \in A \cap B$. Since $\lim z_n = z^*$, there exists an $n_1 \in \mathbb{N}$ such that $z_n \in B$ for all $n \geq n_1$. The inequality

$$|g_m(z_m) - g_n(z_n)| \leq |g_m(z_m) - g_m(a)| + |g_m(a) - g_n(a)| + |g_n(z_n) - g_n(a)|$$

always holds; hence $|g_m(z_m) - g_n(z_n)| \leq 2\varepsilon + |g_m(a) - g_n(a)|$ for all m, $n \geq n_1$. Since $\lim g_n(a)$ exists, there is an n_2 such that $|g_m(a) - g_n(a)| \leq \varepsilon$ for all m, $n \geq n_2$. We have proved that $|g_m(z_m) - g_n(z_n)| \leq 3\varepsilon$ for all m, $n \geq \max(n_1, n_2)$; thus the sequence $g_n(z_n)$ is a Cauchy sequence and therefore convergent. □

We comment on the history of Montel's theorem in 2.3. The theorem is often used in the following form:

3. Montel's convergence criterion. *Let f_0, f_1, $f_2 \ldots$ be a sequence of functions $f_n \in \mathcal{O}(D)$ that is locally bounded in D. If every subsequence of the sequence f_n that converges compactly in D converges to $f \in \mathcal{O}(D)$, then f_n converges compactly in D to f.*

Proof. If not, there would exist a compact set $K \subset D$ such that $|f_n - f|_K$ did not converge to zero. There would then exist an $\varepsilon > 0$ and a subsequence g_j of the sequence f_n such that $|g_j - f|_K \geq \varepsilon$ for all j. Since the sequence g_j would also be locally bounded, by Theorem 1 it would have a subsequence h_k converging compactly in D. But $|h_k - f|_K \geq \varepsilon$ for all k, so f would not be the limit of this sequence. But this is a contradiction. □

As a first application, we prove

4. Vitali's theorem. *Let G be a domain in $\mathbb{C}$, and let f_0, f_1, $f_2, \ldots$ be a*

[2] A sequence $h_n \in \mathcal{C}(D)$ converges *continuously* to $h : D \to \mathbb{C}$ if $\lim h_n(z_n) = h(z^*)$ for every sequence $z_n \in D$ with $\lim z_n = z^* \in D$. Continuous convergence of the sequence h_n in D is equivalent to *compact* convergence in D (the reader may either prove this or refer to I.3.1.5*). If a sequence $h_n \in \mathcal{C}(D)$ is such that, for every sequence $z_n \in D$ with $\lim z_n = z^* \in D$, the sequence of complex numbers $h_n(z_n)$ is a Cauchy sequence, then obviously the sequence h_n converges continuously in D to the limit function defined by $h(z) := \lim h_n(z)$, $z \in D$.

sequence of functions $f_n \in \mathcal{O}(G)$ that is locally bounded in G. Suppose that the set

$$A := \{w \in G : \lim f_n(w) \text{ exists in } \mathbb{C}\}$$

has at least one accumulation point in G. Then the sequence $f_0, f_1, f_2, \ldots$ converges compactly in G.

Proof. Because of Montel's criterion 3, it suffices to prove that all the compactly convergent subsequences of the sequence f_n have the same limit. But this is clear by the identity theorem, for any two such limits must agree on the set A, which has accumulation points in G. □

Imitating this proof gives

Blaschke's convergence theorem. *Let $f_n \in \mathcal{O}(\mathbb{E})$ be a sequence bounded in $\mathbb{E}$. Suppose there exists a countable set $A = \{a_1, a_2, \ldots\}$ in $\mathbb{E}$, with $\sum(1 - |a_\nu|) = \infty$, such that $\lim_n f_n(a_j)$ exists for every point $a_j \in A$. Then the sequence f_n converges compactly in $\mathbb{E}$.*

Proof. Let $f, \widetilde{f} \in \mathcal{O}(\mathbb{E})$ be limits of two compactly convergent subsequences of the sequence f_n; then $f|A = \widetilde{f}|A$. Now f, $\widetilde{f}$ are both bounded in $\mathbb{E}$. By the identity theorem 4.3.2, it follows that $f = \widetilde{f}$. Montel's convergence criterion yields the assertion. □

For the history of Vitali's and Blaschke's theorems, see 3.3.

We will encounter other compelling applications of Montel's theorem in the proof of the Riemann mapping theorem in 8.2.4 and the theory of automorphisms of bounded domains in Chapter 9.

5*. Pointwise convergent sequences of holomorphic functions. In the theorems of Montel and Vitali, can the hypothesis of local boundedness be dropped if the sequence is assumed to converge at *all* points? The answer is negative: in 12.3.1, we will construct sequences of holomorphic functions that converge *pointwise but not compactly* and whose limit functions are not holomorphic. Such limit functions must, however, be holomorphic *almost everywhere*, as we now prove.

Theorem (Osgood, 1901, [O], p. 33). *Let $f_0, f_1, f_2, \ldots$ be a sequence of functions holomorphic in D that converges pointwise in D to a function f. Then this sequence converges compactly on a dense subset D' of D; in particular, f is holomorphic in D'.*

We base the proof on the following lemma.

Lemma. *Let $\mathcal{F}$ be a family of continuous functions $f : D \to \mathbb{C}$ that is pointwise bounded in D (i.e., every set $\{f(z) : f \in \mathcal{F}\}$, $z \in D$, is bounded* *in* $\mathbb{C}$*). Then there exists a nonempty subregion* D' *of* D *such that the restriction of the family to* D', $\{f|D' : f \in \mathcal{F}\}$, *is locally bounded in* D'.

Proof (by contradiction). Suppose the assertion is false. We recursively construct a sequence $g_0, g_1, \ldots$ in $\mathcal{F}$ and a descending sequence $K_0 \supset K_1 \supset \cdots$ of compact discs $K_n \subset D$ such that $|g_n(z)| > n$ for all $z \in K_n$. Let $g_0 \neq 0$ be in $\mathcal{F}$, and let $K_0 \subset D$ be a compact disc such that $0 \notin g_0(K_0)$. Suppose that g_{n-1} and K_{n-1} have already been constructed. By hypothesis, there exist a $g_n \in \mathcal{F}$ and a point z_n in the interior of K_{n-1} such that $|g_n(z_n)| > n$. Since g_n is continuous, there exists a compact disc $K_n \subset K_{n-1}$ about z_n such that $|g_n(z_n)| > n$ for all $z \in K_n$.

The set $\cap K_n$ is nonempty. For each of its points z^*, we have $|g_n(z^*)| > n$ for all n, contradicting the pointwise boundedness of $\mathcal{F}$. □

Osgood's theorem now follows by applying the lemma to the family $\{f_0, f_1, f_2, \ldots\}$ and *all* subregions of D: We obtain a *dense* subregion D' such that the sequence $f_n|D'$ is locally bounded. By Vitali, it then converges compactly in every connected component of D'; the limit function is holomorphic in D' by Weierstrass.

§2. Normal Families

Montel's accumulation-point principle carries over immediately from sequences to families. In the classical literature, the concept of a "normal family" took shape in this setting; it is still widely used today.

1. Montel's theorem for normal families. A family $\mathcal{F} \subset \mathcal{O}(D)$ is called *normal in D* if every sequence of functions in $\mathcal{F}$ has a subsequence that converges compactly in D. We immediately make an obvious

Remark. Any family $\mathcal{F} \subset \mathcal{O}(D)$ *that is normal in* D *is locally bounded in* D.

Proof. It must be shown that the number $\sup\{|f|_K : f \in \mathcal{F}\}$ is finite for every compact set K. If this failed to hold for some compact set $L \subset D$, then there would exist a sequence $f_n \in \mathcal{F}$ with $\lim_{n\to\infty} |f_n| = \infty$. This sequence f_n would have no subsequence converging compactly in D, since for its limit $f \in \mathcal{O}(D)$ we would have $|f|_L \geq |f_n|_L - |f - f_n|_L$. Contradiction!□

The converse of the remark above is the general form of

Montel's theorem. *Every family* $\mathcal{F} \subset \mathcal{O}(D)$ *that is locally bounded in* D *is normal in* D.

This statement follows immediately from Montel's theorem 1.1 for sequences. □

By what has been shown, the expressions "normal family" and "locally bounded family" are equivalent. A colleague reports that in the early forties, after proving this equivalence in his lectures, he received an inquiry from a higher-level administrative department in Berlin asking whether he would also consider the latest findings of racial theory.

Examples of normal families. 1) The family of all holomorphic mappings from D to a (given fixed) *bounded* region D' is normal in D.

2) For every $M > 0$, the family

$$\mathcal{F}_M := \{f = \sum a_\nu z^\nu : |a_\nu| \leq M \text{ for all } \nu \in \mathbb{N}\}$$

is normal in $\mathbb{E}$: For every $r \in (0,1)$ and every $f \in \mathcal{F}_M$, we have $|f(z)| \leq M(1-r)^{-1}$ for all $z \in B_r(0)$, whence $\mathcal{F}_M$ is locally bounded in $\mathbb{E}$.

3) *If $\mathcal{F}$ is a normal family in D, then every family $\{f^{(k)} : f \in \mathcal{F}\}$, $k \in \mathbb{N}$, is also normal in D.*

Proof. Since $\mathcal{F}$ is locally bounded in D, for every disc $B = B_{2r}(c)$ with $\overline{B} \subset D$ there exists an $M > 0$ such that $|f|_B \leq M$ for all $f \in \mathcal{F}$. Using the Cauchy estimates for derivatives and setting $\widehat{B} := B_r(c)$, we have (cf. I.8.3.1)

$$|f^{(k)}|_{\widehat{B}} \leq 2(M/r^k) \cdot k! \quad \text{for all } f \in \mathcal{F} \text{ and all } k \in \mathbb{N}.$$

The family $\{F^{(k)} : f \in \mathcal{F}\}$ is thus bounded in $\widehat{B}$ for fixed $k \in \mathbb{N}$. Hence it is locally bounded and therefore normal in D.

Another example of a normal family can be found in 4*.

Remark. In the literature, the concept of a normal family is often defined more generally than here: *The subsequences are also allowed to converge compactly to* ∞. This formulation is especially useful if one wants to include meromorphic functions as well.

2. Discussion of Montel's theorem. Montel's theorem — in contrast to that of Vitali — is not, strictly speaking, a theorem of function theory, as it can easily be subsumed under a classical theorem of real analysis. We need a new concept. A family $\mathcal{F}$ of functions $f : D \to \mathbb{C}$ is called *equicontinuous in D* if for every $\varepsilon > 0$ there exists a $\delta > 0$ such that, for all $f \in \mathcal{F}$,

$$|f(w) - f(z)| \leq \varepsilon \quad \text{for all } w, z \in D \text{ with } |w - z| \leq \delta.$$

The family $\mathcal{F}$ is called *locally equicontinuous in D* if every point $z \in D$ has a neighborhood $U \subset D$ such that the restriction of the family to U, $\mathcal{F}|U$, is equicontinuous in U. Every function in a family that is locally equicontinuous in D is locally uniformly continuous in D. Our next result follows from carrying the expression "normal family" over verbatim to arbitrary families of functions.

Theorem (Arzelà-Ascoli). *A family of complex-valued functions in D is normal in D whenever the following conditions are satisfied.*

1) *$\mathcal{F}$ is locally equicontinuous in D.*
2) *For every $w \in D$, the set $\{f(w) : f \in \mathcal{F}\} \subset \mathbb{C}$ is bounded in $\mathbb{C}$.*

This theorem, which is not about holomorphic functions, contains Montel's theorem: If a family $\mathcal{F} \subset \mathcal{O}(D)$ is locally bounded in D, then 1) holds because of Lemma 1.1, while 2) is trivial. The reader should realize that in Subsection 1.2 we actually proved the Arzelà-Ascoli theorem for families of continuous functions.

The Arzelà-Ascoli theorem plays an important role in real analysis and functional analysis; regions in $\mathbb{R}^n$ replace regions in $\mathbb{C}$.

3. On the history of Montel's theorem. A process of selecting convergent subsequences from sets of functions was first used in 1899, by David Hilbert, to construct the desired potential function in his proof of the Dirichlet principle ([Hi], pp. 13–14). Hilbert does not yet use the concept of local equicontinuity, introduced in 1884 by G. Ascoli, nor the "Arzelà-Ascoli theorem," discovered in 1895 by C. Arzelà (1847–1912).

Paul Montel was the first to recognize the great significance for function theory of the principle of selecting convergent subsequences. He published the theorem named after him in 1907, in his thesis ([Mo_1], pp. 298–302). Montel reduces his theorem to the Arzelà-Ascoli selection theorem by proving that local boundedness implies local equicontinuity in the holomorphic case (Lemma 1.1). Independently of Montel, Paul Koebe discovered and proved the theorem in 1908 ([K], p. 349); Koebe says that he drew the basic ideas of the proof from Hilbert's fourth communication on the "Theorie der linearen Integralgleichungen," *Gött. Nachr.* 1906, p. 162. In the literature, Montel's theorem is occasionally also called the Stieltjes-Osgood theorem, e.g. in the book by S. Saks and A. Zygmund, ([SZ], p. 119); see also 3.4.

The handy expression "normal family" was introduced by Montel in 1912; cf. *Ann. Sci. Éc. Norm. Sup.* 24 (1912), p. 493. He devoted the work of half a lifetime to these families; in 1927 he published a coherent theory in the monograph [Mo_2].

4*. Square-integrable functions and normal families. For every function $f \in \mathcal{O}(G)$, we set

$$\|f\|_G^2 := \iint_G |f(z)^2|\, do \in [0, \infty] \quad (do := \text{Euclidean surface element}).$$

Example. Let $f = \sum a_\nu (z-c)^\nu \in \mathcal{O}(B_R(c))$ and let $B := B_r(c)$, $0 < r < R$. Then

(1) $$\|f\|_B^2 = \pi \sum \frac{|a_\nu|^2}{\nu+1} r^{2\nu+2}; \quad \text{in particular, } |f(c)| \leq (\sqrt{\pi} r)^{-1} \|f\|_B.$$

Proof. In polar coordinates $z - c = \rho e^{i\varphi}$, we have $do = \rho\, d\rho\, d\varphi$ and

$$|f(z)|^2 = \sum_{\mu,\nu=0}^{\infty} a_\mu \overline{a}_\nu \rho^{\mu+\nu} e^{i(\mu-\nu)\varphi}, \quad z \in B.$$

Thus

$$\|f\|_B^2 = \int_0^r \int_0^{2\pi} |f(z)|^2 \rho\, d\rho\, d\varphi = \sum_{\mu,\nu=0}^{\infty} a_\mu \overline{a}_\nu \int_0^r \rho^{\mu+\nu+1} d\rho \int_0^{2\pi} e^{i(\mu-\nu)\varphi} d\varphi.$$

The integrals on the right-hand side vanish if $\mu \neq \nu$. □

We call $f \in \mathcal{O}(G)$ *square integrable in* G if $\|f\|_G < \infty$. The set $H(G)$ of all functions that are square integrable in G is a $\mathbb{C}$-vector subspace of $\mathcal{O}(G)$ because for all f, $g \in \mathcal{O}(G)$,

$$|af(z) + bg(z)|^2 \leq 2(|a|^2|f(z)|^2 + |b|^2|g(z)|^2), \quad z \in G.$$

If G is bounded, then $H(G)$ contains all functions that are bounded and holomorphic in G. Since $2u \cdot \overline{v} = |u+v|^2 + i|u+iv|^2 - (1+i)(|u|^2 + |v|^2)$, we have

$$\langle f, g \rangle := \iint_G f(z)\, \overline{g(z)}\, do \in \mathbb{C} \quad \text{for all } f,\, g \in H(G).$$

A verification shows that $\langle f, g \rangle$ is a positive-definite Hermitian form on $H(G)$. It always holds that

$$\|f\|_G \leq \sqrt{\operatorname{vol} G} \cdot |f|_G, \text{ where } \operatorname{vol} G := \iint_G do = \text{Euclidean surface area of } G.$$

The next result is more important.

Bergman's inequality. *If K is a compact subset of $G \neq \mathbb{C}$ and d denotes the Euclidean distance from K to ∂G, then*

(2) $$|f|_K \leq (\sqrt{\pi} d)^{-1} \|f\|_G \quad \textit{for all } f \in H(G).$$

Proof. Let $c \in K$ and let $r \in (0, d)$. Then $\|f\|_B \leq \|f\|_G$ since $B := B_r(c) \subset G$. It follows from (1) that, in the limit, $|f(c)| \leq (\sqrt{\pi} d)^{-1} \|f\|_G$ for all $c \in K$. □

The next result follows immediately from Bergman's inequality.

Proposition. *Every ball $\{f \in H(G) : \|f\|_G < r\}$ in the unitary space $H(G)$ is a normal family.*

Proof. This is clear by Montel since, in view of Bergman's inequality, every ball in $H(G)$ is a locally bounded family in $\mathcal{O}(G)$. □

Remark. The results of this subsection can be expressed in a more modern way by using the language of functional analysis. One first establishes that $\mathcal{O}(G)$, with respect to the topology of compact convergence, is a Fréchet space and that $H(G)$, with respect to the scalar product $\langle f, g\rangle$, is a Hilbert space. One can then state:

The injection $H(G) \to \mathcal{O}(G)$ is continuous and compact (in other words, bounded sets in $H(G)$ are relatively compact in $\mathcal{O}(G)$).

Exercises. 1. a) Find an $f \in H(\mathbb{E})$ such that $f' \notin H(\mathbb{E})$.
b) Prove that $H(\mathbb{C}) = \{0\}$.

2) (Schwarz's lemma for square-integrable functions). For all $f \in H(\mathbb{E})$ and all r with $0 < r < 1$,

$$\|f\|_{B_r(0)} \leq r^n \|f\|_{\mathbb{E}}, \quad \text{where } n := o_0(f).$$

§3*. Vitali's Theorem

Man kann die Fortpflanzung der Konvergenz mit der Ausbreitung einer Infektion vergleichen. (The propagation of convergence can be compared to the spread of an infection.) — G. Polyà and G. Szegö, 1924

If a power series $\sum a_\nu z^\nu$ converges at a point $a \neq 0$, then it converges normally in a disc of radius $|a|$ about 0. This elementary convergence criterion is the simplest example of the *propagation of convergence.* The phenomenon also occurs in more general situations: the convergence of sequences of holomorphic functions is frequently *contagious;* it can spread from subsets to the entire domain of definition. An impressive example of this is Vitali's convergence theorem, which was proved in Section 1. Vitali's theorem can be understood especially well by seeing its similarity to the identity theorem. Just as a holomorphic function f in a *domain* G is completely determined once its values are known at infinitely many points of G that have an accumulation point in G, a locally bounded sequence $f_j \in \mathcal{O}(G)$ is compactly convergent in G once it converges at infinitely many points of G that have an accumulation point in G.

1. Convergence lemma. Let $B = B_r(c)$, $r > 0$, and let $f_n \in \mathcal{O}(B)$, $n \in \mathbb{N}$, be a sequence that is bounded in B. Then the following statements are equivalent.

i) *The sequence f_n is compactly convergent in B.*
ii) *For every $k \in \mathbb{N}$, the sequence of numbers $f_0^{(k)}(c), f_1^{(k)}(c), \ldots$ is convergent.*

Proof. Since the sequences $f_n^{(k)}$ of all the derivatives converge compactly in B whenever the sequence f_n does, only the implication ii) $\Rightarrow$ i) needs to be verified. We may assume that $B = \mathbb{E}$ and $|f_n|_{\mathbb{E}} \leq 1$, $n \in \mathbb{N}$. We consider the Taylor series

$$f_n(z) = \sum a_{n\nu} z^\nu, \quad \text{where} \quad a_{n\nu} = \frac{1}{\nu!} f_n^{(\nu)}(0).$$

By hypothesis, all the limits $a_\nu := \lim_n a_{n\nu}$, $\nu \in \mathbb{N}$, exist. Since we always have $|a_{n\nu}| \leq 1$ by Cauchy's inequalities, it follows that $|a_\nu| \leq 1$ for all $\nu \in \mathbb{N}$ and hence that $f(z) = \sum a_\nu z^\nu \in \mathcal{O}(\mathbb{E})$. We fix ρ with $0 < \rho < 1$. For all $z \in \mathbb{C}$ with $|z| \leq \rho$ and all $l \in \mathbb{N}$, $l \geq 1$, we have

$$|f_n(z) - f(z)| \leq \sum_{\nu=0}^{l-1} |a_{n\nu} - a_\nu|\rho^\nu + 2\rho^l/(1-\rho), \quad n \in \mathbb{N}.$$

Now let $\varepsilon > 0$ be arbitrary. Since $\rho < 1$, we can first choose l such that $2\rho^l/(1-\rho) \leq \varepsilon$. Since $\lim_n \sum_{\nu=0}^{l-1} |a_{n\nu} - a_\nu|\rho^\nu = 0$, there now exists an n_0 such that this sum of l terms is less than ε for $n \geq n_0$. It follows that $|f_n(z) - f(z)| \leq 2\varepsilon$ for all $n \geq n_0$ and all z with $|z| \leq \rho$. Since $\rho < 1$ can be chosen arbitrarily close to 1, the sequence f_n converges compactly to f in $\mathbb{E}$. □

If boundedness is not assumed, the implication ii) $\Rightarrow$ i) is false in general: the sequence $f_n(z) := n^n z^n \in \mathcal{O}(\mathbb{E})$ does not converge at any point $z \in \mathbb{E} \setminus \{0\}$, even though $\lim_{n\to\infty} f_n^{(k)}(0) = 0$ for all $k \in \mathbb{N}$.

2. Vitali's theorem (final version). We put the theorem in a form whose similarity to the identity theorem I.8.1.1 is obvious.

Vitali's theorem. *Let G be a domain, and let $f_0, f_1, f_2, \ldots \in \mathcal{O}(G)$ be a sequence of functions that is locally bounded in G. Then the following statements are equivalent.*

i) *The sequence f_n is compactly convergent in G.*
ii) *There exists a point $c \in G$ such that for every $k \in \mathbb{N}$ the sequence of numbers $f_0^{(k)}(c)$, $f_1^{(k)}(c)$, $f_2^{(k)}(c), \ldots$ converges.*
iii) *The set $A := \{w \in G : \lim f_n(w) \text{ exists in } \mathbb{C}\}$ has an accumulation point in G.*

Proof. i) $\Rightarrow$ ii) is clear, since for every k the sequence $f_n^{(k)}$, $n \in \mathbb{N}$, is compactly convergent in G. In order to prove ii) $\Rightarrow$ iii), let B be a disc about c with $\overline{B} \subset G$. Then the sequence $f_n|B$ is bounded in B and hence, by the convergence lemma 1, compactly convergent in B. It follows that $B \subset A$; thus iii) holds.

iii) $\Rightarrow$ i). This was proved in 1.4. □

Vitali's theorem trivially implies Montel's theorem 1.1 *for sequences.* For if f_n is a sequence of functions holomorphic in G that is locally bounded in G, the diagonal process can be used (as in 1.2) to obtain a subsequence g_n that converges pointwise on a countable dense subset of G. By Vitali, this sequence converges compactly in G.

The following exercise shows how contagious convergence can be.

Exercise. Let $g \in \mathcal{O}(G)$ and suppose there exists a point $c \in G$ such that the series

$$g(z) + g'(z) + g''(z) + \cdots + g^{(n)}(z) + \cdots$$

converges (absolutely) at c. Then g is an entire function and the series converges compactly (normally) in all of $\mathbb{C}$.

3. On the history of Vitali's theorem. In 1885 C. Runge observed that sequences of holomorphic functions that converge compactly on the boundaries of domains always converge compactly in the domains themselves: "Wenn ein Ausdruck von der Form $\lim g_n(x)$ auf einer geschlossenen Curve von endlicher Länge gleichmässig convergirt, so ist er auch im Innern derselben gleichmässig convergent." (If an expression of the form $\lim g_n(x)$ converges uniformly on a closed curve of finite length, then it also converges uniformly in its interior.) ([Run], p. 247) This *inward extension of convergence* was also quite familiar to Weierstrass. Runge's observation is the first link in a chain of theorems which, from successively weaker hypotheses, yield the same result: the proof that sequences of holomorphic functions $f_n \in \mathcal{O}(G)$ converge compactly in G. The Dutch mathematician T.-J. Stieltjes saw the *principle of propagation of convergence* clearly in 1894. In his paper [St], he proves Vitali's theorem under the stronger hypothesis that the sequence f_n converges compactly in a subdomain of G; in a letter of 14 February 1894 to Hermite, Stieltjes expresses his surprise at his result: "... ayant longuement réfléchi sur cette démonstration, je suis sûr qu'elle est bonne, solide et valable. J'ai dû l'examiner avec autant plus de soin qu'*a priori* il me semblait que le théorème énoncé *ne pouvait pas exister et devait être faux*" (... having thought at length about this proof, I am sure that it is good, solid, and valid. I had to examine it all the more carefully because *a priori* it seemed to me that the theorem stated *could not exist and had to be false*); cf. [HS], p. 370.

In 1901, W. F. Osgood substantially weakened the hypothesis of Stieltjes; Osgood gets by with pointwise convergence in a subset of G that is dense in a subregion of G ([O], p. 26). In 1903, G. Vitali finally reduced the convergence hypotheses to the minimum ([V], p. 73). The American M. B. Porter (1869–1960) rediscovered Vitali's theorem in 1904 [P]. Montel was still unaware of the Vitali-Porter theorem in 1907; he cites only the work of Osgood and Stieltjes.

The hypothesis of local boundedness in Vitali's theorem can be replaced by the assumption that there exist two different complex constants a and b such that all functions f_n in G omit both the values a and b. This was shown in 1911 by C. Carathéodory and E. Landau in their paper [CL], which also contains a number of historical remarks.

A proof of Vitali's theorem by means of the Schwarz lemma, without recourse to Montel's theorem, can be found in the 1914 Berlin dissertation of R. Jentzsch, which was first published in 1917 ([J], pp. 223–326). The idea is already contained in a 1913 paper of Lindelöf [Li]. Jentzsch's proof was reproduced in the first edition of this book.

W. Blaschke proved his convergence theorem, discussed in 1.4, in 1915 by means of Vitali's theorem and that of Montel-Koebe; cf. [Bl], where, incidentally, only Koebe is cited. A direct proof, which even yields Vitali's original theorem for $\mathbb{E}$, was given in 1923 by K. Löwner and T. Radó (cf. [LR] and also [Bu], p. 219).

§4*. Applications of Vitali's theorem

Vitali's theorem is often viewed as an appendage of Montel's theorem and considered a curiosity. Vitali's theorem is, however, very useful: it often yields easy proofs that complicated analytic expressions are holomorphic. We illustrate this by classical examples; another beautiful application is given in 11.1.3.

Using his theorem, Stieltjes justified the compact convergence of a continued fraction in the slit plane $\mathbb{C}^-$ by its compact convergence in the right half-plane $\mathbb{T}$; he writes to Hermite ([HS], p. 371): "L'utilité que pourra avoir mon théorème, ... ce sera de permettre de reconnaître plus aisément la possibilité de continuation analytique de certaines fonctions définies d'abord dans un domaine restreint." (The usefulness of my theorem may well lie ... in making it easier to tell whether certain functions that are initially defined in a restricted domain can be continued analytically.)

1. Interchanging integration and differentiation. In I.8.2.2 we showed, as an application of Morera's theorem, that

$$F(z) := \int_\gamma f(\zeta, z)d\zeta, \quad z \in D,$$

is holomorphic in D if f is continuous in $|\gamma| \times D$ and, for each fixed $\zeta \in |\gamma|$, holomorphic in D. (We recall the notation used there. If γ is a path with domain $[a,b] \subset \mathbb{R}$, then $|\gamma| := \gamma([a,b])$.) Osgood observed in 1902 that this statement and more follow directly from Vitali's theorem. We denote by $\gamma : [0,1] \to \mathbb{C}$ a continuously differentiable path in $\mathbb{C}$ and claim the following:

Theorem. *Let $f(\zeta, z) : |\gamma| \times D \to \mathbb{C}$ be locally bounded (e.g., continuous). For every point $\zeta \in |\gamma|$, let $f(\zeta, z)$ be holomorphic in D, and assume in addition that every (Riemann) integral $\int_\gamma f(\zeta, z)d\zeta$, $z \in D$, exists. Then* *the function*

$$F(z) := \int_\gamma f(\zeta, z)d\zeta, \quad z \in D,$$

is holomorphic in D. All the integrals $\int_\gamma \frac{\partial f}{\partial z}(\zeta, z)d\zeta$, $z \in D$, *also exist, and*

$$F'(z) = \int_\gamma \frac{\partial f}{\partial z}(\zeta, z), \quad z \in D \quad \textit{(interchange rule)}.$$

Proof ([O], pp. 33–34). We may assume that f is bounded, say $|f|_{|\gamma| \times D} \leq M$. If we set $g(t,z) := f(\gamma(t), z)\gamma'(t)$, then, for every point $z \in D$, every sequence of Riemann sums

$$\sum_{\nu=1}^{n} g(\zeta_\nu^{(n)}, z)(t_{\nu+1}^{(n)} - t_\nu^{(n)})$$

converges by hypothesis to $F(z)$. The functions $S_n(z)$ are holomorphic in D; moreover, $|S_n|_D \leq M \cdot |\gamma'|_I$. Hence, by Vitali, the sequence S_n converges compactly in D and the limit function is therefore holomorphic in D; furthermore, the sequence

$$S_n'(z) = \sum_{\nu=1}^{n} \frac{\partial g}{\partial z}(\zeta_\nu^{(n)}, z)(t_{\nu+1}^{(n)} - t_\nu^{(n)}),$$

as the sequence of derivatives of S_n, converges compactly to F' by Weierstrass. Since the $S_n'(z)$ form arbitrary sequences of Riemann sums for $\frac{\partial g}{\partial z}(t, z)$, everything has been proved. □

2. Compact convergence of the Γ-integral. Let $0 < a < b < \infty$. Since $e^{-t}t^{z-1}$ is continuous in $[a, b] \times \mathbb{C}$ and, for fixed t, holomorphic in $\mathbb{C}$, Theorem 1 implies that

$$f(z, a, b) := \int_a^b t^{z-1}e^{-t}dt \in \mathcal{O}(\mathbb{C}).$$

If we assume the existence of the real Γ-integral $h(x) := \int_0^\infty t^{x-1}e^{-t}dt$, $x > 0$ (pointwise convergence), the next statement follows trivially.

The family $\{f(z, a, b) : a, b \in \mathbb{R} \text{ with } 0 < a < b\}$ *is locally bounded in* $\mathbb{T} = \{z \in \mathbb{C} : \operatorname{Re} z > 0\}$; *for all* $z = x + iy$ *with* $0 < c \leq x \leq d < \infty$, *we have*

$$|f(z, a, b)| \leq \int_0^1 t^{c-1}e^{-t}dt + \int_1^\infty t^{d-1}e^{-t}dt.$$

The next statement follows immediately from Vitali.

For every choice of real sequences a_n, b_n *with* $0 < a_n < b_n$, $\lim a_n = 0$, $\lim b_n = \infty$, *the sequence* $f(z, a_n, b_n)$ *converges compactly in* $\mathbb{T}$ *to a function holomorphic in* $\mathbb{T}$.

Since the limit function is independent of the choice of the sequences a_n and b_n (it equals $h(x)$ on $(0,\infty)$!), we see that

The Γ*-integral* $\int_0^\infty t^{z-1}e^{-t}dt$ *exists in* $\mathbb{T}$ *and is holomorphic there.*

This proof that Γ is holomorphic uses *only the existence* of the real Γ-integral; we need no information about the Γ-function itself.

Analogously, the compact convergence in $\mathbb{T}\times\mathbb{T}$ of the beta integral

$$\int_0^1 t^{x-1}(1-t)^{y-1}dt \quad \text{for } x>0,\ y>0$$

follows from its pointwise convergence. Interested readers may construct their own proof.

3. Müntz's theorem. By the Weierstrass approximation theorem, every real-valued function h that is continuous in $I:=[0,1]$ can be approximated uniformly by real polynomials, e.g. by the sequence of *Bernstein polynomials*

$$q_n(x)=\sum_{\nu=0}^{n}\binom{n}{\nu}h\left(\frac{\nu}{n}\right)x^\nu(1-x)^{n-\nu}, \quad n\in\mathbb{N}$$

for h (cf., for instance, M. Barner and F. Flohr: *Analysis I*, De Gruyter, 1991, p. 324).

Corollary. *Let* h *be continuous on* I *and satisfy*

$$(*)\qquad \int_0^1 h(t)t^n dt=0 \quad \textit{for all } n\in\mathbb{N}.$$

Then h *vanishes identically on* I.

Proof. By $(*)$, $\int_0^1 h(t)q(t)dt=0$ for all polynomials $q\in\mathbb{R}[t]$; thus

$$\int_0^1 [h(t)]^2dt=\int_0^1 h(t)[h(t)-q(t)]dt \quad \text{for all } q\in\mathbb{R}[t].$$

This implies the estimate $\int_0^1 h(t)^2dt\le|h-q|_I\int_0^1|h(t)|dt$ for all $q\in\mathbb{R}[t]$. Since $\inf\{|h-q|_I : q\in\mathbb{R}[t]\}=0$, it follows that $\int_0^1 h(t)^2dt=0$ and hence that $h\equiv 0$. □

We now prove by function-theoretic methods that *not all* powers of t are needed to force $h\equiv 0$ in $(*)$.

Müntz's identity theorem. *Let* k_ν *be a real sequence such that* $0<k_1<\cdots<k_n<\cdots$ *and* $\sum 1/k_\nu=\infty$. *Then a function* h *that is continuous on* I *must be identically zero if*

$$\int_0^1 h(t)t^{k_n}dt=0 \quad \text{for all } n=1,2,\ldots.$$

Proof. Set $f(t,z) := h(t)t^z$ for $t > 0$ and $f(0,z) := 0$; then f is continuous in $I \times \mathbb{T}$. Since $f(t,z)$ is always holomorphic in $\mathbb{T}$ for fixed $t \in I$ and since $|f|_{I\times\mathbb{T}} \leq |h|_I$, Theorem 1 implies that $F(z) := \int_0^1 f(t,z)dt$ is holomorphic in $\mathbb{T}$. Since $|F|_{\mathbb{T}} \leq |h|_I$, F is bounded in $\mathbb{T}$. Since $F(k_n) = 0$ for all $n \geq 1$ and $\sum 1/k_\nu = \infty$, F vanishes identically in $\mathbb{T}$ by 4.3.4b). In particular,

$$F(n+1) = \int_0^1 t \cdot h(t) \cdot t^n dt = 0 \quad \text{for all } n \in \mathbb{N}.$$

Hence, by the corollary, $t \cdot h(t)$ is identically zero in I. □

Historical remark. In his 1914 paper [Mü], C. H. Müntz discovered the following generalization of the Weierstrass approximation theorem:

Let k_n be a real sequence such that $0 < k_1 < \cdots k_n < \cdots$ and $\sum 1/k_\nu = \infty$. Then every continuous function on $[0,1]$ can be approximated uniformly by functions of the form $\sum_1^n a_\nu x^{k_\nu}$.

Müntz derived his identity theorem from this (arguing as in the proof of the corollary above). The function-theoretic proof given here is due to T. Carleman ([Ca], especially p. 15). The converses of Müntz's identity theorem and approximation theorems are true; cf. [Rud], pp. 312–315. — Elementary proofs of Müntz's approximation theorem can be found in the papers of L. C. G. Rogers [Ro] and M. v. Golitschek [G].

§5. Consequences of a Theorem of Hurwitz

In this section we gather some properties of limit functions of sequences of holomorphic functions that will be needed later. There is no connection with the theorems of Montel and Vitali; rather, the following lemma, obtained in I.8.5.5, is our starting point.

Lemma (Hurwitz). *Let the sequence $f_n \in \mathcal{O}(G)$ converge compactly in G to a nonconstant function $f \in \mathcal{O}(G)$. Then for every point $c \in G$ there exist an index $n_c \in \mathbb{N}$ and a sequence $c_n \in G$, $n \geq n_c$, such that*

$$\lim c_n = c \quad \textit{and} \quad f_n(c_n) = f(c), \quad n \geq n_c.$$

This lemma, which is a special case of a more general theorem of Hurwitz (cf. I.8.5.5), has important consequences. The following is well known.

Corollary. *Let the sequence $f_n \in \mathcal{O}(G)$ converge compactly in G to $f \in \mathcal{O}(G)$. If all the functions f_n are nonvanishing in G and f is not identically zero, then f is nonvanishing in G.*

Proof. We may assume that f is not constant. Then if f had a zero $c \in G$, by Hurwitz almost all the f_n would have zeros $c_n \in G$. □

Another corollary follows from this one.

Corollary. *If the sequence $f_n \in \mathcal{O}(G)$ converges compactly in G to a nonconstant function $f \in \mathcal{O}(G)$, then the following statements hold.*

(1) *If all the images $f_n(G)$ are contained in a fixed set $A \subset \mathbb{C}$, then $f(G) \subset A$.*
(2) *If all the maps $f_n : G \to \mathbb{C}$ are injective, then so is $f : G \to \mathbb{C}$.*
(3) *If all the maps $f_n : G \to \mathbb{C}$ are locally biholomorphic, then so is $f : G \to \mathbb{C}$.*

Proof. ad (1). Let $b \in \mathbb{C}\backslash A$. Since $f_n(G) \subset A$, every function $f_n - b$ is nonvanishing in G. Since $f - b \not\equiv 0$, the corollary implies that $f - b$ is nonvanishing in G. This means that $b \notin f(G)$. It follows that $f(G) \subset A$.

ad (2). Let $c \in G$. By the injectivity of all the f_n, the functions $f_n - f_n(c)$ are all nonvanishing in $G\backslash\{c\}$. Since $f - f(c) \not\equiv 0$, the corollary implies that $f - f(c)$ is nonvanishing on $G\backslash\{c\}$. Hence $f(z) \neq f(c)$ for all $z \in G\backslash\{c\}$. Since $c \in G$ was chosen arbitrarily, the injectivity of f follows.

ad (3). The sequence f_n' of derivatives converges compactly in G to f'. Since f is not constant, f' is not the zero function. By the local biholomorphy criterion I.9.4.2, all the f_n' are nonvanishing in G. By the corollary, f' is also nonvanishing in G; hence — again by I.9.4.2 — the map $f : G \to \mathbb{C}$ is locally biholomorphic. □

Remark. If A is a domain such that $\mathring{\overline{A}} = A$ (a disc, for example), then statement (1) follows directly from the open mapping theorem I.8.5.1. (Prove this.)

The following formulation of statements (1) and (2) is used in the proof of the Riemann mapping theorem:

Hurwitz's injection theorem. *Let G and G' be domains, and let $f_n : G \to G'$ be a sequence of holomorphic injections that converges compactly in G to a nonconstant function $f \in \mathcal{O}(G)$. Then $f(G) \subset G'$ and the induced map $f : G \to G'$ is injective.*

We also note a supplement to statement (2), which will be used in 9.1.1.

(2′) *If all the maps f_n are injective and $\lim f_n(b_n) = f(b)$, where $b_n, b \in G$, then $\lim b_n = b$. In particular,*

$$\lim f_n^{-1}(a) = f^{-1}(a) \quad \textit{for every } a \in f(G) \cap \bigcap_{n \geq 0} f_n(G).$$

Proof. If $\lim b_n$ were not equal to b, there would exist $\varepsilon > 0$ and a subsequence $b_{n'}$ of the sequence b_n with $b_{n'} \notin B := B_\varepsilon(b)$. By the injectivity of all the f_n, the sequence $f_{n'} - f_{n'}(b_{n'})$ would then be nonvanishing in B. Its limit $f - f(b)$ would therefore also be nonvanishing in B; but this is impossible. Thus $b_n = b$. □

Bibliography

[Bl] BLASCHKE, W.: Eine Erweiterung des Satzes von Vitali über Folgen analytischer Funktionen, *Ber. Verh. Königl. Sächs. Ges. Wiss. Leipzig* 67, 194–200 (1915); *Ges. Werke* 6, 187–193.

[Bu] BURCKEL, R. B.: *An Introduction to Classical Complex Analysis*, vol. 1, Birkhäuser, 1979.

[Ca] CARLEMAN, T.: Über die Approximation analytischer Funktionen durch lineare Aggregate von vorgegebenen Potenzen, *Ark. för Mat. Astron. Fys.* 17, no. 9 (1923).

[CL] CARATHÉODORY, C. and E. LANDAU: Beiträge zur Konvergenz von Funktionenfolgen, *Sitz. Ber. Königl. Preuss. Akad. Wiss., Phys.-math. Kl.* 26, 587–613 (1911); CARATHÉODORY's *Ges. Math. Schriften* 3, 13–44; LANDAU's *Coll. Works* 4, 349–375.

[G] GOLITSCHEK, M. V.: A short proof of Müntz's theorem, *Journ. Approx. Theory* 39, 394–395 (1983).

[HS] HERMITE, C. and T.-J. STIELTJES: *Correspondance d'Hermite et de Stieltjes*, vol. 2, Gauthier-Villars, Paris, 1905.

[Hi] HILBERT, D.: Über das Dirichletsche Prinzip, *Jber. DMV* 8, 184–188 (1899); *Ges. Abh.* 3, 10–14.

[J] JENTZSCH, R.: Untersuchungen zur Theorie der Folgen analytischer Funktionen, *Acta Math.* 41, 219–251 (1917).

[K] KOEBE, P.: Ueber die Uniformisierung beliebiger analytischer Kurven, Dritte Mitteilung, *Nachr. Königl. Ges. Wiss. Göttingen, Math. phys. Kl.* 1908, 337–358.

[Li] LINDELÖF, E.: Démonstration nouvelle d'un théorème fondamental sur les suites de fonctions monogènes, *Bull. Soc. Math. France* 41, 171–178 (1913).

[LR] LÖWNER, K. and T. RADÓ: Bemerkung zu einem Blaschkeschen Konvergenzsatze, *Jber. DMV* 32, 198–200 (1923).

[Mo$_1$] MONTEL, P.: Sur les suites infinies de fonctions, *Ann. Sci. Éc. Norm. Sup.* 24, 233–334 (1907).

[Mo$_2$] MONTEL, P.: *Leçons sur les familles normales de fonctions analytiques et leurs applications*, Gauthier-Villars, Paris, 1927; reissued 1974 by Chelsea Publ. Co., New York.

[Mü] MÜNTZ, C. H.: Über den Approximationssatz von Weierstraß, *Math. Abh. H. A. Schwarz gewidmet*, 303–312, Julius Springer, 1914.

[O] OSGOOD, W. F.: Note on the functions defined by infinite series whose terms are analytic functions of a complex variable; with corresponding theorems for definite integrals, *Ann. Math.*, 2nd ser., 3, 25–34 (1901–1902).

[P] PORTER, M. B.: Concerning series of analytic functions, *Ann. Math.*, 2nd ser., 6, 190–192 (1904–1905).

[Ro] ROGERS, L. C. G.: A simple proof of Müntz's theorem, *Math. Proc. Cambridge Phil. Soc.* 90, 1–3 (1981).

[Rud] RUDIN, W.: *Real and Complex Analysis*, 3rd ed., McGraw-Hill, New York, 1987.

[Run] RUNGE, C.: Zur Theorie der analytischen Functionen, *Acta Math.* 6, 245–248 (1885).

[St] STIELTJES, T.-J: Recherches sur les fractions continues, *Ann. Fac. Sci. Toulouse* 8, 1–22 (1894).

[SZ] SAKS, S. and A. ZYGMUND: *Analytic Functions*, 3rd ed., Elsevier, Warsaw, 1971.

[V] VITALI, G.: Sopra le serie de funzioni analitiche, *Rend. Ist. Lombardo*, 2. Ser. 36, 772–774 (1903) and *Ann. Mat. Pur. Appl.*, 3. Ser. 10, 65–82 (1904).

8

The Riemann Mapping Theorem

Zwei gegebene einfach zusammenhängende ebene Flächen können stets so auf einander bezogen werden, daß jedem Punkt der einen Ein mit ihm stetig fortrückender Punkt der andern entspricht und ihre entsprechenden kleinsten Theile ähnlich sind. (Two given simply connected planar surfaces can always be related to each other in such a way that every point of one corresponds to one point of the other, which varies continuously with it, and their corresponding smallest parts are similar.) — B. Riemann, 1851

Since Riemann, the problem of determining all domains in the plane that are biholomorphically (= conformally) equivalent to each other has been one of the main interests of geometric function theory. Existence and uniqueness theorems make it possible to study interesting and important holomorphic functions without knowing closed analytic expressions (such as integral formulas or power series) for them. Furthermore, analytic properties of the mapping functions can be obtained from geometric properties of the given domains.

The Riemann mapping theorem — this chapter's epigraph — solves the problem of when *simply connected* domains can be mapped biholomorphically onto each other. In order to understand this theorem, we familiarize ourselves in Section 1 with the *topological* concept of a "simply connected domain." Intuitively, these are domains without holes; that is, domains in which every closed path can be continuously contracted to a point (i.e. is *null homotopic*). We discuss two integral theorems; the result that will be crucial is the following:

Simply connected domains G in $\mathbb{C}$ are homologically simply connected: $\int_\gamma f d\zeta = 0$ for all $f \in \mathcal{O}(G)$ and all piecewise continuously differentiable paths γ in G.

Readers primarily interested in the Riemann mapping theorem may skip Section 1 on a first reading and think of the concepts "simply connected" and "homologically simply connected" as equivalent.

It took many years and the greatest efforts to prove Riemann's assertion.[1] Such mathematicians as C. Neumann, H. A. Schwarz, H. Poincaré, D. Hilbert, P. Koebe, and C. Carathéodory worked on it. Finally, in 1922, the Hungarian mathematicians L. Fejér and F. Riesz gave their ingenious proof by means of an extremal principle. In Section 2 we reproduce Carathéodory's variant of the Fejér-Riesz proof.

In Section 3 we give a detailed account of the history of the mapping theorem. Section 4 contains supplements to the mapping theorem, including a Schwarz lemma for simply connected domains.

§1. Integral Theorems for Homotopic Paths

In star-shaped domains, integrals of holomorphic functions over paths with fixed initial and terminal points do not depend on the choice of path. This path independence remains valid for arbitrary domains as long as the path of integration is "only continuously deformed." Precisely what this means will be explained in this section.

Two notions of homotopy, which will be introduced in Subsections 1 and 2, are fundamental. To each notion of homotopy corresponds a version of the Cauchy integral theorem. The proofs of these integral theorems are elementary but rather technical; all they use from function theory is that holomorphic functions in discs have antiderivatives.

In Subsection 3 we show that null-homotopic paths are always null homologous, but that the converse is not true in general. The basic concept of a "simply connected domain" is introduced and discussed in detail in Subsection 4.

1. Fixed-endpoint homotopic paths. Two paths γ, $\widetilde{\gamma}$ with the same initial point a and terminal point b in a metric (more generally, topological) space X are called *fixed-endpoint homotopic in X* if there exists a

[1] Incidentally, L. Ahlfors writes: "Riemann's writings are full of almost cryptic messages to the future. For instance, Riemann's mapping theorem is ultimately formulated in terms which would defy any attempt of proof, even with modern methods."

continuous map $\psi : I \times I \to X$, $(s,t) \mapsto \psi(s,t)$, such that, for all $s, t \in I$,

$(*)$ $\psi(0,t) = \gamma(t)$ and $\psi(1,t) = \widetilde{\gamma}(t)$, $\psi(s,0) = a$ and $\psi(s,1) = b$.

The map ψ is called a *homotopy between* γ *and* $\widetilde{\gamma}$. For every $s \in I$, $\gamma_s : I \to X$, $t \mapsto \psi(s,t)$, is a path in X from a to b; the family $(\gamma_s)_{s \in I}$ is a "deformation" of the path $\gamma = \gamma_0$ into the path $\gamma_1 = \widetilde{\gamma}$. We note (omitting the simple proof) that "being fixed-endpoint homotopic in X" is an equivalence relation on the set of all paths in X from a to b.

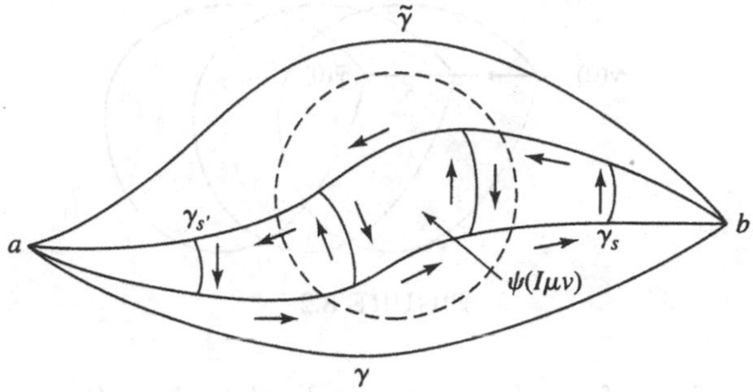

FIGURE 8.1.

The importance of the concept of homotopy just introduced is shown by the

Cauchy integral theorem *(first homotopy version). Let γ, $\widetilde{\gamma}$ be piecewise continuously differentiable fixed-endpoint homotopic paths in the domain $G \subset \mathbb{C}$.*
Then

$$\int_\gamma f\, d\zeta = \int_{\widetilde{\gamma}} f\, d\zeta \quad \textit{for all} \quad f \in \mathcal{O}(G).$$

Observe that the paths γ_s, $0 < s < 1$, need *not* be piecewise continuously differentiable.

The *idea of the proof* can be explained quickly. The rectangle $I \times I$ is subdivided into rectangles $I_{\mu\nu}$ in such a way that the images $\psi(I_{\mu\nu})$ lie in discs $\subset G$ (Figure 8.1). By the integral theorem for discs, the integrals of f along all the boundaries $\partial\psi(I_{\mu\nu})$ are zero. Hence the integrals of f are equal along sufficiently close paths γ_s, $\gamma_{s'}$ (see figure 8.1). — The technically rather tedious details are given in Subsections 5 and 6; they can be skipped on a first reading.

2. Freely homotopic closed paths. Two closed paths γ, $\widetilde{\gamma}$ in X are said to be *freely homotopic in X* if there exists a continuous map $\psi : I \times I \to X$

with the following properties:

$$(*)\qquad \begin{array}{ll} \psi(0,t)=\gamma(t) \quad\text{and}\quad \psi(1,t)=\widetilde{\gamma}(t) & \text{for all } t\in I,\\ \psi(s,0)=\psi(s,1) & \text{for all } s\in I. \end{array}$$

Then all the paths $\gamma_s : I \to X$, $t \mapsto \psi(s,t)$ are closed; their initial points trace the path $\delta : I \to X$, $t \mapsto \psi(t,0)$ in X (Figure 8.2). The paths γ and $\delta+\widetilde{\gamma}-\delta$ have the same initial and terminal points. The following statement is intuitively clear.

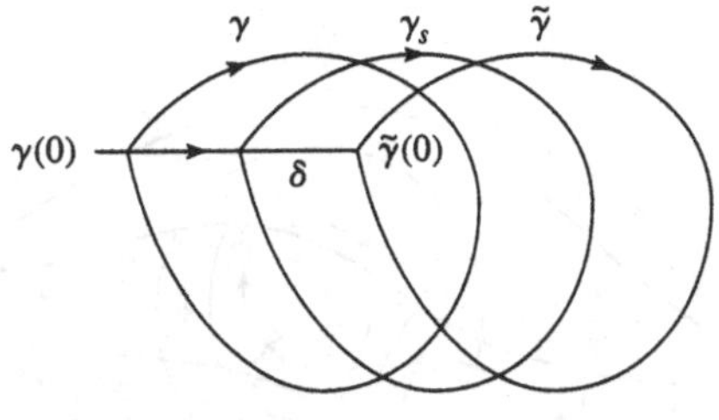

FIGURE 8.2.

(1) *If γ and $\widetilde{\gamma}$ are freely homotopic in X, then the paths γ and $\delta+\widetilde{\gamma}-\delta$ are fixed-endpoint homotopic in X.*

Proof. For every $s \in I$, set $\chi_s := \delta|[0,s] + \gamma_s - \delta|[0,s]$. A parametrization can be chosen (!) such that $\chi : I \times I \to X$, $(s,t) \mapsto \chi_s(t)$, is continuous and hence a fixed-endpoint homotopy between γ and $\delta+\widetilde{\gamma}-\delta$. □

Cauchy integral theorem *(second homotopy version). Let γ, $\widetilde{\gamma}$ be piecewise continuously differentiable closed paths in the domain $G \subset \mathbb{C}$ that are freely homotopic in G. Then*

$$\int_\gamma f\,d\zeta = \int_{\widetilde{\gamma}} f\,d\zeta \quad \textit{for all} \quad f \in \mathcal{O}(G).$$

The proof follows trivially from (1) and Theorem 1 if, *in addition*, δ is piecewise continuously differentiable; for then

$$\int_\gamma f\,d\zeta = \int_{\delta+\widetilde{\gamma}-\delta} f\,d\zeta = \int_\delta f\,d\zeta + \int_{\widetilde{\gamma}} f\,d\zeta - \int_\delta f\,d\zeta = \int_{\widetilde{\gamma}} f\,d\zeta.$$

The proof of the general case is completely analogous to that of Theorem 1 (cf. Subsections 5* and 6*, where ψ is now a "free homotopy" between γ and $\widetilde{\gamma}$).

3. Null homotopy and null homology. A closed path γ in G is said to be *null homotopic in G* if it is freely homotopic to a constant path (point path). By 2(1), this holds if and only if γ is fixed-endpoint homotopic in

G to the constant path $t \mapsto \gamma(0)$. The next statement follows immediately from Theorem 2.

Proposition. *Every piecewise continuously differentiable closed path γ that is null homotopic in G is null homologous in G:*

$$\int_\gamma f d\zeta = 0 \quad \textit{for all} \quad f \in \mathcal{O}(G).$$

The *interior* $\operatorname{Int}\gamma$ of every such path also lies in G; cf. I.9.5.2. Null homology is a consequence of null homotopy, but the converse is false. In $G := \mathbb{C}\backslash\{-1, 1\}$, for instance, consider the boundaries $\gamma_1, \gamma_2, \gamma_3, \gamma_4$ of the discs $B_1(-1)$, $B_1(1)$, $B_2(-2)$, $B_2(2)$, each with 0 as initial and terminal point (Figure 8.3). Then one can prove:

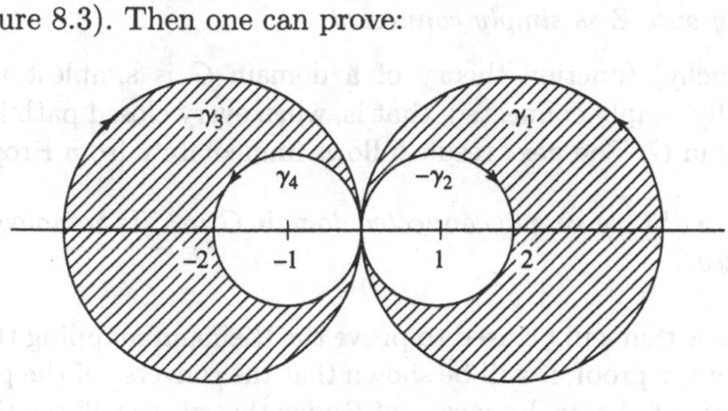

FIGURE 8.3.

The closed path $\gamma := \gamma_1 - \gamma_3 - \gamma_2 + \gamma_4$ is null homologous but not null homotopic in G.

It is easy to see that γ is null homologous in G:

$$\operatorname{Int}\gamma = [B_2(-2)\backslash\overline{B_1(-1)}] \cup [B_2(2)\backslash\overline{B_1(1)}] \subset G \quad \text{(Figure 8.3).}$$

It is also immediately clear from the figure that γ is not null homotopic in G: every deformation ψ from γ to a point must certainly pass through the omitted points 1 or -1. To give a clean argument is harder; one can argue, for example, as follows. Choose a function f holomorphic at 0 which can be continued holomorphically along *every path in G* but whose continuation along γ does not lead back to f (one such function is $f(z) = \sqrt{\log \frac{1}{2}(1+z)}$, where $f(0) := i\sqrt{\log 2}$). Then, by the monodromy theorem, γ is not null homologous in G. The argument shows that the fundamental group of the twice-punctured plane is *not abelian*.

4. Simply connected domains. A path-connected space X is called *simply connected* if every closed path in X is null homotopic. Clearly, this occurs if and only if two arbitrary paths in X with the same initial and terminal points are always fixed-endpoint homotopic in X.

(1) *Every star-shaped domain G in $\mathbb{C}$ (or $\mathbb{R}^n$) is simply connected.*

Proof. Let $c \in G$ be a center of G. If γ is a closed path in G, the continuous map

$$\psi : I \times I \to G, \quad (s,t) \mapsto \psi(s,t) := (1-s)\gamma(t) + sc$$

is a free homotopy between γ and the constant path c. □

In particular, all *convex* domains, of which the plane $\mathbb{C}$ and the unit disc $\mathbb{E}$ are special cases, are simply connected. — Simple connectivity is a topological invariant:

(2) *If $X \to X'$ is a homeomorphism, then X' is simply connected if and only if X is. In particular, every domain in C that can be mapped topologically onto $\mathbb{E}$ is simply connected.*

The (Cauchy) function theory of a domain G is simplest when G is homologically simply connected; that is, when every closed path in G is null homologous in G. The next result follows immediately from Proposition 3.

Proposition. *Every simply connected domain G in $\mathbb{C}$ is homologically simply connected.*

This proposition will be used to prove the Riemann mapping theorem. In the course of the proof, it will be shown that the converse of the proposition also holds; cf. 2.6. Later, by means of Runge theory, we will see that simply connected domains G in $\mathbb{C}$ are also characterized by having no holes (= compact components of $\mathbb{C}\backslash G$); cf. 13.2.4.

Historical note. Riemann introduced the concept of simple connectivity in 1851 and recognized its great significance for many function-theoretic problems; he defines it as follows ([Rie], p. 9): *"Eine zusammenhängende Fläche heißt, wenn sie durch jeden Querschnitt in Stücke zerfällt, eine einfach zusammenhängende."* (A connected surface is called simply connected if it is divided into pieces by every cross cut.) By cross cuts he means, in this context, *"Linien, welche von einem Begrenzungspunkte das innere einfach — keinen Punkt mehrfach — bis zu einem Begrenzungspunkte durchschneiden"* (lines which cut across the interior simply — no point more than once — from a boundary point to a boundary point). It is intuitively clear that this definition does in fact describe simply connected domains.

5*. Reduction of the integral theorem 1 to a lemma. We first prove:

(1) *If $\psi : I \times I \to G$ is continuous, there exist numbers $s_0, s_1, \dots, s_m, t_0, t_1, \dots, t_n$, with $0 = s_0 < s_1 < \cdots < s_m = 1$ and $0 = t_0 < t_1 < \cdots < t_n = 1$, such that for every rectangle $I_{\mu\nu} := [s_\mu, s_{\mu+1}] \times [t_\nu, t_{\nu+1}]$ the ψ-image $\psi(I_{\mu\nu})$ is contained in an open disk $B_{\mu\nu} \subset G$, $0 \leq \mu < m$, $0 \leq \nu < n$.*

Proof. We cover G by open discs B_j, $j \in J$. By the continuity of ψ, the family $\{\psi^{-1}(B_j), j \in J\}$ is an open cover of $I \times I$ (where $I \times I \subset \mathbb{R}^2$ is equipped with the relative topology). There exists a cover of $I \times I$ by open rectangles R parallel to the axes such that each $\overline{R}$ lies in a set $\psi^{-1}(B_j)$. Since $I \times I$ is compact, it is already covered by *finitely many* such rectangles $R_1, \ldots, R_k$. Each rectangle $\overline{R}_\kappa$ is of the form $[\sigma, \sigma'] \times [\tau, \tau']$, with $0 \le \sigma < \sigma' \le 1$, $0 \le \tau < \tau' \le 1$. Arranging and enumerating all the σ and τ appearing here, we obtain numbers $s_0, \ldots, s_m$, $t_0, \ldots, t_n$, with $0 = s_0 < s_1 < \cdots < s_m = 1$ and $0 = t_0 < t_1 < \cdots < t_n = 1$, such that every rectangle $I_{\mu\nu}$ lies in a set $\psi^{-1}(B_j)$. □

Now let $\psi : I \times I \to G$ be a fixed-endpoint homotopy between γ and $\widetilde{\gamma}$. We choose the rectangles $I_{\mu\nu}$ as in (1). Every $f \in \mathcal{O}(G)$ has antiderivatives on $B_{\mu\nu}$, which are determined up to additive constants. A clever choice of these constants gives the following result.

Lemma. *For every $f \in \mathcal{O}(G)$, there exist a continuous function $\varphi : I \times I \to \mathbb{R}$ and antiderivatives $F_{\mu\nu} \in \mathcal{O}(B_{\mu\nu})$ of $f|B_{\mu\nu}$ such that*

$$\varphi|I_{\mu\nu} = F_{\mu\nu} \circ (\psi|I_{\mu\nu}) \quad \textit{for} \quad 0 \le \mu < m,\ 0 \le \nu < n.$$

The functions $\varphi(s, 0)$ and $\varphi(s, 1)$, $s \in I$, are constant; in particular,

$$\varphi(0,0) = \varphi(1,0) \quad \textit{and} \quad \varphi(0,1) = \varphi(1,1).$$

This lemma immediately implies the integral theorem 1: Let $\gamma_0 + \cdots + \gamma_{n-1}$ and $\widetilde{\gamma}_0 + \cdots + \widetilde{\gamma}_{n-1}$ be the partitions of the paths γ and $\widetilde{\gamma}$, respectively, into subpaths that correspond to the interval partition $0 = t_0 < t_1 < \cdots < t_n = 1$. Then

$$\int_\gamma f\, d\zeta = \sum_{\nu=0}^{n-1} \int_{\gamma_\nu} f\, d\zeta \quad \text{and} \quad \int_{\widetilde{\gamma}} f\, d\zeta = \sum_{\nu=0}^{n-1} \int_{\widetilde{\gamma}_\nu} f\, d\zeta.$$

Since $\gamma(t) = \psi(0, t) \subset B_{0\nu}$ for $t \in [t_\nu, t_{\nu+1}]$ and $F_{0\nu}$ is an antiderivative of $f|B_{0\nu}$, it follows from the lemma that

$$\begin{aligned} \int_{\gamma_\nu} f\, d\zeta &= F_{0\nu}(\psi(0, t_{\nu+1})) - F_{0\nu}(\psi(0, t_\nu)) \\ &= \varphi(0, t_{\nu+1}) - \varphi(0, t_\nu), \quad 0 \le \nu < n. \end{aligned}$$

Hence $\int_\gamma f d\zeta = \varphi(0,1) - \varphi(0,0)$. Similarly, $\int_{\widetilde{\gamma}} f d\zeta = \varphi(1,1) - \varphi(1,0)$. The lemma shows that the two integrals are equal.

Remark. For the reader familiar with the theory of path liftings, the lemma is a special form of the *monodromy theorem.* A germ of an antiderivative F_1 of f at a can be continued holomorphically along every path γ_s. These continuations are a lifting of the homotopy ψ to the sheaf space $\mathcal{O}$. By the monodromy theorem for path liftings, the continuations of F_a along all the paths γ_s determine the *same* terminal germ F_b.

6*. Proof of Lemma 5*. Let $F_{\mu\nu} \in \mathcal{O}(G)$ be any antiderivative of $f|B_{\mu\nu}$, $0 \le \mu < m$, $0 \le \nu < n$. Let μ be fixed. The region $B_{\mu\nu} \cap B_{\mu,\nu+1}$ is simply connected and nonempty (it contains $\psi([s_\mu, s_{\mu+1}] \times \{t_{\nu+1}\})$); hence $F_{\mu\nu}$ and $F_{\mu,\nu+1}$ differ there only by a constant. By successively adding constants to $F_{\mu 1}, F_{\mu 2}, \ldots, F_{\mu,\nu-1}$, we can arrange that $F_{\mu\nu}$ and $F_{\mu,\nu+1}$ agree on $B_{\mu\nu} \cap B_{\mu,\nu+1}$, $0 \le \nu < n-1$. Now set

$$(\circ) \qquad \varphi_\mu(s,t) := F_{\mu\nu}(\psi(s,t)) \quad \text{for} \quad (s,t) \in I_{\mu\nu},\ 0 \le \nu < n;$$

then the function φ_μ is continuous on $[s_\mu, s_{\mu+1}] \times I$. Let this construction be carried out for all $\mu = 0, 1, \ldots, m-1$. The sets on which φ_μ and $\varphi_{\mu+1}$ are defined have intersection $\{s_{\mu+1}\} \times I$. We claim:

($*$) *There exists a $c_{\mu+1} \in \mathbb{C}$ such that $\varphi_\mu(s_{\mu+1},t) - \varphi_{\mu+1}(s_{\mu+1},t) = c_{\mu+1}$ for all $t \in I$, $0 \le \mu < m$.*

By the definition of φ_μ,

$$\varphi_\mu(s_{\mu+1},t) - \varphi_{\mu+1}(s_{\mu+1},t) = F_{\mu\nu}(\psi(s_{\mu+1},t)) - F_{\mu+1,\nu}(\psi(s_{\mu+1},t))$$

for all $t \in [t_\nu, t_{\nu+1}]$. Since $\psi(\{s_{\mu+1}\} \times [t_\nu, t_{\nu+1}]) \subset B_{\mu\nu} \cap B_{\mu+1,\nu}$ and $F_{\mu\nu}$ and $F_{\mu+1,\nu}$ are antiderivatives of f in this domain, their difference is constant there. Hence there exists a $c_{\mu\nu} \in \mathbb{C}$ such that

$$\varphi_\mu(s_{\mu+1},t) - \varphi_{\mu+1}(s_{\mu+1},t) = c_{\mu\nu} \text{ for } t \in [t_\nu, t_{\nu+1}],\ 0 \le \mu < m,\ 0 \le \nu < n.$$

Since $t_{\nu+1} \in [t_\nu, t_{\nu+1}] \cap [t_{\nu+1}, t_{\nu+2}]$, it follows that $c_{\mu 0} = c_{\mu 1} = \cdots = c_{\mu,n-1}$ for all $\mu = 0, \ldots, m-1$. This proves ($*$).

Passing from φ_μ to $\varphi_\mu + \sum_{\kappa=1}^{\mu} c_\kappa$, $1 \le \mu < m$, and retaining the previous notation yields

$$\varphi_\mu(s_{\mu+1},t) = \varphi_{\mu+1}(s_{\mu+1},t) \quad \text{for} \quad t \in I,\ 0 \le \mu < m.$$

Finally, ($\circ$) remains valid if we replace $F_{\mu\nu}$ by $F_{\mu\nu} + \sum_{\kappa}^{\mu} c_\kappa$, $1 \le \mu < m$. Now, setting

$$\varphi(s,t) := \varphi_\mu(s,t) \quad \text{for} \quad s \in [s_\mu, s_{\mu+1}],\ t \in I,\ 0 \le \mu < m,$$

defines a continuous function on $I \times I$ for which

$$\varphi|I_{\mu\nu} = F_{\mu\nu} \circ (\psi|I_{\mu\nu}), \quad 0 \le \mu < m,\ 0 \le \nu < n.$$

Since $\psi(s,0) = a$, $\psi(s,1) = b$ for all $s \in I$, it follows for $s \in [s_\mu, s_{\mu+1}]$ that

$$\varphi(s,0) = \varphi_\mu(s,0) = F_{\mu 0}(\psi(s,0)) = F_{\mu 0}(a),$$

$$\varphi(s,1) = \varphi_\mu(s,1) = F_{\mu,n-1}(\psi(s,1)) = F_{\mu,n-1}(b).$$

The functions $\varphi(s,0)$ and $\varphi(s,1)$, $s \in I$, are therefore constant; in particular, $\varphi(0,0) = \varphi(1,0)$ and $\varphi(0,1) = \varphi(1,1)$. □

Remark. The method of proof of the lemma is well known to topologists. It is used to show that the union $G \cup G'$ of simply connected domains G, G' with simply connected intersection $G \cap G'$ is again simply connected. More generally, this method is used to compute the fundamental group $\pi_1(G \cup G')$ of arbitrary domains G, G' if the groups $\pi_1(G)$, $\pi_1(G')$, and $\pi_1(G \cap G')$ are known (a theorem of Seifert and van Kampen; the interested reader may consult W. S. Massey: *Algebraic Topology: An Introduction*, GTM, Springer, 1987, p. 113 ff.).

§2. The Riemann Mapping Theorem

Which domains G in $\mathbb{C}$ can be mapped biholomorphically onto the unit disc $\mathbb{E}$? Certainly $G \neq \mathbb{C}$, since every holomorphic map $\mathbb{C} \to \mathbb{E}$ is constant (Liouville). Since biholomorphic maps are homeomorphisms and $\mathbb{E}$ is simply connected, G must also be simply connected (cf. 1.4(1) and (2)). We claim that G is subject to no further constraints.

Theorem *(Riemann mapping theorem). Every simply connected domain $G \neq \mathbb{C}$ in the plane $\mathbb{C}$ can be mapped biholomorphically onto the unit disc $\mathbb{E}$.*

Note that the possible unboundedness of G plays no role. Thus the upper half-plane $\mathbb{H}$ and the slit plane $\mathbb{C}^-$ are mapped biholomorphically onto the unit disc by the maps

$$\mathbb{H} \xrightarrow{\sim} \mathbb{E},\ z \mapsto \frac{z-i}{z+i} \quad \text{and} \quad \mathbb{E} \xrightarrow{\sim} \mathbb{C}^-,\ z \mapsto \left(\frac{z+1}{z-1}\right)^2.$$

The topological assertion of the mapping theorem is impressive by itself. Since

$$\mathbb{C} \to \mathbb{E}, \quad z \mapsto \frac{z}{\sqrt{1+|z|^2}},$$

is a topological map of $\mathbb{C}$ onto $\mathbb{E}$, we see:

Every simply connected domain G in $\mathbb{C}$ can be mapped topologically onto $\mathbb{E}$. In particular, any two simply connected domains in $\mathbb{R}^2$ are always homeomorphic (that is, they can be mapped topologically onto each other).

It is hardly conceivable that the simply connected domain shown in Figure 8.4 can be mapped topologically, indeed biholomorphically (hence preserving angles and orientation), onto $\mathbb{E}$.

1. Reduction to Q-domains. The first step of the proof consists of replacing the *topological* property of simple connectivity by an *algebraic* property of the ring $\mathcal{O}(G)$.

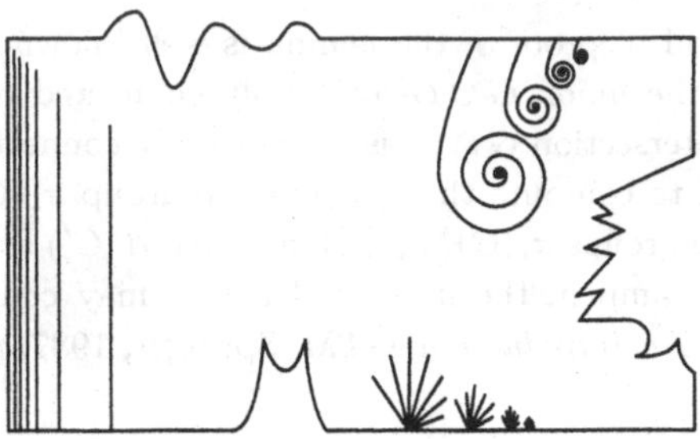

FIGURE 8.4.

Lemma. *If $G \subset \mathbb{C}$ is simply connected, then every unit in $\mathcal{O}(G)$ has a square root in $\mathcal{O}(G)$ (square root property).*

Proof. By Proposition 1.4, G is homologically simply connected. By I.9.3.3, homologically simply connected domains have the square root property. □

In what follows, only the square root property will be used. The topological notion of simple connectivity may be forgotten for the time being. (Of course, once the mapping theorem has been proved, the square root property and simple connectivity turn out to be equivalent; cf. Theorem 2.6.) — We prove the following elementary invariance statement:

(1) *If $f : G \xrightarrow{\sim} \widehat{G}$ is biholomorphic and G has the square root property, then so does $\widehat{G}$.*

Proof. If $\widehat{u}$ is a unit in $\mathcal{O}(\widehat{G})$, then $u := \widehat{u} \circ f$ is a unit in $\mathcal{O}(G)$. If $u = v^2$ with $v \in \mathcal{O}(G)$, then $\widehat{u} = \widehat{v}^2$ with $\widehat{v} := v \circ f^{-1}$. □

For convenience, we call domains G in $\mathbb{C}$ with $0 \in G$ that have the square root property simply *Q-domains*. The Riemann mapping theorem is then contained in the following statement:

Theorem. *Every Q-domain $G \neq \mathbb{C}$ can be mapped biholomorphically onto $\mathbb{E}$.*

The proof of this theorem will be carried out in the next three subsections. The essential tools are

- a "square root trick" due to Carathéodory and Koebe,
- Montel's theorem,
- Hurwitz's injection theorem, and
- the involutory automorphisms $g_c : \mathbb{E} \to \mathbb{E}$, $z \mapsto (z-c)/(\bar{c}z-1)$, $c \in \mathbb{E}$.

2. Existence of holomorphic injections. *For every Q-domain $G \neq \mathbb{C}$ there exists a holomorphic injection $f : G \to \mathbb{E}$ with $f(0) = 0$.*

Proof. Let $a \in \mathbb{C}\backslash G$. Then $z - a$ is a unit in $\mathcal{O}(G)$; hence there exists a $v \in \mathcal{O}(G)$ such that $v(z)^2 = z - a$. The map $v : G \to \mathbb{C}$ is *injective.* It must be true that

$$(*) \qquad v(G) \cap (-v)(G) = \emptyset,$$

since if there were points $b, b' \in G$ with $v(b) = -v(b')$, we would have $b - a = v(b)^2 = v(b')^2 = b' - a$, whence $b = b'$ and thus $v(b) = 0$; but this is impossible since $b \neq a$.

Since $(-v)(G)$ is nonempty and open, there exists by $(*)$ a disc $B = B_r(c)$, $r > 0$, such that $v(G) \subset \mathbb{C}\backslash\overline{B}$. Since

$$g(z) := \frac{1}{2}r \cdot \left(\frac{1}{z-c} - \frac{1}{v(0)-c}\right)$$

maps $\mathbb{C}\backslash\overline{B}$ injectively into $\mathbb{E}$, the function $f := g \circ v$ has the desired properties. □

A substantial generalization of the injection theorem will be given in 13.2.4.

Remark. The statement is unexciting for domains whose complements contain interior points c: in that case, maps $z \mapsto \varepsilon(z-c)^{-1}$, ε small, give an immediate solution. In all cases where $\mathbb{C}\backslash G$ has no interior points (slit regions, e.g. $\mathbb{C}^-$), the "square root trick" yields the simplest conformal maps onto domains which contain discs in their complement. This trick is due to P. Koebe, who noticed as early as 1912 that every holomorphic square root of a unit $(z-a)/(z-b) \in \mathcal{O}(G)$, where $a, b \in \partial G$, $a \neq b$, gives an image domain whose complement has interior points ([Koe$_3$], p. 845).

3. Existence of expansions. If G is a domain with $0 \in G \subset \mathbb{E}$, then every holomorphic injection $\kappa : G \to \mathbb{E}$ for which

$$\kappa(0) = 0 \quad \text{and} \quad |\kappa(z)| > |z| \quad \text{for all} \quad z \in G\backslash\{0\}$$

is called a (proper) *expansion of* G to $\mathbb{E}$ (relative to the origin). We construct expansions as inverses of contractions. We denote by j the square map $\mathbb{E} \to \mathbb{E}$, $z \mapsto z^2$, and begin by making a simple but not obvious observation:

Every map $\psi_c : \mathbb{E} \to \mathbb{E}$, $z \mapsto (g_{c^2} \circ j \circ g_c)(z)$, where $c \in \mathbb{E}$, is a proper contraction:

$$(1) \qquad \psi_c(0) = 0, \quad |\psi_c(z)| < |z| \quad \textit{for all } z \in \mathbb{E}\backslash\{0\}.$$

Proof. This is clear by Schwarz: ψ_c is not a rotation about 0 since $j \notin \operatorname{Aut} \mathbb{E}$. □

Lemma (square root method). *Let $G \subset \mathbb{E}$ be a Q-domain and let $c \in \mathbb{E}$ with $c^2 \notin G$. Let $v \in \mathcal{O}(G)$ be the square root of $g_{c^2}|G \in \mathcal{O}(G)$ with $v(0) = c$. Then the map $\kappa : G \to \mathbb{E}$, $z \mapsto g_c(v(z))$, is an expansion of G. Moreover,*

$$(2) \qquad id_G = \psi_c \circ \kappa.$$

Proof. Since g_{c^2} is nonvanishing in G and $g_{c^2}(0) = c^2$, v is well defined; since $v(G) \subset \mathbb{E}$, so is κ. We have $\kappa(G) \subset \mathbb{E}$ and $\kappa(0) = g_c(v(0)) = g_c(c) = 0$. It follows from $g_c \circ g_c = id_{\mathbb{E}}$ and $j \circ v = g_{c^2}$ that

$$\psi_c \circ \kappa = g_{c^2} \circ j \circ g_c \circ g_c \circ v = g_{c^2} \circ g_{c^2} = id_G.$$

Therefore $\kappa : G \to \mathbb{E}$ is injective; by (1), $|z| = |\psi_c(\kappa(z))| < |\kappa(z)|$ for all $\kappa(z) \neq 0$, hence for all $z \in G \backslash \{0\}$. □

Remark. Since square roots are used in the construction of expansions, it is hardly surprising that $x \mapsto \sqrt{x}$ is a simple expansion of the interval $[0, 1)$.

The "auxiliary function" ψ_c can be given explicitly:

$$(3) \qquad \psi_c(z) = z\frac{z-b}{\bar{b}z-1}, \quad \text{where} \quad b := \psi_c'(0) = \frac{2c}{1+|c|^2} \in \mathbb{E}.$$

The calculation is left to the reader. By (2), it follows in particular that

$$(4) \qquad \kappa'(0) = \frac{1+|c|^2}{2c}.$$

ψ_c is a *finite* map of $\mathbb{E}$ onto itself of *mapping degree* 2; cf. 9.3.2 and 9.4.4. □

Historical note. The contraction ψ_c was introduced geometrically by Koebe in 1909 and used as a majorant ([Koe$_2$], p. 209). Carathéodory writes it explicitly in 1912 ([Ca$_1$], p. 401); he normalizes it by $\psi_c'(0) > 0$. We will come across the function ψ_c again in the appendix to this chapter, in "Carathéodory-Koebe theory."

4. Existence proof by means of an extremal principle. If $G \neq \mathbb{C}$ is a Q-domain, the family

$$\mathcal{F} := \{f \in \mathcal{O}(G) : f \text{ maps } G \text{ injectively into } \mathbb{E},\ f(0) = 0\}$$

is nonempty by 2. Each $f \in \mathcal{F}$ maps G biholomorphically onto $f(G) \subset \mathbb{E}$ (biholomorphy criterion I.9.4.1). The functions in $\mathcal{F}$ with $f(G) = \mathbb{E}$ can be characterized surprisingly simply by an extremal property.

Proposition. *Let $G \neq \mathbb{C}$ be a Q-domain and let $p \neq 0$ be a fixed point in G. Then $h(G) = \mathbb{E}$ for every function $h \in \mathcal{F}$ such that*

$$(*) \qquad |h(p)| = \sup\{|f(p)| : f \in \mathcal{F}\}.$$

Proof. By 1(1), $h(G) \subset \mathbb{E}$ is a Q-domain whenever G is. Suppose that $h(G) \neq \mathbb{E}$; then by Lemma 3 there exists an expansion $\kappa : h(G) \to \mathbb{E}$, and $g := \kappa \circ h \in \mathcal{F}$. Since $h(p) \neq 0$ by the injectivity of h, it follows that $|g(p)| = |\kappa(h(p))| > |h(p)|$. Contradiction! □

At this point Theorem 1 can be proved quickly. Let $p \in G\backslash\{0\}$ be fixed. Since $\mathcal{F}$ is nonempty, $\mu := \sup\{|f(p)| : f \in \mathcal{F}\} > 0$. We choose a sequence $f_0, f_1, \ldots$ in $\mathcal{F}$ with $\lim |f_n(p)| = \mu$. Since $\mathcal{F}$ is bounded, Montel implies that a subsequence h_j of the sequence f_n converges compactly to a function $h \in \mathcal{O}(G)$. We have $h(0) = 0$ and $|h(p)| = \mu$. Since $\mu > 0$, h is not constant; Hurwitz's injection theorem 7.5.1 now implies that $h : G \to \mathbb{E}$ is an injection. It follows that $h \in \mathcal{F}$. By the proposition above, $h(G) = \mathbb{E}$. Thus $h : G \to \mathbb{E}$ is biholomorphic. This proves Theorem 1 and hence also the Riemann mapping theorem. □

Propaedeutic hint. The extremal principle used here can be better understood by observing that, in general, *biholomorphic* maps $f : G \to \mathbb{E}$ with $f(0) = 0$ are characterized by the following extremal property (apply Schwarz's lemma to $f \circ h^{-1}$):

$|h(z)| \geq |f(z)|$ *for all* $z \in G$. *If equality holds at a point* $p \neq 0$, *then* $f : G \xrightarrow{\sim} \mathbb{E}$ *biholomorphically.*

Thus, once one believes that biholomorphic maps $G \xrightarrow{\sim} \mathbb{E}$ exist, one *must* look among *all* holomorphic maps $f : G \to \mathbb{E}$ with $f(0) = 0$ for those with $|f(p)|$ maximal.

5. On the uniqueness of the mapping function. The following theorem, due to Poincaré, can be obtained from Schwarz's lemma alone.

Uniqueness theorem. *Let h and $\widehat{h}$ be biholomorphic maps from a domain G onto $\mathbb{E}$, and suppose there exists a point $a \in G$ such that $h(a) = \widehat{h}(a)$ and $h'(a)/\widehat{h}'(a) > 0$. Then $h = \widehat{h}$.*

Proof. Let $b := h(a)$. If $b = 0$, then $f := \widehat{h} \circ h^{-1} \in \operatorname{Aut}\mathbb{E}$ with $f(0) = 0$, $f'(0) = \widehat{h}'(a)/h'(a) > 0$. It follows from Schwarz that $f = id$; hence $\widehat{h} = h$.

If $b \neq 0$, set $g := -g_b$, $h_1 := g \circ h$, and $\widehat{h}_1 := g \circ \widehat{h}$. This is the case already treated: it follows that $\widehat{h}_1 = h_1$, hence that $\widehat{h} = h$. □

We now have the following theorem:

Existence and uniqueness theorem. *If $G \neq \mathbb{C}$ is simply connected, then for every point $a \in G$ there exists exactly one biholomorphic map $h : G \xrightarrow{\sim} \mathbb{E}$ with $h(a) = 0$ and $h'(a) > 0$.*

Proof. Only the existence of h needs to be proved. By Riemann, there exists a biholomorphic map $h_1 : G \xrightarrow{\sim} \mathbb{E}$. For $h_2 := g_c \circ h_1$ with $c := h_1(a)$, it follows that $h_2(a) = 0$. For $h := e^{i\varphi}h_2$ with $e^{i\varphi} := |h_2'(a)|/h_2'(a)$, we then have $h(a) = 0$ and $h'(a) > 0$. □

6. Equivalence theorem. *The following statements about a domain G in $\mathbb{C}$ are equivalent:*

i) *G is homologically simply connected.*
ii) *Every function holomorphic in G is integrable in G.*
iii) *For all $f \in \mathcal{O}(G)$ and every closed path γ in G,*

$$\operatorname{ind}_\gamma(z)f(z) = \frac{1}{2\pi i}\int_\gamma \frac{f(\zeta)}{\zeta - z}d\zeta, \quad z \in G\backslash|\gamma|.$$

iv) *The interior* $\operatorname{Int}\gamma$ *of every closed path γ in G lies in G.*
v) *Every unit in $\mathcal{O}(G)$ has a holomorphic logarithm in G.*
vi) *Every unit in $\mathcal{O}(G)$ has a holomorphic square root in G.*
vii) *Either $G = \mathbb{C}$ or G can be mapped biholomorphically onto $\mathbb{E}$.*
viii) *G can be mapped topologically onto $\mathbb{E}$.*
ix) *G is simply connected.*

Proof. Equivalences i) through vi) are known from I.9.5.4. — vi) ⇒ vii). This is Theorem 1 (where it is now unnecessary that $0 \in G$). — vii) ⇒ viii) ⇒ ix). Trivial (note the introduction to this section and 1.4(1) and (2)). — ix) ⇒ i). This is Proposition 1.4. □

This theorem is an aesthetic peak of function theory. The following turn out to be equivalent:

- *topological* statements (simple connectivity);
- *analytic* statements (Cauchy integral formula);
- *algebraic* statements (existence of a square root).

Each of these statements implies that one is actually looking at $\mathbb{C}$ or $\mathbb{E}$.

The list of nine equivalences can be considerably and nontrivially lengthened. Thus one can add:

x) *G has no holes.*
xi) *Every function in $\mathcal{O}(G)$ can be approximated compactly in G by polynomials. (G is a Runge domain.)*
xii) *G is homogeneous relative to* $\operatorname{Aut} G$ *and $G \not\approx \mathbb{C}^\times$.*
xiii) *There exists a point $a \in G$ with infinite isotropy group* $\operatorname{Aut}_a G$.
xiv) *The monodromy theorem holds for G.*

That ix), x), and xi) are equivalent is proved in 13.2.4. The equivalence xii) ⇔ ix) is proved in 9.1.3 for bounded domains "with smooth boundary components." The last two equivalences will not be pursued further.

§3. On the History of the Riemann Mapping Theorem

The names of many mathematicians are inseparably linked to the history of the Riemann mapping theorem:

Carathéodory, Courant, Fejér, Hilbert, Koebe,

Riemann, Riesz, Schwarz, Weierstrass.

There are three different ways to prove the theorem, using

- the Dirichlet principle,
- methods of potential theory, or
- the Fejér-Riesz extremal principle.

We describe the principal steps in the development of each approach.

1. Riemann's dissertation. Riemann stated the mapping theorem in his dissertation in 1851, and sketched a proof tailored to bounded domains with piecewise smooth boundaries ([Rie], p. 40). He uses a method of proof that ties the problem of the existence of a biholomorphic map $G \xrightarrow{\sim} \mathbb{E}$ to the Dirichlet boundary value problem for harmonic functions. Riemann solves the boundary value problem with the aid of the *Dirichlet principle*, which characterizes the desired function as that function $\varphi(x, y)$ with given boundary values for which the Dirichlet integral

$$\int\int_G (\varphi_x^2 + \varphi_y^2)dxdy$$

has the smallest possible value. Riemann's revolutionary ideas are not accepted by his contemporaries; the time is not yet ripe. An awareness of the mapping theorem as an existence theorem is missing. Only after his death does Riemann find public relations agents in H. A. Schwarz, L. Fuchs, and, above all, F. Klein; cf. F. Klein: "Riemann und seine Bedeutung für die Entwicklung der modernen Mathematik," a lecture given in 1894 (*Ges. Math. Abh.* 3, pp. 482–497).

What did the philosophy faculty of the venerable Georgia Augusta at Göttingen think of Riemann's dissertation? E. Schering reports on this in a eulogy, "Zum Gedächniss an B. Riemann," delivered before the Göttingen Academy on 1 December 1866 but not published until 1909, in Schering's *Ges. Math. Werke* 2, p. 375, Verlag Mayer und Müller Berlin. This eulogy remained almost unknown until it appeared recently in the new edition of Riemann's *Werke* edited by R. Narasimhan, pp. 828–847, Springer und Teubner 1990; cf. p. 836 in particular.

In Figure 8.5, we reproduce a copy of parts of the Riemann-Akte Nr. 135, with the kind permission of the Göttingen University archives. The dean of the faculty,

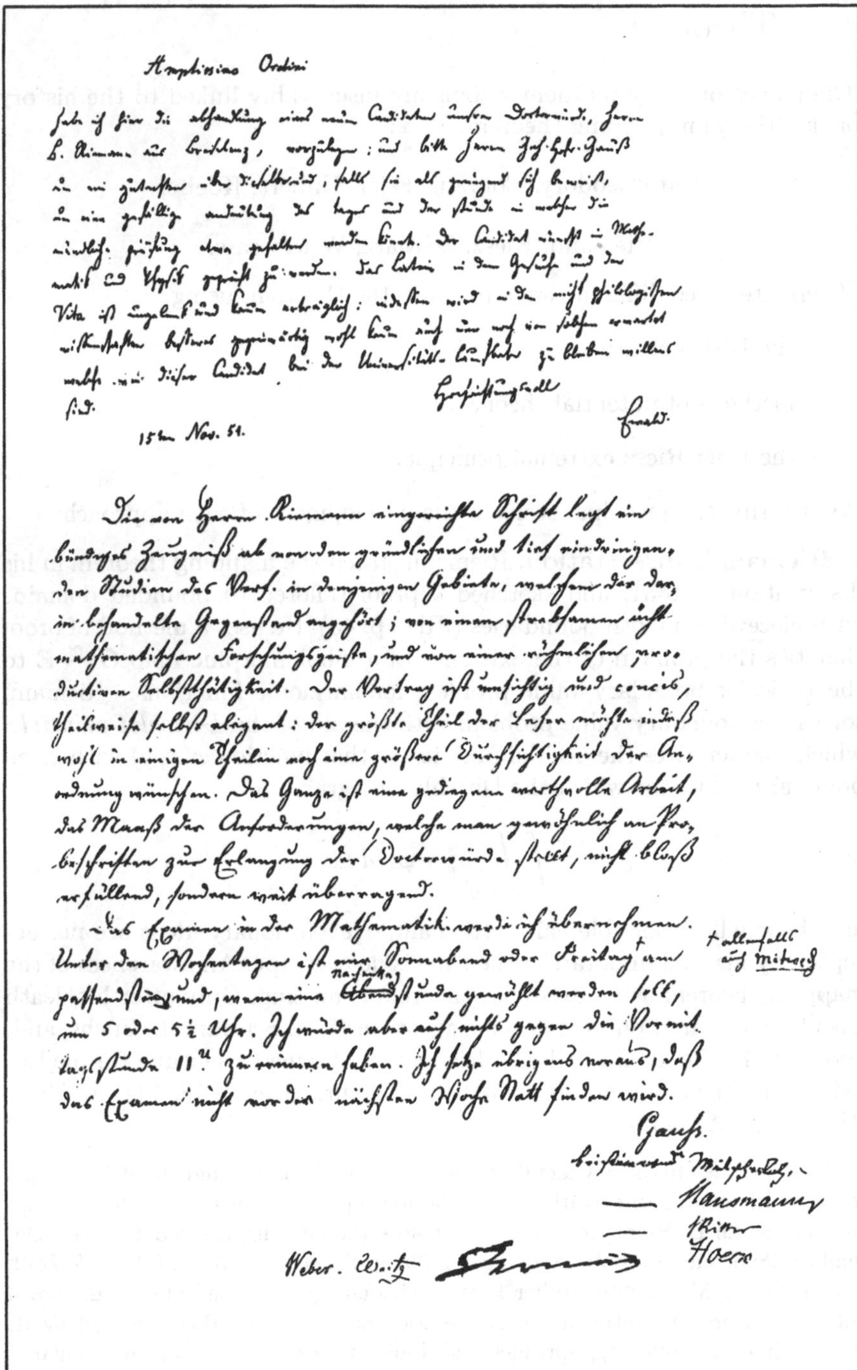

Amplissimo Ordini

habe ich hier die Abhandlung eines neuen Candidaten unserer Doctorwürde, Herrn B. Riemann aus Breselenz, vorzulegen: und bitte Herrn Geh. Hofr. Gauss um ein Gutachten über dieselbe, falls sie als geeignet sich erweist, um eine gefällige Andeutung des Tages und der Stunde in welcher die mündliche Prüfung etwa gehalten werden könnte. Der Candidat wünscht in Mathematik und Physik geprüft zu werden. Das Latein in dem Gesuch und der Vita ist ungelenk und kaum erträglich: indessen wird in den nicht philologischen Wissenschaften bisweilen gegenwärtig wohl kaum auch nur ein solches erwartet, welche wie dieser Candidat bei der Universitäts-Laufbahn zu bleiben willens sind.

Ehrerbietungsvoll

Ewald.

15ten Nov. 51.

Die von Herrn Riemann eingereichte Schrift legt ein bündiges Zeugniß ab von den gründlichen und tief eindringenden Studien des Verf. in demjenigen Gebiete, welchem der darin behandelte Gegenstand angehört; von einem strebsamen ächt mathematischen Forschungsgeiste, und von einer rühmlichen productiven Selbstthätigkeit. Der Vortrag ist umsichtig und concis, theilweise selbst elegant: der größte Theil der Leser möchte indeß wohl in einigen Theilen noch eine größere Durchsichtigkeit der Anordnung wünschen. Das Ganze ist eine gediegene werthvolle Arbeit, das Maaß der Anforderungen, welche man gewöhnlich an Probeschriften zur Erlangung der Doctorwürde stellt, nicht bloß erfüllend, sondern weit überragend.

Das Examen in der Mathematik werde ich übernehmen. Unter den Wochentagen ist mir Sonnabend oder Freitag[+] am passendsten, und, wenn eine Nachmittagsstunde gewählt werden soll, um 5 oder 5½ Uhr. Ich würde aber auch nichts gegen die Vormittagsstunde 11U zu erinnern haben. Ich setze übrigens voraus, daß das Examen nicht vor der nächsten Woche statt finden wird.

+allenfalls auch Mittwoch

Gauß.

beistimmend Mitscherlich.

— Hausmann

— Ritter

Hoeck

Weber. Waitz Ewald

Ewald (1803–1875, an evangelical theologian, orientalist, and politician, one of the Göttingen Seven), asks Gauss for an expert opinion; the dean finds "das Latein in dem Gesuche und der Vita [von Riemann] ungelenk und kaum erträglich" (the Latin in the application and vita [of Riemann] clumsy and almost unbearable). Gauss, in his terse commentary, does not say a word about the content of the work. The reviewer, known to be sparing of praise, does speak of "gründliche und tief eindringende Studien in demjenigen Gebiete, welchem der darin behandelte Gegenstand angehört" (profound and deeply penetrating studies in that area to which the subject treated there belongs), of "strebsame ächt mathematische Forschungsgeiste" (industrious and ambitious genuine mathematical spirit of research), and of "rühmliche productive Selbstthätigkeit" (commendable productive creativity). He thinks that "der größter Theil der Leser möchte wohl in einigen Theilen noch eine größere Durchsichtigkeit der Anordnung wünschen" (most readers may well wish in some places for greater clarity in the presentation); he summarizes, however, by saying: "Das Ganze ist eine gediegene werthvolle Arbeit, das Maaß der Anforderungen, welche man gewöhnlich an Probeschriften zur Erlangung der Doctorwürder stellt, nicht bloß erfüllend, sondern weit überragend." (The whole is a solid work of high quality, not merely fulfilling the requirements usually set for a doctoral thesis, but far surpassing them.) Gauss did not suggest a grade; a third of his letter concerns a time and date that would be convenient for him, not too early in the afternoon, for the oral exam.

2. Early history. In his 1870 paper [Wei], Weierstrass temporarily pulled the rug out from under Riemann's proof with his criticism of the Dirichlet principle, by showing through examples that the existence of a minimal function is by no means certain. Around the turn of the century Hilbert weakened this criticism by a rigorous proof of the Dirichlet principle to the extent required by Riemann ([Hi], pp. 10–14 and 15–37). Since then it has regained its place among the powerful tools of classical analysis; cf. also [Hi], pp. 73–80, and the 1910 dissertation of Courant, [Cou].

In the meantime, other methods were developed. C. Neumann and H. A. Schwarz devised the so-called *alternating method*; see the encyclopedia articles [Lich] and [B] of L. Lichtenstein and L. Bieberbach, especially [Lich], §48. The alternating method allows the potential-theoretic boundary value problem to be solved for domains that are the union of domains for which the boundary value problem is already known to be solvable. With this method, which also uses the Poisson integral and the Schwarz reflection principle, Schwarz arrived at results which finally culminated in the following theorem ([Schw], vol. 2, passim):

If G is a simply connected domain bounded by finitely many real analytic curves which intersect in angles not equal to 0, then there exists a topological map from $\overline{G}$ onto $\overline{\mathbb{E}}$ that maps G biholomorphically onto $\mathbb{E}$.

In Schwarz's time, statements of this kind were considered the hardest in all analysis. — Simply connected domains with arbitrary boundary were first treated in 1900 by W. F. Osgood [Osg]; Osgood's research is based on earlier developments by Schwarz and Poincaré.

3. From Carathéodory-Koebe to Fejér-Riesz. The proof in 2.2–4 is an amalgam of ideas of C. Carathéodory, P. Koebe, L. Fejér, and F. Riesz. All the methods of proof known up to 1912 use the detour via the solution of the (real) boundary value problem for the potential equation $\Delta u = 0$. In 1912 Carathéodory had the felicitous idea, for a given domain G, of mapping the unit disc $\mathbb{E}$ onto a sequence of Riemann surfaces whose "kernel" converges to G; the sequence f_n itself then converges compactly to a biholomorphic map $f : \mathbb{E} \xrightarrow{\sim} G$ (cf. [Ca$_1$], pp. 400–405, and Satz VI, p. 390). Thus for the first time, with relatively simple, purely function-theoretic means, "die Abbildungsfunktion durch ein *rekurrentes Verfahren* gewonnen [wurde], das bei jedem Schritt nur die Auflösung von Gleichungen ersten und zweiten Grades verlangt" (the mapping function [was] obtained by an iteration process, which at each step required only the solution of first- and second-degree equations) (loc. cit., p. 365). Koebe could immediately, to a large extent, eliminate Carathéodory's auxiliary Riemann surfaces [Koe$_{3,4}$]: thus there emerged a very transparent *constructive* proof for the *fundamental theorem of conformal mapping.* The construction of expansions described in Lemma 2.3 plays the central role ([Koe$_4$], pp. 184–185). We will present this beautiful Carathéodory-Koebe theory in detail in the appendix to this chapter and, in doing so, will also go more deeply into the "competition" between these two mathematicians.

In 1922 L. Fejér and F. Riesz realized that the desired Riemann mapping function could be obtained as the solution of an extremal problem for derivatives. They had their stunningly short proof published by T. Radó in the recently founded Hungarian journal *Acta Szeged.* Radó needed one full page to present it ([Ra], pp. 241–242); the Carathéodory-Koebe square root transformation was godfather to this "existence proof of the first water."

Fejér and Riesz consider bounded Q-domains G and show:

There exist $\rho > 0$ *and a biholomorphic map* $h : G \xrightarrow{\sim} B_\rho(0)$ *such that* $h(0) = 0$ *and* $h'(0) = 1$.

The key to the proof is furnished by the (nonempty) family

$$\mathcal{H} := \{f \in \mathcal{O}(G) : f \text{ is bounded and injective, } f(0) = 0,\ f'(0) = 1\}.$$

Corresponding to $\rho := \inf\{|f|_G,\ f \in \mathcal{H}\} < \infty$, there exists a sequence $f_j \in \mathcal{H}$ that is bounded in G and satisfies $\lim |f_j|_G = \rho$. By Montel, a subsequence converges compactly in G to some $h \in \mathcal{O}(G)$. We have $h(0) = 0$, $h'(0) = 1$, and $|h|_G = \rho$. By Hurwitz, $h : G \to B_\rho(0)$ is *injective*; in particular, it follows that $h \in \mathcal{H}$ and $\rho > 0$. The *surjectivity* of h is now the crucial point: assuming that $h(G) \neq B_\rho(0)$, Fejér and Riesz use the Carathéodory-Riesz square root method to construct a function $\widehat{h} \in \mathcal{H}$ with $|\widehat{h}|_G < \rho$; this contradicts the minimality of ρ.

4. Carathéodory's final proof. For the proof that $\widehat{h}'(0) = 1$, Fejér and Riesz must compute explicit derivatives; cf. [Ra], pp. 241–242. In 1929 A.

Ostrowski published a variant of the Fejér-Riesz proof in which "sämtliche Rechnungen — auch die Berechnungen der Nullpunktsableitung der Abbildungsfunktion — vermieden werden" (all computations — even the calculation of the derivative at zero of the mapping function — are avoided) ([Ost], pp. 17–19).

Ostrowski works with the family $\mathcal{F}$ of 2.4 and begins by observing (substitute this for Proposition 2.4):

If $G \neq \mathbb{C}$ is a Q-domain, then $h(G) = \mathbb{E}$ for every function $h \in \mathcal{F}$ with

$$(*). \qquad |h'(0)| = \sup\{|f'(0)| : f \in \mathcal{F}\}.$$

Indeed, for every $g \in \mathcal{F}$ with $f(G) \neq \mathbb{E}$ there exists by Lemma 2.3 an expansion $\kappa : g(G) \to \mathbb{E}$. Since $|\kappa'(0)| > 1$ (see 1.1 of the appendix to this chapter), it follows that $|\widehat{g}'(0)| > |g'(0)|$ for $\widehat{g} := \kappa \circ g \in \mathcal{F}$. — The existence of a function $h \in \mathcal{F}$ satisfying $(*)$ now follows again from the fact that, by Montel, there exists a sequence $h_j \in \mathcal{F}$ with $\lim |h_j'(0)| = \mu := \sup\{|f'(0)| : f \in \mathcal{F}\}$ that converges compactly to some $h \in \mathcal{O}(G)$. Then $|h'(0)| = \mu$. Since $\mathcal{F} \neq \{0\}$ implies that $\mu > 0$, it follows from Hurwitz that $h \in \mathcal{F}$.

When Ostrowski published his proof, he was unaware that not long before, in 1929, in the almost unavailable *Bulletin of the Calcutta Mathematical Society*, Carathéodory had presented a variant completely free of derivatives: "... durch eine geringe Modifikation in der Wahl des Variationsproblems [kann man] den *Fejér-Rieszschen Beweis* noch wesentlich vereinfachen" (... through a minor modification in the choice of the variational problem, [one can] further simplify the *Fejér-Riesz proof* considerably) ([Ca$_1$], pp. 300–301). We have presented Carathéodory's version in 2.2–4. It is the most elegant proof of the Riemann mapping theorem. In the prevailing literature, however, up to the present, Ostrowski's version has won out over that of Carathéodory; the 1985 textbook [N] of R. Narasimhan is an exception.

In 1928 Carathéodory wrote the following on the history of the proof of the mapping theorem ([Ca$_1$], p. 300): "Nachdem die Unzulänglichkeit des ursprünglichen *Riemannschen* Beweises erkannt worden war, bildeten für vielen Jahrzehnte die wunderschönen, aber sehr umständlichen Beweismethoden, die *H.A. Schwarz* entwickelt hatte, den einzigen Zugang zu diesem Satz. Seit etwa zwanzig Jahren sind dann in schneller Folge eine Große Reihe von neuen kürzeren und besseren Beweisen [von ihm selbst und von Koebe] vorgeschlagen worden; es war aber den ungarischen Mathematikern *L. Fejér* und *F. Riesz* vorbehalten, auf den Grundgedanken von *Riemann* zurückzukehren und die Lösung des Problems der konformen Abbildung wieder mit der Lösung eines Variationsproblems zu verbinden. Sie wählten aber nicht ein Variationsproblem, das, wie das *Dirichletsche* Prinzip, außerordentlich schwer zu behandeln ist, sondern ein solches, von dem die Existenz einer Lösung feststeht. Auf diese Weise entstand ein Beweis, der nur wenige Zeilen lang ist, und der auch sofort in allen neueren

Lehrbüchern aufgenommen worden ist." (After the inadequacy of *Riemann*'s original proof was recognized, the exquisite but very intricate methods of proof developed by *H. A. Schwarz* were for many decades the only approach to this theorem. Then, since about twenty years ago, a great number of new shorter and better proofs [his own and Koebe's] have been proposed in rapid succession; but it remained for the Hungarian mathematicians *L. Fejér* and *F. Riesz* to return to *Riemann's* basic ideas and once again tie the solution of the conformal mapping problem to the solution of a variational problem. They did not, however, choose a variational problem that, like the *Dirichlet* principle, is extremely hard to handle, but rather one that is certain to have a solution. In this way a proof emerged that is only a few lines long and that was also immediately adopted by all new textbooks.) In fact, the Fejér-Riesz proof appears as early as 1927 in Bieberbach's *Lehrbuch der Funktionentheorie*, vol. 2, p. 5. The old proofs were forgotten: *The better is the enemy of the good.*

5. Historical remarks on uniqueness and boundary behavior. In his dissertation, Riemann not only asserted the existence of a conformal map $f : G_1 \xrightarrow{\sim} G_2$ between simply connected domains G_1, G_2, but stated that f can moreover be chosen to be a topological map $\overline{G}_1 \xrightarrow{\sim} \overline{G}_2$: that, in particular, f maps the boundaries ∂G_1, ∂G_2 topologically onto each other (for Riemann, all boundaries are piecewise smooth). Riemann also had precise ideas about when f is uniquely determined ([Rie], p. 40):

> *Zu Einem innern Punkte und zu Einem Begrenzpunkte [kann] der entsprechende beliebig gegeben werden; dadurch aber ist für alle Punkte die Beziehung bestimmt.* (The points corresponding to an interior point and to a boundary point [can] be assigned arbitrarily; doing this, however, determines the relationship for all points.)

In 1884 Poincaré proved a uniqueness theorem that assumes nothing about the existence of the map on the boundary of G; cf. *Lemme fondamental* ([Po], p. 327). Poincaré's lemma is — in modern language — nothing but the uniqueness theorem 2.5; our proof is essentially the same as Poincaré's (loc. cit., pp. 327–328). — The uniqueness problem played an important role in the history of conformal mapping theory. In 1912 Carathéodory, by determining that the uniqueness theorem is ultimately based on Schwarz's lemma, put the elegant finishing touches on these matters.

Schwarz, in 1869, sharply separated the problem of mapping a domain conformally onto a disc from the problem of extending this map continuously to the boundary. Carathéodory studied the extension problem from 1913 on and developed his penetrating *theory of prime ends*; cf. the first three papers in the fourth volume of his *Gesammelte Mathematische Schriften*.

In order to formulate the high points of extension theory, we consider a biholomorphic map $f : \mathbb{E} \to G$ onto a *bounded* domain G. The following lemma is the starting point.

Extension lemma. *The map f can be extended to a continuous map from $\overline{\mathbb{E}}$ to $\overline{G}$ if and only if the boundary of G is a closed path (in other words, if there exists a continuous map $\varphi : \partial\mathbb{E} \to \mathbb{C}$ with $\varphi(\partial\mathbb{E}) = \partial G$).*

This lemma is used to prove

Carathéodory's theorem. *The map $f : \mathbb{E} \to G$ can be extended to a topological map from $\overline{\mathbb{E}}$ onto $\overline{G}$ if and only if the boundary of G is a closed Jordan curve (in other words, if there exists a topological map $\varphi : \partial\mathbb{E} \to \mathbb{C}$ with $\varphi(\partial\mathbb{E}) = \partial G$).*

A simple corollary is Schoenflies's theorem on Jordan curves, which has nothing to do with function theory:

Every topological map from one Jordan curve onto another can be extended to a topological map from $\mathbb{C}$ onto itself.

Details of this theory can be found in the book [Pom].

6. Glimpses of several variables. There is no obvious generalization of the Riemann mapping theorem to simply connected domains in $\mathbb{C}^n$, $n > 1$, even in the case $n = 2$. The *polydisc* $\{(w, z) \in \mathbb{C}^2 : |w| < 1, |z| < 1\}$ and the *ball* $\{(w, z) \in \mathbb{C}^2 : |w|^2 + |z|^2 < 1\}$ are natural analogues of the unit disc; both domains are topologically cells, thus certainly simply connected. But Poincaré proved in 1907:

There is no biholomorphic map from the ball onto the polydisc.

Simple proofs are given in [Ka], p. 8, and [Ran], p. 24. There even exist *families of bounded domains of holomorphy G_t, $t \in \mathbb{R}$, with boundaries ∂G_t that are real analytic everywhere*, such that all the domains G_t are *diffeomorphic* to the 4-dimensional cell but two domains G_t, $G_{\tilde{t}}$ are biholomorphically equivalent only if $t = \tilde{t}$.

Positive statements can be obtained if the automorphisms of the domain are brought into play. Thus, for example, É. Cartan proved in 1935 [Car_1]:

Every bounded homogeneous domain in $\mathbb{C}^2$ can be mapped biholomorphically onto either the ball or the bidisc.

For $n \geq 3$ the situation is more complicated, even for bounded homogeneous domains.

§4. Isotropy Groups of Simply Connected Domains

The automorphism group $\operatorname{Aut} G$ of all the biholomorphic maps of a domain G onto itself contains important information about the function theory of G. Two domains G, G' can be mapped biholomorphically onto each other only if their groups $\operatorname{Aut} G$, $\operatorname{Aut} G'$ are isomorphic. In addition to automorphisms, we study *inner* maps of G; these are *holomorphic maps from G to itself.* The set $\operatorname{Hol} G$ of all the inner maps of G, with composition as group operation, is a *semigroup* with $\operatorname{Aut} G$ as a *subgroup.*

For every point $a \in G$, the set $\operatorname{Hol}_a G$ of all the inner maps of G with *fixed point* a is a *subsemigroup* of $\operatorname{Hol} G$. The set $\operatorname{Aut}_a G$ of the automorphisms of G with fixed point a is a *subgroup* of $\operatorname{Hol}_a G$; $\operatorname{Aut}_a G$ is also called the *isotropy group* of G at a. The map

$$\sigma : \operatorname{Hol}_a G \to \mathbb{C}, \quad f \mapsto f'(a)$$

is fundamental for the study of $\operatorname{Hol}_a G$ and $\operatorname{Aut}_a G$. It is *multiplicative* (by the chain rule):

$$(f \circ g)'(a) = f'(a)g'(a), \quad f,\ g \in \operatorname{Hol}_a G;$$

in particular, σ induces a homomorphism $\operatorname{Aut}_a G \to \mathbb{C}^\times$ of the group $\operatorname{Aut}_a G$ into the *multiplicative group* $C^\times$.

In Subsection 1, we describe σ for four special domains. In Subsection 2, σ is studied for simply connected domains $\neq \mathbb{C}$. The tools for doing this are the Riemann mapping theorem and the Schwarz lemma, which we use in the following form:

(S) $|g'(0)| \leq 1$ *for* $g \in \operatorname{Hol}_0 \mathbb{E}$. *Moreover,* $\operatorname{Aut}_0 \mathbb{E} = \{g \in \operatorname{Hol}_0 \mathbb{E} : |g'(0)| = 1\}$. *The map* $\operatorname{Aut}_0 \mathbb{E} \to S^1$, $g(z) \mapsto g'(0)$, *is a group isomorphism.*

1. Examples. 1) Since $\operatorname{Aut}_a \mathbb{C} = \{z \mapsto uz + a(1-u) : u \in \mathbb{C}^\times\}$, $\sigma : \operatorname{Aut}_a \mathbb{C} \to \mathbb{C}^\times$ is an *isomorphism.*

2) Since $\operatorname{Aut}_a \mathbb{C}^\times = \{\mathrm{id}_{\mathbb{C}^\times}, z \mapsto a^2/z\}$, $\sigma : \operatorname{Aut}_a \mathbb{C}^\times \to \mathbb{C}^\times$ is *injective; the image is the cyclic group* $\{1, -1\}$ *of order* 2.

3) By (S), $\sigma : \operatorname{Aut}_0 \mathbb{E} \to \mathbb{C}^\times$ is *injective; the image is the circle group* S^1.

4) Let $\zeta := \exp(2\pi i/m)$, with $m \in \mathbb{N}\backslash\{0\}$; let $\widehat{\mathbb{E}} := \mathbb{E}\backslash\{\frac{1}{2}\zeta, \frac{1}{2}\zeta^2, \ldots, \frac{1}{2}\zeta^m\}$. Then $\operatorname{Aut}_0 \widehat{\mathbb{E}} = \{z \mapsto \zeta^\mu z,\ \mu \in \mathbb{N}\}$. Thus $\sigma : \operatorname{Aut}_0 \widehat{\mathbb{E}} \to \mathbb{C}^\times$ is *injective; the image is the cyclic group* $\{1, \zeta, \ldots, \zeta^{m-1}\}$ *of order* n.

In these examples σ is always *injective*; when $G \neq \mathbb{C}$, the image is always a *subgroup of* S^1, and in addition to S^1 all *finite cyclic* groups occur as isotropy groups (of bounded domains). With the aid of uniformization theory it can be shown that the examples are already characteristic: the only groups other than $\mathbb{C}$ that have infinite isotropy groups are those that can be mapped biholomorphically onto $\mathbb{C}$.

2. The group $\mathrm{Aut}_a G$ for simply connected domains $G \neq \mathbb{C}$. By the Riemann mapping theorem, for every point $a \in G$ there exists a biholomorphic map $u : \mathbb{E} \to G$ with $u(0) = a$. The next statement can be readily verified.

(∗) *The correspondence* $\imath : \mathrm{Hol}_a\, G \to \mathrm{Hol}_0\, \mathbb{E}$, $f \mapsto g := u^{-1} \circ f \circ u$, *is bijective and a semigroup homomorphism. Moreover,* $\sigma(f) = g'(0)$.
$\mathrm{Aut}_a\, G$ *is mapped biholomorphically onto* $\mathrm{Aut}_0\, \mathbb{E}$ *by* $\imath$.

A Schwarz lemma for simply connected domains now follows immediately from (∗) and (S) of the introduction:

(1) *If $G \neq \mathbb{C}$ is simply connected, then $|f'(a)| \leq 1$ for every $f \in \mathrm{Hol}_a\, G$, $a \in G$. Moreover,* $\mathrm{Aut}_a\, G = \{f \in \mathrm{Hol}_a\, G : |f'(a)| = 1\}$.

Since σ is the composition of the isomorphism $\imath : \mathrm{Aut}_a\, G \to \mathrm{Aut}_0\, \mathbb{E}$ with the isomorphism $\mathrm{Aut}_0\, \mathbb{E} \to S^1$, $g \mapsto g'(0)$, we have another result:

Proposition. *If $G \neq \mathbb{C}$ is simply connected, then $\sigma : \mathrm{Aut}_a\, G \to \mathbb{C}$, $f \mapsto f'(a)$, maps the group $\mathrm{Aut}_a\, G$ isomorphically onto the circle group S^1, $a \in G$.*

One corollary of this is a uniqueness theorem.

Uniqueness theorem. *Let $G \neq \mathbb{C}$. Suppose that either G is simply connected or $G \simeq \mathbb{C}^\times$. Let $f \in \mathrm{Aut}_a\, G$ satisfy $f'(a) > 0$. Then $f = \mathrm{id}_G$.*

Proof. The case $G \simeq \mathbb{C}^\times$ is clear (Example 1.2). In the other case, $|f'(a)| = 1$; hence $f'(a) = 1$ since $f'(a) > 0$. Since σ is injective, it follows that $f = \mathrm{id}_G$. □

This yields the following uniqueness theorem.

Let $G \neq \mathbb{C}$ be simply connected and let $g, h \in \mathrm{Aut}_a G$, $a \in G$. Then $g = h$ if g and h "have the same direction at a"; that is, if $g'(a)/|g'(a)| = h'(a)/|h'(a)|$.

Proof. For $f := g^{-1} \circ h \in \mathrm{Aut}_a\, G$, verify that $f'(a) = |h'(a)|/|g'(a)| > 0$. □

Iteration theory will be used in 9.2.3 to prove the results of this subsection for *arbitrary bounded* domains. Uniformization theory can be used to show that the results are actually valid for *all* domains $\neq \mathbb{C}$.

3*. Mapping radius. Monotonicity theorem. If $G \neq \mathbb{C}$ is simply connected, then for every point $a \in G$ there exists *exactly one* biholomorphic map f from G onto a disc $B_\rho(0)$ with $f(a) = 0$, $f'(a) = 1$: namely $f := (1/h'(a))h$, where $h : G \xrightarrow{\sim} \mathbb{E}$, with $h(a) = 0$ and $h'(a) > 0$, is chosen as in 2.5. The number $\rho = \rho(G, a)$ is called the *mapping radius of G relative to a*. Thus

(1) $$\rho(G, a) = 1/h'(a), \quad h := \text{mapping function of 2.5.}$$

We also set $\rho(\mathbb{C}, a) := \infty$ for all $a \in \mathbb{C}$.

Monotonicity theorem. *If $\widehat{G}$, G are simply connected and $\widehat{G} \subset G$, then*

(2) $$\rho(\widehat{G}, a) \leq \rho(G, a) \quad \textit{for all} \quad a \in \widehat{G}.$$

If there exists a point $b \in \widehat{G}$ such that $\rho(\widehat{G}, b) = \rho(G, b)$, then $\widehat{G} = G$.

Proof. Let $G \neq \mathbb{C}$ and let $h : G \xrightarrow{\sim} \mathbb{E}$, $\widehat{h} : \widehat{G} \xrightarrow{\sim} \mathbb{E}$ be chosen as in 2.5. Then $g := \widehat{h}^{-1} \circ h \in \text{Hol}_a G$. Since $g'(a) = h'(a)/\widehat{h}'(a) > 0$, 2(1) implies that $g'(a) \leq 1$. Hence $h'(a) \leq \widehat{h}'(a)$, and (2) follows from (1).

It follows from $\rho(\widehat{G}, b) = \rho(G, b)$ that $h'(b) = \widehat{h}'(b)$; hence $g'(b) = 1$ for the maps corresponding to b. Now 2(1) and Theorem 2 imply that $\widehat{h}^{-1} \circ h = \text{id}_G$; hence $\widehat{h}^{-1}(\mathbb{E}) = G$, i.e. $\widehat{G} = G$. □

Exercises

1) Compute $\rho(G, a)$, $a \in G$, in the following cases:

a) $G := B_r(c)$, $c \in \mathbb{C}$, $r > 0$;
b) $G := \mathbb{H}$ = upper half-plane;
c) $G := \{z = re^{i\varphi} \in \mathbb{C} : r > 0 \text{ arbitrary}, 0 < \varphi < \varphi_0, \text{ where } \varphi_0 \in (0, 2\pi]\}$.

2) If G is simply connected and $g : G \to \mathbb{C}$ is a holomorphic injection, then

$$\rho(g(G), g(a)) = |g'(a)|\rho(G, a) \quad \text{for all} \quad a \in G.$$

3) Let $G \neq \mathbb{C}$ be simply connected and let $a \in G$. Prove:

a) $|f|_G \geq \rho(G, a)$ for every $f \in \mathcal{O}(G)$ with $f(a) = 0$, $f'(a) = 1$.
b) Equality holds in a) if and only if f maps G biholomorphically onto $B_\rho(0)$ ("minimum-maximorum principle").

Bibliography for Chapter 8: See p. 201

Appendix to Chapter 8: Carathéodory-Koebe Theory

The Fejér-Riesz-Carathéodory proof is *not constructive*: No rule is given (in 8.2.4) telling how the sequence f_n with $\lim_n |f_n(p)| = \mu$ is to be constructed, and absolutely nothing is said about how to find the subsequence h_j of the sequence f_n. A proof free of these defects was given in 1914 by P. Koebe, who applied ideas of Carathéodory: By *expansion*, the domain is mapped successively onto subdomains on $\mathbb{E}$ in such a way that these subdomains exhaust the unit disc. Koebe obtains the expansion maps by elementary means, solving a quadratic equation and determining a boundary point with *minimal* distance from the origin; his *expansion sequences* converge — though slowly — to the desired biholomorphic map $G \xrightarrow{\sim} \mathbb{E}$; *no passage to subsequences is necessary.*

In Section 1 we discuss the expansions used by Koebe; in Section 2 we describe the Carathéodory-Koebe algorithm and apply it to the special expansion family $\mathcal{K}_2$. In Section 3 we construct other families suitable for the algorithm.

§1. Simple Properties of Expansions

We begin with a simple expansion lemma (Subsection 1), which is fundamental for the considerations of this appendix. In Subsection 2 "admissible" expansions are discussed. In Subsection 3, as an example of such expansions, we study the "crescent expansion."

1. Expansion lemma. If G is a domain in $\mathbb{C}$ with $0 \in G$, then

$$r(G) := \sup\{t \in \mathbb{R} : B_t(0) \subset G\} = d(0, \partial G)$$

is called the *inner radius of G (relative to the origin)*. We have $0 < r(G) \leq \infty$; when $r(G) \neq \infty$, there always exist boundary points $a \in \partial G$ with $|a| = r(G)$. The following *monotonicity property of inner radii* with respect to holomorphic maps is crucial for our purposes in this appendix.

Expansion lemma. *Let f, $g \in \mathcal{O}(G)$ be nonconstant and let the map $g : G \to \mathbb{C}$ be injective. Assume moroever that $f(0) = 0$ and $|f(z)| \geq |g(z)|$ for all $z \in G$. Then $r(f(G)) \geq r(g(G))$.*

Proof. Let $B = B_s(0) \subset g(G)$, $s < \infty$. It suffices to prove that $B \subset f(g^{-1}(B))$. This will follow if we prove that $|b| \geq s$ *for every point* $b \in \partial f(g^{-1}(B))$.

We set $h := f \circ g^{-1}$. There exists a sequence $a_n \in B$ with $a := \lim a_n \in \overline{B}$ and $b = \lim h(a_n)$. Then $|a| \geq s$, since otherwise $a \in B$; thus $b = h(a) \in h(B)$. Now the inequality $|h(w)| = |f(g^{-1}(w))| \geq |g(g^{-1}(w))| = |w|$ holds for all $w \in B$. Hence $|b| = \lim |h(a_n)| \geq \lim |a_n| = |a| \geq s$. □

Now let $G \subset \mathbb{E}$ and $0 \in G$. Then $r(G) \leq 1$. According to 8.2.3, holomorphic injections $\kappa : G \to \mathbb{E}$ with $\kappa(0) = 0$ and $|\kappa(z)| > |z|$, $z \in G\backslash\{0\}$, are called *expansions of* G. It is trivial that

(1) *$\widehat{\kappa} \circ \kappa : G \to \mathbb{E}$ is an expansion if $\kappa : G \to \mathbb{E}$ and $\widehat{\kappa} : \widehat{G} \to \mathbb{E}$ are expansions with $\widehat{G} \supset \kappa(G)$.*

The following is important:

Proposition. *If $\kappa : G \to \mathbb{E}$ is an expansion, then $|\kappa'(0)| > 1$ and $r(\kappa(G)) \geq r(G)$.*

Proof. We have $\kappa(z) = zf(z)$, with $f \in \mathcal{O}(G)$. Since $|f(z)| > 1$ in $G\backslash\{0\}$, we also have $|f(0)| > 1$ (minimum principle); hence $|\kappa'(0)| = |f(0)| > 1$. — The inequality $r(\kappa(G)) \geq r(G)$ follows immediately from the expansion lemma. □

Warning. The reader should not think that $\kappa(G) \supset G$ for all expansions. The crescent expansions are instructive counterexamples; see Subsection 3.

One often speaks of expansions if, besides $\kappa(0) = 0$, only $|\kappa(z)| \geq |z|$ holds. In that case, the proposition is true with $|\kappa'(0)| \geq 1$.

2. Admissible expansions. The square root method. A Q-domain G is called a *Koebe domain* if $G \Subset \mathbb{E}$ (this means that $G \subset \mathbb{E}$ but $G \neq \mathbb{E}$). Then $0 < r(G) < 1$. An expansion κ of a Koebe domain is called *admissible* if $\kappa(G)$ is again a Koebe domain. The next statement follows trivially from 8.2.1(1).

(1) *Every expansion $\kappa : G \to \mathbb{E}$ of a Koebe domain such that $\kappa(G) \neq \mathbb{E}$ is admissible.*

Admissible expansions have already appeared in Lemma 8.2.3. We state matters more precisely here:

Theorem *(square root method). Let G be a Koebe domain and let $c \in \mathbb{E}$, $c^2 \notin G$. Let $v \in \mathcal{O}(G)$ be the square root of $g_{c^2}|G$ with $v(0) = c$. Let $\vartheta : \mathbb{E} \to \mathbb{E}$ be a rotation about 0. Then $\kappa := \vartheta \circ g_c \circ v$ is an admissible expansion of G and*

$$|\kappa'(0)| = \frac{1 + |c|^2}{2|c|}. \tag{2}$$

Proof. By Lemma 8.2.3, $g_c \circ v : G \to \mathbb{E}$ is an expansion; hence so is $\kappa : G \to \mathbb{E}$. If we had $\kappa(G) = \mathbb{E}$, it would follow that $v(G) = \mathbb{E}$ and (because $g_{c^2} = v^2$) that $g_{c^2}(G) = \mathbb{E}$, i.e. $G = \mathbb{E}$. But this is impossible. By (1), κ is therefore admissible. Equation (2) is equation 8.2.3(4). □

In particular, for every Koebe domain G the square root method yields an admissible expansion $\kappa : G \to \mathbb{E}$ with $\kappa'(0) > 1$; these "normalized" expansions play a leading role in Carathéodory-Koebe theory; cf. 2.3. □

The following representation of the square root v is useful in computations:

(3) *Set* $G^* := g_{c^2}(G)$. *Then*

$$v = q \circ g_{c^2}, \quad \textit{where } q \in \mathcal{O}(G^*) \textit{ with } q^2 = z|G^*,\ q(c^2) = c.$$

For all expansions κ constructed in the theorem above,

(4) $r(G) < r(\kappa(G))$ *(sharpened version of Proposition 1)*

(5) $\mathbb{E}\backslash\kappa(G)$ *always has interior points in* $\mathbb{E}$.

Proof. ad (4). Let $b \in \mathbb{E} \cap \partial(\kappa(G))$ with $|b| = r(\kappa(G))$. There exists a sequence $z_n \in G\backslash\{0\}$ with $\lim \kappa(z_n) = b$ and $a := \lim z_n \in \partial G$. Then $|a| \geq r(G)$. By Lemma 8.2.3, $id = \psi_c \circ \kappa$, where $\psi_c \in \mathcal{O}(\mathbb{E})$ and $|\psi_c(z)| < |z|$ for $z \in \mathbb{E}\backslash\{0\}$. It follows that $z_n = \psi_c(\kappa(z_n))$; hence $a = \psi_c(b)$, so $|a| < |b|$ and $r(G) < r(\kappa(G))$.

ad (5). Since ϑ, g_c, $g_{c^2} \in \operatorname{Aut}\mathbb{E}$, it suffices by (3) to show that $\mathbb{E}\backslash q(G^*)$ has interior points in $\mathbb{E}$. But this is clear: $(-q)(G^*) \subset \mathbb{E}$ and $q(G^*) \cap (-q)(G^*) = \emptyset$ (the second statement because $0 \notin G^*$ — compare the proof of 8.2.2).

3*. The crescent expansion. All slit domains $G_t := \mathbb{E}\backslash[t^2, 1)$, $0 < t < 1$, are Koebe domains. The effect of the admissible expansion $\kappa := g_t \circ v$ on G_t is surprising:

(1) *The image domain* $\kappa(G_t)$ *is the "crescent"* $\mathbb{E}\backslash K$, *where* K *is the closed disc about* $\rho := (1 + t^2)/2t$ *with* $t \in \partial K$ (Figure 8.6). *In particular,* $\kappa(G_t) \not\supset G_t$.

Proof. By 2(3), $\kappa : G \xrightarrow{\sim} \kappa(G)$ can be factored as follows:

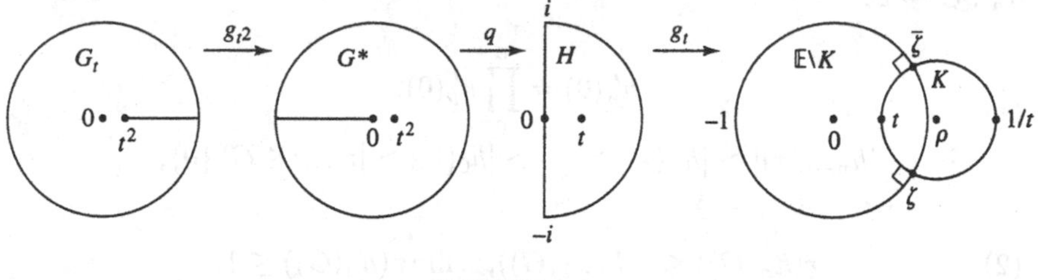

FIGURE 8.6.

Here $G^* = g_{t^2}(G_t) = \mathbb{E}\backslash(-1, 0]$ and $H := q(G^*) = \{z \in \mathbb{E} : \operatorname{Re} z > 0\}$. Since $g_t(t) = 0$, it thus suffices to show that $g_t(\partial H)$ is the boundary of $\mathbb{E}\backslash K$. Since $g_t(\partial\mathbb{E}) = \partial\mathbb{E}$, the functions g_t map the semicircle in ∂H onto the arc in $\partial\mathbb{E}$ that goes from $\zeta := g_t(i)$ through $-1 = g_t(1)$ to $\overline{\zeta} := g_t(-i)$ (note that $\text{Im}\,\zeta < 0$). The image under g_t of the line $i\mathbb{R}$ is the circle L through $t = g_t(0)$ (punctured at $1/t$) that intersects $\partial\mathbb{E}$ orthogonally at ζ and $\overline{\zeta}$ (conformality). The equation for L is thus $|z - m|^2 = m^2 - 1$, with $m > 1$ (by Pythagoras; the tangents to $\partial\mathbb{E}$ at $\zeta, \overline{\zeta}$ are perpendicular to the radii and intersect each other on $\mathbb{R}$). Since $t \in L$, it follows that $m = \rho$; hence $L = (\partial K)\backslash\{t^{-1}\}$. This proves that $g_t(\partial H) = \partial(\mathbb{E}\backslash K)$. □

Remark. The crescent expansion is discussed from a computational point of view in [PS], Problem 90, Part IV.

§2. The Carathéodory-Koebe Algorithm

If a *nonempty* set $\mathcal{D}(G)$ of *admissible* expansions is assigned (according to some rule) to every Koebe domain G, we call the union $\mathcal{D} := \bigcup \mathcal{D}(G)$, where G runs through all Koebe domains, an *expansion family.* With the aid of such families, "expansion sequences" can be assigned to every Koebe domain G in various ways. We set $G_0 := G$ and choose $\kappa_0 \in \mathcal{D}(G_0)$. Since κ_0 is admissible, $G_1 := \kappa_0(G_0)$ is a Koebe domain; we can therefore choose $\kappa_1 \in \mathcal{D}(G_1)$. Since $G_2 := \kappa_1(G_1)$ is again a Koebe domain, we can continue and (recursively) determine Koebe domains G_n and expansions $\kappa_n \in \mathcal{D}(G_n)$ with $\kappa_n(G_n) = G_{n+1}$, $n \in \mathbb{N}$. Then, by 1(1),

$$h_n := \kappa_n \circ \kappa_{n-1} \circ \cdots \circ \kappa_0 : G \to \mathbb{E}, \quad h_n = \kappa_n \circ h_{n-1}$$

is an admissible expansion. The procedure just described is called the *Carathéodory-Koebe algorithm* and the sequence h_n thus constructed an *expansion sequence (for G relative to $\mathcal{D}$).* It will be shown that "correctly chosen" expansion sequences converge to biholomorphic maps $h : G \xrightarrow{\sim} \mathbb{E}$.

1. Properties of expansion sequences. *For all expansion sequences $h_n : G \to \mathbb{E}$:*

$$(1) \qquad \begin{gathered} h_n'(0) = \prod_0^n \kappa_\nu'(0), \\ |h_{n+1}(z)| > |h_n(z)| > \cdots > |h_0(z)| > |z|, \; z \in G\backslash\{0\}, \end{gathered}$$

$$(2) \qquad r(h_n(G)) \leq r(h_{n+1}(G)), \quad \lim r(h_n(G)) \leq 1,$$

$$(3) \qquad \lim |\kappa_n'(0)| = 1.$$

Proof. Statements (1) and (2) follow immediately from the equation $h_n = \kappa \circ h_{n-1}$, where Proposition 1.1 is used in the proof of (2). To prove (3), let $B_t(0) \subset G$, $t > 0$. Then every map $\mathbb{E} \to \mathbb{E}$, $z \mapsto h_n(tz)$, is holomorphic. Since $h_n(0) = 0$, it follows from Schwarz that $|h_n'(0)| \le 1/t$. The sequence $|h_n'(0)|$ is thus *bounded.* Since $|\kappa_\nu'(0)| > 1$ for all ν by Proposition 1.1, the sequence is also *monotone increasing* by (1). Hence $a := \lim |h_n'(0)| = \prod_0^\infty |\kappa_\nu'(0)|$ exists. Since $a \neq 0$ (indeed, $a > 1$), it follows that $\lim |\kappa_n'(0)| = 1$ (cf. 1.1.1). □

Koebe, in 1915, called the algorithm a *Schmiegungsverfahren* (osculation process). The nth *osculation operation* is the choice of the expansion κ_n, which produces the domain G_{n+1} from G_n ([Koe$_4$], pp. 183–185). The *osculation effect* is expressed by (2). Equation (3) is the key to forcing the desired equation $\lim r(h_n(G)) = 1$ by means of "cleverly" chosen sequences κ_n; see Subsection 3.

2. Convergence theorem. We seek expansion sequences $h_n : G \to \mathbb{E}$ that converge to biholomorphic maps $G \to \mathbb{E}$. The following convergence theorem shows when this occurs. An expansion sequence $h_n : G \to \mathbb{E}$ is called an *osculation sequence* if $h_n'(0)$ is always *positive* and $\lim r(h_n(G)) = 1$.

Convergence theorem. 1) *Let $h_n : G \to \mathbb{E}$ be an expansion sequence that converges compactly to a function h. Then $h : G \to \mathbb{E}$ is an expansion, and $r(h(G)) \ge \lim r(h_n(G))$.*

2) *Every osculation sequence $h_n : G \to \mathbb{E}$ converges compactly in G to a biholomorphic expansion $h : G \xrightarrow{\sim} \mathbb{E}$.*

Proof. ad 1). By 1(1), $|h(z)| > |h_n(z)| > |z|$ for all $z \in G\backslash\{0\}$ and all $n \in \mathbb{N}$. Since $h(0) = 0$, it follows that h is not constant; hence, since all the h_n are injections, h is injective by Hurwitz. Therefore $h : G \to \mathbb{E}$ is an expansion of G. Furthermore, it follows from the expansion lemma 1.1 that $r(h(G)) \ge \lim r(h_n(G))$ for all n. Hence $r(h(G)) \ge \lim r(h_n(G))$.

ad 2). Since subsequences of expansion sequences are again expansion sequences, 1) implies that every limit function h of a subsequence of the sequence h_n is an expansion and hence maps G biholomorphically onto $h(G) \subset \mathbb{E}$. Since moreover $r(h(G)) \ge 1$ by 1), we have $h(G) = \mathbb{E}$; that is, $h : G \xrightarrow{\sim} \mathbb{E}$ is biholomorphic. — To see that the sequence h_n converges compactly, it suffices since $h_n(G) \subset \mathbb{E}$ to show that all its compactly convergent subsequences have the same limit (Montel's convergence criterion 7.1.3). If h, $\widehat{h}$ are such limits, then, by what has already been proved, they furnish biholomorphic maps $G \xrightarrow{\sim} \mathbb{E}$ with $h(0) = \widehat{h}(0) = 0$. Since $h_{n+1}'(0) \ge h_n'(0) > 0$ for all n, it follows that $h'(0) > 0$, $\widehat{h}'(0) > 0$; hence $h = \widehat{h}$ by the uniqueness theorem 8.2.5. □

Observe that 1(3) was not used in this proof and that the limit map in 2) is, by 8.2.5, the uniquely determined map $G \xrightarrow{\sim} \mathbb{E}$. □

Since simply connected domains $\neq \mathbb{C}$ can be mapped biholomorphically onto Koebe domains (cf. 8.2.2), the Riemann mapping theorem follows from 2) once an osculation sequence h_n has been constructed for G. Special expansion families are used in this construction. In the next subsection we describe how Koebe constructed such a family. Since he wants $\lim r(h_n(G)) = 1$ and knows that $\lim |\kappa_n'(0)| = 1$ (by 1(3)), he "artificially" produces a relationship between inner radii and derivatives of expansions.

3. Koebe families and Koebe sequences. τ will denote a *continuous real-valued* function on $(0,1)$ for which $\tau(x) > 1$, $x \in (0,1)$. An expansion family $\mathcal{K}$ is called a *Koebe family for* τ if

$$\text{(a)} \qquad \kappa'(0) = \tau(r(G)) \quad \text{for all} \quad \kappa : G \to \mathbb{E} \quad \text{in} \quad \mathcal{K}.$$

An expansion sequence $h_n = \kappa_n \circ \kappa_{n-1} \circ \cdots \circ \kappa_0 : G \to \mathbb{E}$ relative to a Koebe family $\mathcal{K}$ is called a *Koebe sequence.* If τ is the function corresponding to $\mathcal{K}$, it follows from (a) that

$$\text{(a*)} \qquad \tau(r(h_n(G))) = \kappa_{n+1}'(0), \quad n \in \mathbb{N}.$$

From (a*) and 1(3), we now immediately obtain the crucial

Osculation lemma. *Every Koebe sequence is an osculation sequence.*

Proof. By 1(2), $r := \lim r(h_n(G)) \in (0,1]$. If r were less than 1, the continuity of τ in r, together with (a*), would imply that $\lim \kappa_n'(0) = \tau(r)$. But this contradicts 1(3) since $\tau(r) > 1$. Thus $r = 1$. Since $\kappa_\nu'(0) > 0$, 1(1) implies that $h_n'(0) > 0$ for all n; hence h_n is an osculation sequence. □

To complete our study of Carathéodory-Koebe theory, the "only" thing we need to prove is the existence of Koebe families. For every Koebe domain G, we use the square root method of 1.2 to construct *all expansions* $\kappa : G \to \mathbb{E}$ *with* $\kappa'(0) > 0$, *where* $c \in \mathbb{E}$ *is always chosen in such a way that* c^2 *is a boundary point of* G *with minimum distance to the origin.* We let G run through *all* Koebe domains and denote by $\mathcal{K}_2$ the set of all expansions obtained in this way. We set

$$\tau_2 := \frac{1}{2}\frac{x - x^{-1}}{x^{\frac{1}{2}} - x^{-\frac{1}{2}}} = \frac{1+x}{2\sqrt{x}}, \qquad x \in (0,1);$$

this function is continuous in $(0,1)$, and $\tau_2(x) > 1$ for all $x \in (0,1)$.

Proposition. *The family $\mathcal{K}_2$ is a Koebe family for τ_2.*

Proof. By Theorem 1.2, $\mathcal{K}_2$ is an expansion family. For all $\kappa \in \mathcal{K}_2$, $\kappa'(0) = \tau_2(r(G))$ by 1.2(2), since $\kappa'(0) > 0$ and $|c|^2 = r(G)$ by the choice of c. □

Other Koebe families are constructed in Section 3.

4. Summary. Quality of convergence. Carathéodory-Koebe theory contains the following as a special case.

Koebe's main theorem. *For every Koebe domain G there exist Koebe domains G_n, with $G_0 = G$, and expansions $\kappa_n : G_n \to \mathbb{E}$, with $\kappa_n(G_n) = G_{n+1}$, such that the sequence $\kappa_n \circ \kappa_{n-1} \circ \cdots \circ \kappa_0 : G \to \mathbb{E}$ converges compactly to a biholomorphic expansion $G \xrightarrow{\sim} \mathbb{E}$. Every expansion κ_n can be explicitly constructed by a square root operation: $\kappa_n \in \mathcal{K}_2$.*

By 1.2(4), $r(G) < r(\kappa(G))$ for all expansions $\kappa \in \mathcal{K}_2$. Since $|c|^2 = r(G)$, it can even be proved that

$$r(\kappa(G)) \geq \frac{\sqrt{r}}{1+r}(\sqrt{2(1+r^2)} + r - 1), \quad \text{if} \quad r := r(G), \quad \kappa \in \mathcal{K}_2$$

([Ca$_2$], pp. 285–286; also [PS], Part IV, Problem 91). For numerical approximation of the Riemann mapping function $h : G \xrightarrow{\sim} \mathbb{E}$, the square root method converges *very slowly*. Suppose that the inner radius r_n of the domain $h_n(G)$ is chosen as a measure of the quality of the nth approximation $h_n = \kappa_n \circ \kappa_{n-1} \circ \cdots \circ \kappa_0$. Ostrowski showed in 1929 ([Ost], p. 174):

There exists a constant $M > 0$, depending on $r(G)$, such that $r_n > 1 - M/n$, $n > 0$.

5. Historical remarks. The competition between Carathéodory and Koebe. Square root maps occur early in Koebe's work (e.g. in 1907 in [Koe$_1$], p. 203 and p. 644, and also in 1909 in [Koe$_2$], p. 209 and p. 216). But they were first used systematically by Carathéodory in 1912, in his recursive procedure for constructing the Riemann mapping function. He works explicitly with the function

$$z\frac{(1+r^2)z - 2re^{i\theta}}{2rz - (1+r^2)e^{i\theta}} \quad ([\text{Ca}_1], \text{ p. } 401);$$

in the terminology of 8.2.3, this is the function $e^{-i\theta}\psi_c$ with $c := re^{i\theta}$.

To prove that his sequence converges compactly, Carathéodory used Montel's theorem (loc. cit., pp. 376–378). In 1914, in the Schwarz Festschrift, he explained his method in detail; he formulated the convergence theorem explicitly and proved it directly, without resorting to Montel ([Ca$_1$], pp. 280–284).

Koebe did not let Carathéodory's breakthrough rest in peace. He had dedicated his life to conformal maps and enriched the theory by a vast abundance of papers — from 1905 to 1909 alone he wrote more than 14 papers, some of them quite long — and could immediately (in 1912) let the approximating Riemann surfaces that had appeared in Carathéodory's work "arise automatically." Stimulated by "Mr. C. Carathéodory's interesting work," he takes up his "earlier thoughts again" and "reveals a new elementary method of mapping the most general [schlicht] simply connected region onto the surface of the unit disc," which has "ideal perfection in more than one respect"; cf. [Koe_3], pp. 844–845. Here Koebe sketches the square root algorithm for the first time. He presents his "osculation process" in detail in 1914 ([Koe_4], p. 182):

> Die [Konstruktion] der konformen Abbildung des gegebenen Bereichs auf das Innere des Einheitskreises werden wir durch unendlich viele Quadratwurzeloperationen bewirken, ... die wesentliche Eigenschaft der einzelnen dieser Operationen ... ist, eine Verstärkung der Anschmiegung der Begrenzungslinie des jeweilig abzubildenden Bereichs an die Peripherie des Einheitskreises, und zwar vom Innern her zu bewirken. (We will carry out the [construction] of the conformal map of the given region onto the interior of the unit disc by infinitely many square root operations; ... the crucial property of each of these operations ... is to improve the osculation of the boundary curve of the region to be mapped to the boundary of the unit disc and, more precisely, to do so by working outward from the interior.)

Koebe argues geometrically — as did Carathéodory before him; thus two-sheeted Riemann surfaces are used again. In 1916, at the 4th Scandindavian congress of mathematicians, Lindelöf gave a detailed lecture on Koebe's proof; cf. [Lin].

G. Pólya and G. Szegö, in 1925, split Koebe's proof into nine problems ([PS], Problems 88–96, Part IV, pp. 15–16); they prove the convergence theorem directly (without Montel) — as, of course, does Koebe. Their arguments are free of Riemann surfaces and simpler than in the pioneering work of Carathéodory and Koebe.

§3. The Koebe Families $\mathcal{K}_m$ and $\mathcal{K}_\infty$

Koebe noticed immediately (in 1912) that, in his osculation process, taking square roots "could readily be replaced by taking roots of higher orders or taking logarithms" ([Koe_2], p. 845). We want to construct corresponding Koebe families. In the construction, we use the fact that holomorphic functions that have no zeros in Q-domains always have holomorphic mth roots, $m \in \mathbb{N}$, and holomorphic logarithms. — In what follows, the (large) family

$$\mathcal{E} = \{f : \mathbb{E} \to \mathbb{E} \textit{ holomorphic}, \ f(0) = 0, \ f \notin \operatorname{Aut}\mathbb{E}, \ f(\mathbb{E}) \textit{ is not a Koebe domain}\}$$

plays an important role. G again always denotes a Koebe domain. We first generalize Lemma 8.2.3.

1. A lemma. *Let $\varphi \in \mathcal{E}$, and let $\kappa : G \to \mathbb{E}$ be a holomorphic map such that $\kappa(0) = 0$ and $\varphi \circ \kappa = \mathrm{id}$. Then κ is an admissible expansion of G.*

Proof. Since $\varphi \notin \mathrm{Aut}\,\mathbb{E}$, Schwarz implies that $|\varphi(w)| < |w|$, $w \in \mathbb{E}^\times$. Since $\varphi \circ \kappa = id$, κ is injective; moreover, $|z| = |\varphi(\kappa(z))| < |\kappa(z)|$ if $\kappa(z) \neq 0$, i.e. if $z \in G\backslash\{0\}$. Hence κ is an expansion of G. If $\kappa(G)$ were equal to $\mathbb{E}$, then $\varphi(\mathbb{E}) = G$ would be a Koebe domain, contradicting $\varphi \in \mathcal{E}$. Thus κ is admissible by 1.2(1). □

The trick is now to track down those functions $\varphi \in \mathcal{E}$ for which there exists a κ as in the lemma. We give two examples for which this is possible.

Example 1. Let $m \in \mathbb{N}$, $m \geq 2$, and let $c \in \mathbb{E}^\times$. We denote the map $\mathbb{E} \to \mathbb{E}$, $z \mapsto z^m$, by j_m and claim:

The map $\psi_{c,m} := g_{c^m} \circ j_m \circ g_c : \mathbb{E} \to \mathbb{E}$ is in $\mathcal{E}$ and

$$(1) \qquad \psi'_{c,m}(0) = mc^{m-1}\frac{|c|^2 - 1}{|c|^{2m} - 1} = m\left(\frac{c}{|c|}\right)^{m-1} \cdot \frac{|c| - |c|^{-1}}{|c|^m - |c|^{-m}}.$$

Proof. $\psi_{c,m} \in \mathcal{E}$ since $j_m \notin \mathrm{Aut}\,\mathbb{E}$ and $\psi_{c,m}(\mathbb{E}) = \mathbb{E}$. Equation (1) follows (by the chain rule) since $j'_m(c) = mc^{m-1}$ and $g'_a(0) = 1/g'_a(a) = |a|^2 - 1$, $a \in \mathbb{E}$. □

Example 2. Let $c \in \mathbb{E}^\times$. We choose $b \in \mathbb{H}$ such that $c = e^{ib}$. The map

$$q_b : \mathbb{H} \to \mathbb{E}, \quad z \mapsto \frac{z - b}{z + \bar{b}} \quad \text{with} \quad q_b^{-1} : \mathbb{E} \to \mathbb{H}, \quad z \mapsto \frac{b - \bar{b}z}{1 - z}$$

is biholomorphic (a generalized Cayley map). The function $\varepsilon(z) := e^{iz}$ maps $\mathbb{H}$ onto $\mathbb{E}\backslash\{0\}$. We claim:

The map $\chi_c := g_c \circ \varepsilon \circ q_b^{-1} : \mathbb{E} \to \mathbb{E}$ is in $\mathcal{E}$ and

$$(2) \qquad \chi'_c(0) = \frac{2c \log|c|}{|c|^2 - 1} = 2\frac{c}{|c|}\frac{\log|c|}{|c| - |c|^{-1}}.$$

Proof. Since $\varepsilon : \mathbb{H} \to \mathbb{E}\backslash\{0\}$ is not biholomorphic and $\chi_c(\mathbb{E}) = \mathbb{E}\backslash\{c\}$ is not a Q-domain, we have $\chi_c \in \mathcal{E}$. Equation (2) follows, since $\chi'_c(0) = g'_c(c)\varepsilon'(b)(q_b^{-1})'(0)$ and $g'_c(c) = 1/(|c|^2 - 1)$, $\varepsilon'(b) = ic$, $(q_b^{-1})'(0) = b - \bar{b} = -i\log|c|^2$. □

In the next subsection, for the functions $\psi_{m,c}$, $\chi_c \in \mathcal{E}$, we construct maps κ satisfying the hypotheses of the lemma.

2. The families $\mathcal{K}_m$ and $\mathcal{K}_\infty$. The functions

$$(1) \quad \tau_m(x) := \frac{1}{m}\frac{x - x^{-1}}{x^{1/m} - x^{-1/m}}, \ m = 2, 3, \ldots; \quad \tau_\infty(x) := \frac{x - x^{-1}}{2\log x}, \ x \in (0, 1),$$

are continuous and map $(0, 1)$ to $(1, \infty)$. (Prove this.) For every Koebe domain, we *construct* admissible expansions κ such that $|\kappa'(0)| = \tau_m(r(G))$, $2 \leq m \leq \infty$.

The mth root method, $m \geq 2$. *Let c be chosen in such a way that c^m is a boundary point of G with minimum distance to the origin. Let $v \in \mathcal{O}(G)$ be the mth root of $g_{c^m}|G$ with $v(0) = c$. Then $\kappa := g_c \circ v : G \to \mathbb{E}$ is an admissible* *expansion of G and*

$$(2) \qquad |\kappa'(0)| = \tau_m(r(G)).$$

Proof. $\kappa : G \to \mathbb{E}$ is well defined since $v(G) \subset \mathbb{E}$. We have

$$(*) \qquad \kappa(0) = g_c(c) = 0 \quad \text{and} \quad \psi_{c,m} \circ \kappa = id \quad (\text{because } j_m \circ v = g_{c^m}|G).$$

Example 1.1 shows that $\psi_{c,m} \in \mathcal{E}$; hence κ is an admissible expansion of G by Lemma 1. By $(*)$, $|\kappa'(0)| = 1/|\psi'_{c,m}(0)|$. Thus (2) follows from 1(1) since $|c^m| = r(G)$. □

The logarithm method. *Let c be a boundary point of G with minimum distance to the origin, and let* $i\nu \in \mathcal{O}(G)$ *be a logarithm of* $g_c|G$. *Then* $\kappa := q_b \circ \nu$, *with* $b := \nu(0)$, *is an admissible expansion of G and*

$$(3) \qquad |\kappa'(0)| = \tau_\infty(r(G)).$$

Proof. Since $e^{i\nu(z)} = g_c(z) \in \mathbb{E}$, $z \in G$, we have $\nu(G) \subset \mathbb{H}$; in particular, $b \in \mathbb{H}$. Hence $\kappa : G \to \mathbb{E}$ is well defined. It follows (with $\varepsilon \circ \nu = g_c$) that

$$(*) \qquad \kappa(0) = q_b(b) = 0 \quad \text{and} \quad \chi_c \circ \kappa = id.$$

Example 1.2 shows that $\chi_c \in \mathcal{E}$; hence κ is an admissible expansion of G by Lemma 1. By $(*)$, $|\kappa'(0)| = 1/|\chi'_c(0)|$. Thus (3) follows from 1(2) since $|c| = r(G)$. □

In addition, we "normalize" each expansion obtained above (by multiplying by an $a \in S^1$) so that $\kappa'(0) > 0$; this normalized expansion is again admissible. Now let $\mathcal{K}_m$ and $\mathcal{K}_\infty$ denote, respectively, the families of all normalized expansions constructed by means of the mth root method and the logarithm method. Then 2.3(a) is satisfied by $\tau := \tau_m$ and by $\tau := \tau_\infty$. The next assertion follows.

Proposition. $\mathcal{K}_m$ *is a Koebe family for* τ_m, $m = 2, 3, \ldots$; $\mathcal{K}_\infty$ *is a Koebe family for* τ_∞.

Remark. The family $\mathcal{K}_\infty$ is the "limit" of the families $\mathcal{K}_m$: For fixed G and $c \in \partial G$, each expansion $\kappa \in \mathcal{K}_\infty$ is the limit of a sequence $\kappa_m \in \mathcal{K}_m$. For the functions τ_m and τ_∞, it is immediate that $\lim \tau_m(x) = \tau_\infty(x)$ for all $x \in (0, 1]$.

For the proof of the Riemann mapping theorem, the square root method is generally used. Carathéodory applied the logarithm method in 1928, choosing $c = h \in (0, 1)$ and working with the function

$$\frac{h - \exp\left[(\log h) \cdot \frac{1+z}{1-z}\right]}{1 - h\exp\left[(\log h) \cdot \frac{1+z}{1-z}\right]} \qquad \text{(cf. [Ca}_1\text{], p. 304, formula (6.2)),}$$

which is precisely the function χ_h with $b = -i \log h$. (Prove this.) This says, by the way, that he could just as well have chosen the function

$$\frac{z(2\sqrt{h} - (1+h)z)}{(1+h) - 2\sqrt{h}z}, \qquad \text{thus} \quad \psi_{2,h} \quad \text{(cf. p. 305).}$$

H. Cartan ([Car$_2$], p. 191) also uses the logarithm method. He works with the admissible expansion $\widehat{\kappa} := q_b \circ \nu$, where $e^b = -c$ and $\nu \in \mathcal{O}(G)$ is a logarithm of $-g_c$; it still holds that $\widehat{\kappa}'(0) = \tau_\infty(r(G))$. The reader should carry out the computations and determine a $\widehat{\chi} \in \mathcal{E}$ with $\widehat{\chi} \circ \widehat{\kappa} = id$. □

The families $\mathcal{K}_m$, $\mathcal{K}_\infty$, and others are treated in detail by P. Henrici; see [He], pp. 328–345, where the rate of convergence of the corresponding Koebe algorithms is discussed.

The functions $\psi_{h,m}$, $m \geq 2$, $h \in (0,1)$, and χ_h are studied by Carathéodory in [Ca$_2$], pp. 30–31; it is also shown there that $\lim \psi_{h,m} = \chi_h$.

Bibliography for Chapter 8 and the Appendix

[B] BIEBERBACH, L.: Neuere Untersuchungen über Funktionen von komplexen Variablen, *Encyc. Math. Wiss.* II, 3.1, 379–532, Teubner, 1921.

[Ca$_1$] CARATHÉODORY, C.: *Ges. Math. Schriften* 3.

[Ca$_2$] CARATHÉODORY, C.: *Conformal Representation*, Cambridge University Press 1932, 2nd ed., 1952.

[Car$_1$] CARTAN, E.: Sur les domaines bornés homogènes de l'espace de n variables complexes, *Abh. Math. Sem. Hamburg* 11, 116–162 (1935); *Œuvres* I, 2, 1259–1306.

[Car$_2$] CARTAN, H.: *Elementare Theorie der analytischen Funktionen einer oder mehrerer komplexer Veränderlichen*, Hochschultaschenbücher BI, Mannheim, 1966.

[Cou] COURANT, R.: Über die Anwendung des Dirichletschen Prinzipes auf die Probleme der konformen Abbildung, *Math. Ann.* 71, 145–183 (1912).

[He] HENRICI, P.: *Applied and Computational Complex Analysis*, vol. 3, J. Wilcy and Sons, 1986.

[Hi] HILBERT, D.: *Ges. Math. Abh.* 3, Springer, 1970.

[Ka] KAUP, L. and B. KAUP: *Holomorphic Functions of Several Variables: An Introduction to the Fundamental Theory*, trans. M. BRIDGLAND, de Gruyter, 1983.

[Koe$_1$] KOEBE, P.: Ueber die Uniformisierung beliebiger analytischer Kurven, *Nachr. Königl. Ges. Wiss. Göttingen, Math.-phys. Kl.* 1907, erste Mitteilung 191–210; zweite Mitteilung 633–669.

[Koe$_2$] KOEBE, P.: Ueber die Uniformisierung beliebiger analytischer Kurven I, *Math. Ann.* 67, 145–224 (1909).

[Koe$_3$] KOEBE, P.: Ueber eine neue Methode der konformen Abbildung und Uniformisierung, *Nachr. Königl. Ges. Wiss. Göttingen, Math.-phys. Kl.* 1912, 844–848.

[Koe$_4$] KOEBE, P.: Abhandlungen zur Theorie der konformen Abbildung, I, Die Kreisabbildung des allgemeinsten einfach und zweifach zusammenhängenden schlichten Bereichs und die Ränderzuordnung bei konformer Abbildung, *Journ. reine angew. Math.* 145, 177–223 (1915).

[Lich] LICHTENSTEIN, L.: Neuere Entwicklungen der Potentialtheorie, Konforme Abbildung, *Encykl. Math. Wiss.* II 3.1, 177–377, Teubner, 1919.

[Lin] LINDELÖF, E.: Sur la représentation conforme d'une aire simplement connexe sur l'aire d'un cercle, *Compte rendu du quatrième congrès des mathématiciens scandinaves*, 59–90; ed. G. MITTAG-LEFFLER, 1920.

[N] NARASIMHAN, R.: *Complex Analysis in One Variable*, Birkhäuser, 1985.

[Osg] OSGOOD, W. F.: On the existence of the Green's function for the most general simply connected plane region, *Trans. Amer. Math. Soc.* 1, 310–314 (1900).

[Ost] OSTROWSKI, A.: Mathematische Miszellen XV. Zur konformen Abbildung einfach zusammenhängender Gebiete, *Jber. DMV* 38, 168–182 (1929); *Coll. Math. Papers* 6, 15–29.

[Po] POINCARÉ, H.: Sur les groupes des équations linéaires, *Acta Math.* 4, 201-311 (1884); *Œuvres* 2, 300–401.

[Pom] POMMERENKE, C.: *Boundary Behaviour of Conformal Maps*, Springer, 1991.

[PS] PÓLYA, G. and G. SZEGÖ: *Problems and Theorems in Analysis*, vol. 2, trans. C. E. BILLIGHEIMER, Springer, New York, 1976.

[Ra] RADÓ, T.: Über die Fundamentalabbildung schlichter Gebiete, *Acta Sci. Math. Szeged* 1, 240–251 (1922–1923); see also Fejér's *Ges. Arb.* 2, 841–842.

[Ran] RANGE, R. M.: *Holomorphic Functions and Integral Representations in Several Complex Variables*, Springer, 1986.

[Rie] RIEMANN, B.: Grundlagen für eine allgemeine Theorie der Functionen einer veränderlichen complexen Grösse, Inauguraldissertation Göttingen 1851, *Werke* 3–45.

[Schw] SCHWARZ, H. A.: *Ges. Math. Abhandlungen*, 2 vols., Julius Springer, 1890.

[Wei] WEIERSTRASS, K.: Über das sogenannte Dirichletsche Princip, *Math. Werke* 2, 49–54.

9
Automorphisms and Finite Inner Maps

The group Aut G and the semigroup Hol G, which were already studied in 8.4, are central to Sections 1 and 2. For *bounded* domains G, every sequence $f_n \in \operatorname{Hol} G$ has a convergent subsequence (Montel); this fact has surprising consequences. For example, in H. Cartan's theorem, one can read off from the convergence behavior of the *sequence of iterates* of a map $f : G \to G$ whether f is an automorphism of G. In 2.5, as an application of Cartan's theorem, we give a *homological characterization* of automorphisms.

An obvious generalization of the biholomorphic maps $G \to G'$ is the *finite* holomorphic maps $G \to G'$, for which *all* fibers are finite sets which always have the *same number* of points (branched coverings). Such maps are studied in Sections 3 and 4; we show, among other things, that every finite holomorphic map of a nondegenerate annulus onto itself is *linear* (fractional), hence biholomorphic.

§1. Inner Maps and Automorphisms

We first show that composing inner maps and taking inverses of automorphisms are compatible with compact convergence (Subsection 1). The proofs are easy, since the sequences $f_n \in \mathcal{O}(G)$ that converge *compactly* in G are precisely those that converge *continuously* in G to their limit f, and thus for which

$$\lim f_n(c_n) = f(c) \quad \text{if} \quad \lim c_n = c \in G$$

(see the footnote on p. 150 or I.3.1.5*).

We write $f = \lim f_n$ if the sequence $f_n \in \mathcal{O}(G)$ converges compactly (continuously) to $f \in \mathcal{O}(G)$. If G is *bounded* and $f_n \in \operatorname{Hol} G$, then, by Vitali, $f = \lim f_n$ if the sequence f_n just converges *pointwise* to f.

1. Convergent sequences in Hol G and Aut G. The next statement follows immediately from 7.5(1).

(1) *If* $f_n \in \operatorname{Hol} G$ *and* $\lim f_n = f \in \mathcal{O}(G)$ *is not constant, then* $f \in \operatorname{Hol} G$.

We also note:

(2) *If* $f_n \in \operatorname{Hol} G$, $g_n \in \mathcal{O}(G)$ *and* $\lim f_n = f \in \operatorname{Hol} G$, $\lim g_n = g \in \mathcal{O}(G)$, *then* $\lim(g_n \circ f_n) = g \circ f \in \mathcal{O}(G)$.

Proof. Let $c_n \in G$ be a sequence with limit $c \in G$. Continuous convergence implies that $\lim f_n(c_n) = f(c)$ and also that $\lim g_n(f_n(c_n)) = g(f(c))$. The sequence $g_n \circ f_n$ therefore converges continuously in G to $g \circ f$. □

For the inverse map, we show:

(3) *If* $f_n \in \operatorname{Aut} G$ *and* $\lim f_n = f \in \operatorname{Aut} G$, *then* $\lim f_n^{-1} = f^{-1} \in \operatorname{Aut} G$.

Proof. Let $c_n \in G$ be a sequence with limit $c \in G$. For $b_n := f_n^{-1}(c_n) \in G$, $b := f^{-1}(c) \in G$, we then have $\lim f_n(b_n) = f(b)$. It follows from 7.5(2′) that $\lim f_n^{-1}(c_n) = f^{-1}(c)$ and hence that the sequence f_n^{-1} converges continuously to f^{-1}.

Remark. Behind (2) and (3) lies a general theorem on topologizing transformation groups. If X is a *locally compact* space, then the set of *all homeomorphisms from X onto itself* is a *group* Top X. Consider the sets $\{f \in \operatorname{Top} X : f(K) \subset U\}$, where K is compact and U is open in X. The family of all finite intersections of such sets forms a base for a topology on Top X. In this so-called C-O topology (compact-open topology), the group operation $\operatorname{Top} X \times \operatorname{Top} X \to \operatorname{Top} X$, $(f, g) \mapsto f \circ g$, is *continuous.* If X is moreover *locally connected,* the inverse map $\operatorname{Top} X \to \operatorname{Top} X$, $f \mapsto f^{-1}$ is also continuous. Hence the following holds:

Theorem (Arens [A], 1946). *If X is locally compact and locally connected, then* Top X — *equipped with the C-O topology — is a topological group.*

In the case of domains G in $\mathbb{C}$, Aut G is a *closed* subgroup of Top G; the sequences that converge in the C-O topology are precisely those that converge compactly in G. If G is *bounded,* the topological group Aut G is *locally compact* and in fact a Lie group. These results were proved in 1932 and 1936 by H. Cartan for all bounded domains in $\mathbb{C}^n$, $1 \le n < \infty$ ([C], pp. 407–420 and 474–523).

2. Convergence theorem for sequences of automorphisms. *If G is bounded and $f_n \in \operatorname{Aut} G$ is a sequence with $f = \lim f_n \in \mathcal{O}(G)$, then two cases are possible.*

(1) *If f is not constant, then $f \in \operatorname{Aut} G$.*
(2) *If f is constant, then $f(G)$ is a boundary point of G.*

Proof. By Montel, the sequence of inverses $g_n := f_n^{-1} \in \operatorname{Aut} G$ contains a subsequence that converges compactly in G to some $g \in \mathcal{O}(G)$. We may assume that $g = \lim g_n$ (omit all troublesome g_n and f_n). We claim that

$$(*) \qquad g'(f(w)) \cdot f'(w) = 1 \quad \text{for all } w \in G \text{ with } f(w) \in G.$$

Since $g_n \circ f_n = id$ and therefore $g_n'(f_n(z)) \cdot f_n'(z) = 1$ for all n and all $z \in G$, we need only show that $\lim g_n'(f_n(w)) = g'(f(w))$ for all $w \in G \cap f^{-1}(G)$. But this holds because $\lim f_n(w) = f(w)$ and the sequence g_n' converges continuously to g'.

Now if f is not constant, then $f \in \operatorname{Hol} G$ by 1(1). By $(*)$, g is not constant; it follows from 1(1) that $g \in \operatorname{Hol} G$. Since $g_n \circ f_n = id = f_n \circ g_n$, 1(2) implies that $g \circ f = id = f \circ g$; hence $f \in \operatorname{Aut} G$.

But if f is constant, then $(*)$ implies that $c := f(G)$ cannot be a point in G. It follows that $c \in \partial G$. □

In the degenerate case (2), one can prove the sharper result that $f(G)$ is not an isolated boundary point of G. (Prove this.) All sequences

$$f_n : \mathbb{E} \to \mathbb{E}, \quad z \mapsto \frac{z - c_n}{\overline{c}_n z - 1}, \quad c_n \in \mathbb{E} \text{ with } \lim c_n =: c \in \partial\mathbb{E},$$

are examples of (2). They converge compactly in $\mathbb{E}$ to $f(z) \equiv c$, although $f_n \circ f_n = id$ for all n.

3. Bounded homogeneous domains. The disc $\mathbb{E}$ is *homogeneous*: the group $\operatorname{Aut} \mathbb{E}$ acts *transitively* on $\mathbb{E}$. The following converse holds:

Every homogeneous domain $G \not\approx \mathbb{C}, \mathbb{C}^\times$ can be mapped biholomorphically onto $\mathbb{E}$.

This theorem can be obtained most easily with the aid of the uniformization theorem. Here we prove a special case:

Let G be bounded and homogeneous, and suppose there exist a boundary point $p \in \partial G$ and a neighborhood U of p such that $U \cap G$ is simply connected. Then G can be mapped biholomorphically onto $\mathbb{E}$.

Proof. We fix $c \in G$ and choose a sequence $g_n \in \operatorname{Aut} G$ with $\lim g_n(c) = p$. We may assume that $g := \lim g_n \in \mathcal{O}(G)$ (Montel). Then $g(c) = p \in \partial G$; hence $g(z) \equiv p$ by Theorem 2. Thus for every closed path γ in G there exists an m such that the image path $\gamma_m := g_m \circ \gamma$ lies in $U \cap G$. Since γ_m is null homotopic in $U \cap G$ by hypothesis, $\gamma = g_m^{-1} \circ \gamma_m$ is null homotopic in G. Hence G is simply connected. The assertion follows from the Riemann mapping theorem. □

Boundary points p with the required property always exist when ∂G contains "smooth boundary pieces."

4*. Inner maps of $\mathbb{H}$ and homotheties. *Let $f : \mathbb{H} \to \mathbb{H}$ be holomorphic and suppose there exists a positive real number $\lambda \neq 1$ such that $f(\lambda i) = \lambda f(i)$. Then f is a homothety: $f(z) = \alpha z$ for all $z \in \mathbb{H}$, where $\alpha := |f(i)|$.*

Proof. Let $g := \alpha^{-1} f \in \operatorname{Hol}\mathbb{H}$; then $g(\lambda i) = \lambda g(i)$ and $|g(i)| = 1$. It must be shown that $g = id_{\mathbb{H}}$. By the Schwarz-Pick lemma (cf. I.9.2.5),

$$(*) \qquad \left|\frac{g(w) - g(z)}{g(w) - \overline{g(z)}}\right| \leq \left|\frac{w - z}{w - \overline{z}}\right| \quad \text{for all } w, z \in \mathbb{H}.$$

For $w := \lambda i$, $z := i$, it follows that

$$\left|\frac{\lambda g(i) - g(i)}{\lambda g(i) - \overline{g(i)}}\right| \leq \frac{|\lambda - 1|}{\lambda + 1}; \quad \text{hence} \quad |\lambda g(i) - \overline{g(i)}| \geq \lambda + 1.$$

Hence, for $\beta := \overline{g(i)}/g(i)$, we have $|\lambda - \beta| \geq \lambda + 1$ and $|\beta| = 1$. It follows that $\beta = -1$ and hence that $\overline{g(i)} = -g(i)$; that is, $g(i) \in \mathbb{R}i$. Since $g(i) \in \mathbb{H} \cap S^1$, it follows that $g(i) = i$; therefore $g(\lambda i) = \lambda i$. Thus equality holds in $(*)$ for $w := \lambda i$ and $z := i$. By the Schwarz-Pick lemma, g is an automorphism of $\mathbb{H}$. Since g has the two fixed points i and λi, it follows that $g = id_{\mathbb{H}}$. □

The proof given here is due to E. Mues and H. Köditz. Under the stronger hypothesis that $f(\lambda z) = \lambda f(z)$ for all $z \in \mathbb{H}$, the theorem follows trivially from the

Carathéodory-Julia-Landau-Valiron theorem. *For every function $f \in \operatorname{Hol}\mathbb{H}$ there exists a real constant $\alpha \geq 0$ such that, in every angular sector*

$$S_\varepsilon := \{re^{i\varphi} : r > 0 \text{ and } \varepsilon < \varphi < \pi - \varepsilon\}, \quad \text{where } \varepsilon \in (0, \tfrac{1}{2}\pi),$$

$f(z)/z$ converges uniformly to α as z tends to ∞.

The number α is called the *angular derivative* of f at ∞. Proofs of the theorem for $\mathbb{E}$ instead of $\mathbb{H}$ can be found in C. Carathéodory: *Theory of Functions II*, 28–32, trans. F. Steinhardt, Chelsea, 1954; A. Dinghas: *Vorlesungen über Funktionentheorie*, 236–237, Springer, 1961; and C. Pommerenke: *Univalent Functions*, 306–307, Vandenhoeck & Ruprecht, 1975.

If $f(\lambda z) = \lambda f(z)$ for all $z \in \mathbb{H}$ and some $\lambda > 1$, it follows from the theorem stated that

$$f(z)/z = f(\lambda^n z)/\lambda^n z, \ n \in \mathbb{N}; \quad \text{hence} \quad f(z)/z = \lim_{n\to\infty} f(\lambda^n z)/\lambda^n z = \alpha, \ z \in \mathbb{H}.$$

§2. Iteration of Inner Maps

For every inner map $f \in \operatorname{Hol} G$, the *iterates* $f^{[n]} \in \operatorname{Hol} G$ are defined recursively by

$$f^{[0]} := id_G, \quad f^{[n]} := f \circ f^{[n-1]}, \quad n = 1, 2, \ldots.$$

This sequence contains valuable information about f; for example, $f \in \operatorname{Aut} G$ if $f^{[m]} \in \operatorname{Aut} G$ for just one $m \geq 1$. This trivial observation is sharpened considerably in Subsection 1; as a corollary we obtain, in Subsection 2, a theorem of H. Cartan for bounded domains.

The equation

$$(*) \qquad (f^{[n]})'(a) = f'(a)^n, \quad f \in \operatorname{Hol}_a G, \quad a \in G, \quad n \geq 1,$$

which is clear at once by induction, has surprising consequences when combined with the theorems of Montel and Cartan; we give samples in Subsections 2, 3, and 5. — We often write f_n for $f^{[n]}$.

1. Elementary properties. *Suppose that a subsequence f_{n_k} of the sequence of iterates of $f \in \operatorname{Hol} G$ converges compactly in G to a function $g \in \mathcal{O}(G)$. Then the following statements hold.*

a) *If $g \in \operatorname{Aut} G$, then $f \in \operatorname{Aut} G$.*
b) *If g is not constant, then every convergent subsequence of the sequence $h_k := f_{n_{k+1}-n_k} \in \operatorname{Hol} G$ has the limit function id_G.*

Proof. ad a) *f is injective.* From $f(a) = f(b)$, a, $b \in G$, it follows that $f_n(a) = f_n(b)$ for all n; hence $g(a) = g(b)$, and therefore $a = b$ since $g \in \operatorname{Aut} G$.

f is surjective. We always have $f_{n_k}(G) \subset f(G)$. It follows from 7.5(1) that $g(G) \subset f(G) \subset G$. But $g(G) = G$ because $g \in \operatorname{Aut} G$; hence $f(G) = G$.

ad b) By 1.1(1), $g \in \operatorname{Hol} G$. Suppose h is the limit of a subsequence of the sequence h_k. Then, by 1.1(2), $f_{n_{k+1}} = h_k \circ f_{n_k}$ implies that $g = h \circ g$. Hence h is the identity on $g(G)$. Since $g(G)$ is open in G, it follows that $h = \mathrm{id}_G$. □

Statement b) gives, in particular (with $n_k := k$):

If the sequence $f^{[n]}$ converges compactly in G to a nonconstant function, then $f = \mathrm{id}_G$.

Thus sequences of iterates $f^{[n]}$, $f \neq \mathrm{id}_G$, if they converge at all, have constant limits. We consider an *example*. Let $0 < a < 1$. For $f := -g_a \in \operatorname{Aut} \mathbb{E}$, where $g_a(z) = (z-a)(az-1)$, we have $f^{[n]} = -g_{a_n}$ with $a_1 := a$, $a_n := (a+a_{n-1})/(1+aa_{n-1})$, $n \geq 2$. (Prove this.) Since $a_1 < a_2 < \cdots < 1$, it follows that $\lim a_n = 1$. Hence

$$\lim f^{[n]}(z) = \lim \frac{a_n - z}{a_n z - 1} = \frac{1-z}{z-1} = -1 \quad \text{for all } z \in \mathbb{E}.$$

The sequence $f^{[n]}$ thus converges compactly in $\mathbb{E}$ to the fixed point -1 of f.

The next theorem follows easily from a) and b).

2. H. Cartan's theorem. *Let G be bounded and let $f \in \operatorname{Hol} G$. Suppose there exists a subsequence f_{n_k} of the sequence of iterates of f that converges compactly in G to a nonconstant function. Then $f \in \operatorname{Aut} G$.*

Proof. By Montel, the sequence $h_k := f_{n_{k+1}-n_k}$ has a convergent subsequence. By 1,b) its limit is id_G; by 1,a), $f \in \operatorname{Aut} G$. □

Corollary 1. *If G is bounded and $f \in \operatorname{Hol} G$ has two distinct fixed points $a, b \in G$, then f is an automorphism of G.*

Proof. By Montel, a subsequence f_{n_k} converges in G to a function g. Since $f_n(a) = a$ and $f_n(b) = b$ for all n, it follows that $g(a) = a \neq b = g(b)$. Hence g is not constant; by Cartan, $f \in \operatorname{Aut} G$. □

Corollary 2. (Cf. 8.4.2(1).) *If G is bounded and $a \in G$, then $|f'(a)| \leq 1$ for all $f \in \operatorname{Hol}_a G$. Moreover, $\operatorname{Aut}_a G = \{f \in \operatorname{Hol}_a G : |f'(a)| = 1\}$.*

Proof. Again let $g = \lim f_{n_k} \in \mathcal{O}(G)$ (Montel). By $(*)$ of the introduction, $\lim f'(a)^{n_k} = g'(a)$. This is possible only if $|f'(a)| \leq 1$. In the case $|f'(a)| = 1$, we have $|g'(a)| = 1$. Then g is not constant, and Cartan implies that $f \in \operatorname{Aut} G$.

Conversely, if $f \in \operatorname{Aut}_a G$, then $f^{-1} \in \operatorname{Aut}_a G$. Thus, since $|f'(a)| \leq 1$, we also have $|1/f'(a)| = |(f^{-1})'(a)| \leq 1$, i.e. $|f'(a)| = 1$. □

Historical note. H. Cartan published his theorem in 1932; he considers arbitrary bounded domains in $\mathbb{C}^n$, $1 \leq n < \infty$ (cf. [C], pp. 417–418).

Exercises. 1) Let G be bounded, $a \in G$, and $f \in \operatorname{Hol}_a G$ but $f \notin \operatorname{Aut} G$. Prove that the sequence $f^{[n]}$ converges in G to $g(z) \equiv a$.

2) Prove that the sequence $f_{n_{k+1}-n_k}$ in Cartan's theorem converges.

Remark. The proofs of Cartan's theorem and of both corollaries work because the sequences h_k and $f^{[n]}$ have convergent subsequences. It can be shown by means of uniformization theory that this occurs for all domains $\not\approx \mathbb{C}, \mathbb{C}^\times$.

3. The group $\operatorname{Aut}_a G$ for bounded domains. *If G is bounded and $a \in G$, then $\sigma : \operatorname{Aut}_a G \to \mathbb{C}^\times$, $f \mapsto f'(a)$, maps the group $\operatorname{Aut}_a G$ isomorphically onto either the circle group S^1 or a finite cyclic subgroup of S^1* (cf. Proposition 8.4.2).

Proof. a) By Corollary 2.2, $\operatorname{Image} \sigma \subset S^1$. If we show that $\operatorname{Image} \sigma$ is closed in S^1, then $\operatorname{Image} \sigma$ is either S^1 or finite cyclic by Theorem 4 (next subsection). Thus let $c \in S^1$ be the limit of a sequence $c_n \in \operatorname{Image} \sigma$. Choose $h_n \in \operatorname{Aut}_a G$ with $\sigma(h_n) = c_n$. By Montel, a subsequence of h_n converges in G to an $h \in \mathcal{O}(G)$. Since $h(a) = a$, it follows from Theorem 1.2 that $h \in \operatorname{Aut}_a G$. Hence $\sigma(h) = h'(a) = c$.

b) It remains to show that σ is injective, in other words that if $f \in \operatorname{Aut}_a G$ and $f'(a) = 1$, then $f = id_G$. We may assume that $a = 0$. The Taylor series of f about 0 has the form $z + a_m z^m +$ *higher-order terms, where* $m \geq 2$. Then $z + n a_m z^m + \cdots$ is the Taylor series of $f^{[n]}$ about 0 (see $(*)$ below).

Since a subsequence of the sequence $f^{[n]}$ converges, so does a subsequence of $na_m = (f^{[n]})^{(m)}(0)/m!$, $n = 1, 2, \ldots$. This can happen only if $a_m = 0$. Therefore $f(z) \equiv z$. □

The following was used in the proof of (b):

(∗) *Let G be a domain with $0 \in G$, and let $z + a_m z^m + \sum_{\nu>m} a_\nu z^\nu$, with $m \geq 2$, be the Taylor series for $f \in \mathrm{Hol}_0 G$ about 0. Then*

$$z + na_m z^m + \text{terms in } z^\nu \text{ with } \nu > m$$

is the Taylor series for $f^{[n]}$ about 0, $n = 1, 2, \ldots$.

Proof (by induction). We set $f_n = f^{[n]}$. The case $n = 1$ is clear. Suppose the assertion has already been verified for $n = k \geq 1$. Since $f_{k+1} = f \circ f_k$, we have

$$f_{k+1}(z) = f_k(z) + a_m f_k(z)^m + g(z), \quad \text{where} \quad g(z) := \sum_{\nu>m} a_\nu f_k(z)^\nu.$$

Since $f_k(0) = 0$ implies that $o_0(g) > m$, by taking the induction hypothesis into account we see that the Taylor series of f_{k+1} looks like

$$z + ka_m z^m + a_m(z + ka_m z^m + \cdots)^m + \cdots = z + (k+1)a_m z^m + \cdots.$$

□

The theorem contains the following (proof as in 8.4.2):

Uniqueness Theorem. *Let G be bounded, $a \in G$, $f \in \mathrm{Aut}_a G$, $f'(a) > 0$. Then $f = id_G$.*

L. Bieberbach discovered this theorem in 1913 and proved it by iteration, as above ([B], pp. 556–557).

4. The closed subgroups of the circle group. *Every closed subgroup $H \neq S^1$ of S^1 is finite and cyclic.*

We first prove a lemma.

Lemma. *Let L be a closed subgroup of the additive group $\mathbb{R}$ such that $L \neq \{0\}$ and $L \neq \mathbb{R}$. Then $L = r\mathbb{Z}$, where $r := \inf\{x \in L : x > 0\} \in \mathbb{R}$.*

Proof. 1) r is well defined since $L \neq \{0\}$, and $r \geq 0$. If r were equal to 0, then for every $\varepsilon > 0$ there would exist an $s \in L$ with $0 < s < \varepsilon$. In every interval in $\mathbb{R}$ of length 2ε, there would now be an integer multiple of s. Hence, for every $t \in \mathbb{R}$, there would exist an $x \in L$ with $|t - x| < \varepsilon$. Thus for $\varepsilon := 1/n$, $n \geq 1$, there would be an $x_n \in L$ with $|t - x_n| < 1/n$. Since L is closed in $\mathbb{R}$, it follows that $t = \lim x_n \in L$, giving the contradiction $L = \mathbb{R}$.

2) We show that $L = r\mathbb{Z}$. Since L is closed, $r \in L$. The inclusion $r\mathbb{Z} \subset L$ is clear. Let $x \in L$ be arbitrary. Since $r > 0$ there exists an $n \in \mathbb{Z}$ such that

$r(n-1) < x \le rn$. This means that $0 \le rn - x < r$. Since $rn - x \in L$, the minimality of r implies that $x = rn$. □

The theorem on subgroups of S^1 now follows immediately. The "polar coordinate epimorphism" $p : \mathbb{R} \to S^1$, $\varphi \mapsto e^{i\varphi}$, is continuous; hence $L := p^{-1}(H)$ is a *closed* subgroup of the additive group $\mathbb{R}$. Now $L \ne \mathbb{R}$ since $H \ne S^1$; hence $L = r\mathbb{Z}$ with $r \ge 0$ by the lemma. With $\eta := e^{ir}$, we then have $H = p(L) = \{\eta^n : n \in \mathbb{Z}\}$. Since $2\pi \in L$ because $p(2\pi) = 1$, there exists an $m \in \mathbb{N}\backslash\{0\}$ with $rm = 2\pi$. This means that $\eta^m = 1$. Thus $H = \{1, \eta, \eta^2, \ldots, \eta^{m-1}\}$.

5*. Automorphisms of domains with holes. Annulus theorem. We first note a sufficient criterion for a sequence $g_n \in \mathcal{O}(G)$ that converges compactly in a domain G to have a *nonconstant* limit function g.

(1) *Suppose there exists a closed path γ in G such that the intersection of all the sets* $\operatorname{Int}(g_n \circ \gamma)$ *contains at least two points. Then g is not constant.*

Proof. If g were constant, say $g(z) \equiv a$, there would exist $b \in \operatorname{Int}(g_n \circ \gamma)$, $b \ne a$, such that

$$\int_\gamma \frac{g_n'}{g_n - b} d\zeta = \int_{g_n \circ \gamma} \frac{d\eta}{\eta - b} \in 2\pi i\mathbb{Z}\backslash\{0\} \quad \text{for almost all } n.$$

Since the sequence $g_n'/(g_n - b)$ converges compactly to 0 on γ, this gives a contradiction. □

In what follows, we use terminology and results from the theory of domains with holes (see Chapters 13 and 14). We consider *bounded domains G that have no isolated boundary points and have at least one but not infinitely many holes* (and thus are m-connected, $2 \le m < \infty$). For such domains, the following theorem holds.

Theorem ([C], pp. 448–449). *Let $f \in \operatorname{Hol} G$ be such that every closed path in G that is not null homologous in G has an image under f that is not null homologous in G. Then $f \in \operatorname{Aut} G$.*

Proof. Let $g \in \mathcal{O}(G)$ be the limit of a subsequence g_n of the iterates of f. Since G has holes, there is a closed path γ in G that is not null homologous in G (cf. 13.2.4). Hence, since $f^{[n]} \circ \gamma = f \circ (f^{[n-1]} \circ \gamma)$, no path $g_n \circ \gamma$ is null homologous in G. Thus for every n there exists a hole L_n of G such that $L_n \subset \operatorname{Int}(g_n \circ \gamma)$; cf. 13.1.4. Since G has only finitely many holes, we may assume (by passing to a subsequence and renumbering the holes) that $L_1 \subset \operatorname{Int}(g_n \circ \gamma)$ for every n. Since L_1 has at least two points, it follows from (1) that g is not constant. Theorem 2 now implies that $f \in \operatorname{Aut} G$. □

The hypothesis on f means that f induces a *monomorphism* $\widetilde{f} : H(G) \to H(G)$ of the homology group of G; cf. 14.1.2. The next theorem now follows immediately from elementary homology theory.

Annulus theorem. *If* $A := \{z \in \mathbb{C} : r < |z| < s\}$, $0 < r < s < \infty$, *is a (nondegenerate) annulus and* $f \in \operatorname{Hol} A$ *is such that* f *maps at least one closed path that is not null homologous in* A *to another such path, then* $f \in \operatorname{Aut} A$*; hence* $f(z) = \eta z$ *or* $f(z) = \eta r s z^{-1}$, $\eta \in S^1$.

Proof. If Γ denotes a circle about 0 in A, then (cf. 14.2.1)

$$H(A) = \mathbb{Z}\overline{\Gamma} \simeq \mathbb{Z} \quad \text{and} \quad \widetilde{f}(H(A)) \neq 0.$$

Hence $\widetilde{f} : \mathbb{Z} \to \mathbb{Z}$ is injective, and it follows from the theorem that $f \in \operatorname{Aut} A$. By Theorem 3.4, p. 215, f has the asserted form. □

Historical note. The annulus theorem was proved in 1950, without the use of Cartan's theorem, by H. Huber ([Hu], p. 163). There are direct proofs; see for example E. Reich: Elementary proof of a theorem on conformal rigidity, *Proc. Amer. Math. Soc.* 17, 644–645 (1966).

Glimpse. In general, domains with several holes have *finite* automorphism groups. M. H. Heins, in 1946, proved the following for domains G with *exactly* n holes, $2 \leq n < \infty$ [Hei₂]:

The group $\operatorname{Aut} G_n$ *is isomorphic to a subgroup of the group of all linear fractional transformations. The best possible upper bound* $N(n)$ *for the number of elements of* $\operatorname{Aut} G_n$ *is:*

$$N(n) := 2n \quad \text{if} \quad n \neq 4,\ 6,\ 8,\ 12,\ 20;$$
$$N(4) := 12; \quad N(6) := N(8) := 24; \quad N(12) := N(20) := 60.$$

The numbers $2n$, 12, 24, and 60 are the orders of the dihedral, tetrahedral, octahedral, and icosahedral groups, respectively. — (Bounded) domains with infinitely many holes can have infinite groups; for example, $\operatorname{Aut}(\mathbb{C}\backslash\mathbb{Z}) \simeq \operatorname{Aut}(\mathbb{H}\backslash\{i+\mathbb{Z}\}) = \{z \mapsto z+n : n \in \mathbb{Z}\}$.

§3. Finite Holomorphic Maps

A sequence $z_n \in G$ is called a *boundary sequence* in G if it has no accumulation point in G. A holomorphic map $f : G \to G'$ is called *finite* if the following is true.

If z_n *is a boundary sequence in* G, *then* $f(z_n)$ *is a boundary sequence in* G'.

Biholomorphic maps are finite. A map $f : \mathbb{C} \to \mathbb{C}$ is finite if and only if f is a nonconstant polynomial. (Prove this.)

In Subsection 2 we give all the finite holomorphic maps $\mathbb{E} \to \mathbb{E}$. In the remaining subsections we study finite holomorphic maps between annuli. Our tools are the maximum and minimum principles.

1. Three general properties. Finite holomorphic maps $f : G \to G'$ have the following properties:

(1) *Every f-fiber $f^{-1}(w)$, $w \in G'$, is finite.*
(2) *Every compact set L in G' has a compact preimage under f.*
(3) *f is surjective: $f(G) = G'$.*

Proof. ad (1). First, f is not constant. Hence every f-fiber is *locally finite* in G. If there were an infinite fiber F, then there would be a boundary sequence $z_n \in F$ in G. But then the constant image sequence $f(z_n)$ would not be a boundary sequence in G'. Contradiction.

ad (2). We show that every sequence $z_n \in K := f^{-1}(L)$ has an accumulation point in K. Since $f(z_n) \in L$ is not a boundary sequence in G', z_n is not a boundary sequence in G. Hence it has an accumulation point $\widehat{z} \in G$. Since K is closed, it follows that $\widehat{z} \in K$.

ad (3). If $f(G)$ were not equal to G', then $f(G)$ would have a boundary point $p \in G'$. Choose a sequence $z_n \in G$ with $\lim f(z_n) = p$. Then z_n is not a boundary sequence in G, and therefore has an accumulation point $\widehat{z} \in G$. Hence $p = f(\widehat{z}) \in f(G)$. Contradiction. □

Other general statements about finite holomorphic maps can be found in §4.

2. Finite inner maps of $\mathbb{E}$. Our first assertion is clear.

An inner map $f : \mathbb{E} \to \mathbb{E}$ is finite if and only if

$$(1) \qquad \lim_{|z| \to 1} |f(z)| = 1 \quad \textit{(boundary rule)}.$$

The next lemma follows immediately from the maximum and minimum principles.

Lemma. *Let G be bounded, and let g be a unit in $\mathcal{O}(G)$. Suppose that $\lim_{z \to \partial G} |g(z)| = 1$. Then g is constant.*

This tool easily gives our next result.

Theorem. *The following statements about a function $f \in \mathcal{O}(\mathbb{E})$ are equivalent:*

i) *f is a finite map $\mathbb{E} \to \mathbb{E}$.*
ii) *There exist finitely many points $c_1, \dots, c_d \in \mathbb{E}$, $d \geq 1$, and an $\eta \in S^1$ such that*

$$f(z) = \eta \prod_1^d \frac{z - c_\nu}{\bar{c}_\nu z - 1} \quad \textit{(finite Blaschke product).}$$

Proof. i) $\Rightarrow$ ii). The set $f^{-1}(0) \subset \mathbb{E}$ is finite and nonempty by 1(1) and (3). Let $c_1, \ldots, c_d \in \mathbb{E}$ be the zeros of f in $\mathbb{E}$, where c_ν occurs according to the order of f at c_ν. Set $f_\nu := (z - c_\nu)/(\overline{c}_\nu z - 1) \in \mathcal{O}(\mathbb{E})$; then $g := f/(f_1 f_2 \cdots f_d)$ is a unit in $\mathbb{E}$. Since $\lim_{|z|\to 1} |f_\nu(z)| = 1$, $1 \le \nu \le d$, and $\lim_{|z|\to 1} |f(z)| = 1$ by (1), it follows that $\lim_{|z|\to 1} |g(z)| = 1$. The lemma now implies that $g(z) = \eta \in S^1$, i.e. $f = \eta f_1 f_2 \cdots f_d$.

ii) $\Rightarrow$ i). $f(\mathbb{E}) \subset \mathbb{E}$ because $|(z - c_\nu)/(\overline{c}_\nu z - 1)| < 1$ for all $z \in \mathbb{E}$. Since $\lim_{|z|\to 1} |f(z)| = 1$, f is finite. □

Corollary. *Let q be a polynomial. Suppose there exists an $R \in (0, \infty)$ such that the region $\{z \in \mathbb{C} : |q(z)| < R\}$ has a disc $B_r(c)$, $0 < r < \infty$, as a connected component. Then $q(z) = a(z - c)^d$, where $d \ge 1$ and $|a| = R/r^d$.*

Proof. The induced map $B_r(c) \xrightarrow{q} B_R(0)$ is finite (!). The polynomial $p(z) := q(rz + c)/R$ induces a finite map $\mathbb{E} \to \mathbb{E}$. By the theorem, it follows that $p(z) = \eta z^d$ with $\eta \in S^1$, $d \ge 1$. This is the assertion. □

The theorem shows that $\mathbb{E}$ admits *many finite inner maps* that are not automorphisms. The simplest such maps with $f(0) = 0$ are given by $f(z) = z(z - b)/(\overline{b} z - 1)$, $b \in \mathbb{E}$. The derivative f' vanishes at the unique point c of $\mathbb{E}$ that satisfies $b = 2c/(1 + |c|^2)$! These maps were denoted by ψ_c in 8.2.3. — In general, bounded domains have *no* finite inner maps other than automorphisms; the standard examples are annuli; cf. Theorem 4.

Historical note. P. Fatou (French mathematician, 1878–1929) initiated the theory of finite holomorphic maps in 1919. Using the Schwarz reflection principle and nontrivial theorems on the boundary behavior of bounded holomorphic functions in $\mathbb{E}$, he showed that finite inner maps of $\mathbb{E}$ are given by rational functions ([F$_1$], pp. 209–212). He observed in 1923 ([F$_2$], p. 192) that these functions are finite Blaschke products. Meanwhile Radó, in 1922, had already introduced the concept of finite holomorphic maps; the elegant proof above is his.

With the aid of uniformization theory, one can prove:

If $f : \mathbb{E} \to G$ is holomorphic and finite, then G can be mapped biholomorphically onto $\mathbb{E}$ (where G can be any Riemann surface).

3. Boundary lemma for annuli. $A := A(r, s)$ and $A' := A(r', s')$ always denote annuli in $\mathbb{C}$ with center 0, *inner* radii $r, r' \ge 0$, and *outer* radii s, $s' \le \infty$. In order to study finite maps $A \to A'$ we need a purely topological lemma, which generalizes the boundary rule 2(1) and, intuitively, says that boundary components of A are mapped to boundary components of A', where the inner and outer boundaries may be interchanged.

Boundary lemma. *If $f : A \to A'$ is holomorphic and finite, then either*

$$\lim_{|z|\to r} |f(z)| = r' \quad \textit{and} \quad \lim_{|z|\to s} |f(z)| = s'$$

or

$$\lim_{|z|\to r} |f(z)| = s' \quad \text{and} \quad \lim_{|z|\to s} |f(z)| = r'.$$

Proof. Let $t \in (r', s')$ be fixed, and let $S := \{z \in \mathbb{C} : |z| = t\} \subset A'$. Since f is finite, $f^{-1}(S)$ is compact in A by 1(2) and hence has a positive distance from ∂A. Thus there exist numbers ρ, σ with $r < \rho < \sigma < s$ such that, for the annuli $C := A(r, \rho)$ and $D := A(\sigma, s)$ (as in Figure 9.1),

$$C \cap f^{-1}(S) = \emptyset = D \cap f^{-1}(S).$$

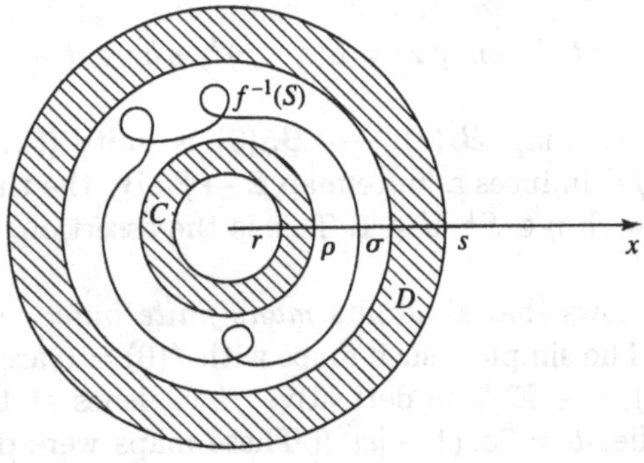

FIGURE 9.1.

This means that

$$f(C) \subset A' \backslash S \quad \text{and} \quad f(D) \subset A' \backslash S.$$

Since f is continuous and C and D are connected, $f(C)$ and $f(D)$ are also connected; it follows that

1) $$f(C) \subset A(r', t) \quad \textit{or} \quad f(C) \subset A(t, s')$$

and

2) $$f(D) \subset A(r', t) \quad \textit{or} \quad f(D) \subset A(t, s').$$

For every sequence z_n in A with $\lim |z_n| = r$, we have $z_n \in C$ for almost all n; hence 1) implies, for *all* such sequences:

Either $r' < |f(z_n)| < t$ for almost all n or $t < |f(z_n)| < s'$ for almost all n. Since $t \in (r', s')$ is arbitrary, we see:

Either $\lim_{|z|\to r} |f(z)| = r'$ *or* $\lim_{|z|\to r} |f(z)| = s'$.

Similarly, applying 2) shows:

Either $\lim_{|z|\to s} |f(z)| = r'$ *or* $\lim_{|z|\to s} |f(z)| = s'$.

The lemma will thus be proved if we also show that $\lim_{|z|\to r}|f(z)|$ cannot equal $\lim_{|z|\to s}|f(z)|$. Since $f(A) = A'$ by 1(3), there always exist sequences z_n and w_n in A with $\lim |f(z_n)| = r'$ and $\lim |f(w_n)| = s'$, where, by the finiteness of f, one of the sequences $|z_n|$ and $|w_n|$ converges to r and the other to s. □

Remark. In the proof of the lemma, all that is used is that f is continuous and surjective and has the boundary sequence property.

4. Finite inner maps of annuli. Every *rotation* $z \mapsto \eta z$, $\eta \in S^1$, is an automorphism of the annulus $A = A(r,s)$. If A is *nondegenerate*, i.e. if $0 < r < s < \infty$, then all the *combined rotations and reflections* $z \mapsto \eta r z^{-1}$ are automorphisms of A; they interchange the boundary components. It now follows quickly from Lemma 3 that in the nondegenerate case these are all the finite inner maps.

Theorem. *If A is nondegenerate, then every finite holomorphic map $f: A \to A$ is an automorphism and*

$$f(z) = \eta z \quad \textit{or} \quad f(z) = \eta r s z^{-1},\ \eta \in S^1.$$

Proof. We define a function $g \in \mathcal{O}(A)$ by

$$g(z) := \begin{cases} \dfrac{f(z)}{z} & \text{if } \lim\limits_{|z|\to r} |f(z)| = r, \\[2ex] \dfrac{zf(z)}{rs} & \text{if } \lim\limits_{|z|\to r} |f(z)| = s. \end{cases}$$

This definition makes sense because $rs \neq 0$ and because of Lemma 3 (with $A' = A$). The function g is *nonvanishing* in A. By Lemma 3,

$$\lim_{|z|\to r} |g(z)| = 1 = \lim_{|z|\to s} |g(z)|.$$

It follows from Lemma 2 that $g(z) = \eta \in S^1$. □

Remark. If A is *degenerate*, there exist finite inner maps of A that are not biholomorphic: if $r = 0$ and $s < \infty$, every map

$A(0,s) \to A(0,s)$, $z \mapsto az^d$, where $d \in \mathbb{N}\backslash\{0\}$ and $a \in \mathbb{C}^\times$ with $|a| = s^{1-d}$,

is finite. The reader should show that these are all the finite inner maps of $A(0,s)$.

Exercise. Determine all the finite inner maps of $A(r,\infty)$, $0 < r < \infty$, and of $A(0,\infty) = \mathbb{C}^\times$.

The theorem can be generalized considerably, as Radó showed as early as 1922 [R]:

Let $G \subset \mathbb{C}$ be a domain with exactly n holes, $1 \le n < \infty$; suppose that no hole consists of a single point. Then every finite holomorphic map $G \to G$ is biholomorphic.

An elegant proof was given in 1941 by M. H. Heins [Hei₁]; the statement is false for $n = \infty$.

5. Determination of all the finite maps between annuli. In order to generalize Theorem 4, we prove a lemma.

Lemma. *Let $f \in \mathcal{O}(A)$. If $|f|$ is constant on every circle about 0 in A, then $f(z) = az^m$ with $a \in \mathbb{C}$, $m \in \mathbb{Z}$.*

Proof. We may assume f is nonvanishing. Then $f(z) = e^{g(z)}z^m$, with $g \in \mathcal{O}(A)$ and $m \in \mathbb{Z}$ (lemma on units; cf. Exercise I.12.1.4 as well as Theorem 14.2.4). By the hypothesis on $|f|$, $\operatorname{Re} g$ is constant on all circles about 0 in A. Then, for every $\eta \in S^1$, $\exp[g(z) - g(\eta z)]$ has absolute value 1 everywhere in A. The function is therefore constant; it follows that $g(z) - g(\eta z) = i\delta$, $\delta \in \mathbb{R}$, for all $z \in A$, and hence (by induction) $g(z) - g(\eta^n z) = i\delta n$, $n \ge 1$. Since g is bounded on circles about 0, it follows that $\delta = 0$; thus $g(z) = g(\eta z)$ for all $\eta \in S^1$. Hence g is constant by the identity theorem. □

For every nondegenerate annulus $A = A(r,s)$, the ratio of the radii $\mu(A) := s/r > 1$ is called the *modulus* of A.

Theorem. *The following statements about nondegenerate annuli A, A' are equivalent:*

i) *There exists a finite holomorphic map $f : A \to A'$.*
ii) *There exists a natural number $d \ge 1$ such that $\mu(A') = \mu(A)^d$.*

If ii) *is satisfied, then all finite holomorphic maps $f : A \to A'$ are given by the functions*

$$f(z) = \eta r'(z/r)^d \quad \text{and} \quad f(z) = \eta s'(r/z)^d, \quad \eta \in S^1.$$

Proof. i) ⇒ ii). By Lemma 3 (the boundary lemma) and Lemma 2, all functions $f(z)/f(e^{i\alpha}z)$, $\alpha \in \mathbb{R}$, are constant on A with absolute value 1. In particular, $|f(z)|$ is constant on every circle about 0 in A. Thus, by the lemma, $f(z) = az^m$ with $m \in \mathbb{Z}\backslash\{0\}$, $a \in \mathbb{C}^\times$. Since $f(A) = A'$ by 1(3), it follows that $\mu(A') = \mu(A)^d$ with $d := |m|$; moreover, it is clear that f has the given form.

ii) ⇒ i). The given functions induce finite maps $A \to A'$. □

Corollary. *Two nondegenerate annuli A and A' are biholomorphically equivalent if and only if they have the same modulus. In this case, every fi-*

nite holomorphic map $A \to A'$ *is biholomorphic and of the form* $z \mapsto \eta r' z/r$ *or* $z \mapsto \eta s' r/z$, $\eta \in S^1$.

If one continues to assume that the outer radii s, s' are finite but allows the inner radii to be equal to 0, matters are different. The reader should prove:

If $r = 0$ *and* $r' > 0$, *or if* $r > 0$ *and* $r' = 0$, *then there exists no finite holomorphic map from* A *to* A'. *If* $r = r' = 0$, *then the functions* $f(z) = \eta s' s^{-d} z^d$, $\eta \in S^1$, $d \in \mathbb{N}\backslash\{0\}$, *are precisely the finite maps* $A \to A'$.

Exercise. Discuss the remaining cases of finite maps between degenerate annuli.

The corollary appears in Koebe's 1914 paper [K] (especially pp. 195–200); all the degenerate cases are also treated there.

§4*. Radó's Theorem. Mapping Degree

For all the finite holomorphic maps $\mathbb{E} \to \mathbb{E}$, $A \to A'$ that have been discussed, every point in the image has the same number of points in its preimage (if these are counted according to their multiplicity). This is no coincidence: we will see in Subsection 3 that for *every* finite holomorphic map, all the fibers have the same number of points; this number is called the *mapping degree* In Subsection 1, finite maps are characterized without using boundary sequences. In Subsection 2 we consider *winding maps.* These are the simplest finite maps. Locally, all nonconstant holomorphic maps are winding maps; they are the building blocks of finite maps along each fiber (Proposition 2). — G, G' always denote domains in $\mathbb{C}$.

1. Closed maps. Equivalence theorem. A map $f : X \to Y$ between topological (metric) spaces is called *closed* if every closed set in X has a closed image in Y. For such maps, we have the following result:

(1) *For every open neighborhood* U *in* X *of a fiber* $f^{-1}(y)$, $y \in Y$, *there exists an open neighborhood* V *of* y *in* Y *such that* $f^{-1}(V) \subset U$.

Proof. Since $X\backslash U$ is closed in X, $f(X\backslash U)$ is closed in Y. The set $V := Y\backslash f(X\backslash U)$ is a neighborhood with the desired property. □

(1) will be crucial in the proof of Radó's theorem in Subsection 3. We now prove an

Equivalence theorem. *The following statements about a holomorphic map* $f : G \to G'$ *are equivalent:*

i) f *maps boundary sequences in* G *to boundary sequences in* G' *(finiteness).*
ii) *Every compact set in* G' *has a compact preimage under* f.
iii) f *is nonconstant and closed.*

Proof. i) $\Rightarrow$ ii). This is statement 3.1(2).

ii) $\Rightarrow$ iii). $f(G)$ is certainly not a single point, for then $G = f^{-1}(f(G))$ would be compact. Suppose that A is closed in G. Let $p \in G'$ be the limit of a sequence $f(z_n)$, $z_n \in A$; we show that $p \in f(A)$. Since $L := \{p, f(z_0), f(z_1), \dots\} \subset G'$ is compact, $f^{-1}(L) \subset G$ is also compact. Hence the sequence $z_n \in A \cap f^{-1}(L)$ has a subsequence with a limit $\widehat{z} \in A$. It follows that $p = f(\widehat{z}) \in f(A)$. Thus f is closed.

iii) $\Rightarrow$ i). If there were a boundary sequence $z_n \in G$ whose image $f(z_n)$ had a limit $p \in G'$, then — since $f^{-1}(p)$ is locally finite in G — there would exist a sequence $\widehat{z}_n \in G \backslash f^{-1}(p)$ such that

$$|\widehat{z}_n - z_n| < 1/n \quad \text{and} \quad |f(\widehat{z}_n) - f(z_n)| < 1/n.$$

Then $\widehat{z}_n$ would again be a boundary sequence in G. Since $p = \lim f(\widehat{z}_n)$ and $f(\widehat{z}_n) \neq p$ for all n, the set $\{f(\widehat{z}_n)\}$ would not be closed in G'. But it is the image under f of the set $\{\widehat{z}_n\}$, which is closed in G. Contradiction. □

2. Winding maps. A nonconstant holomorphic map $f : U \to V$ is called a *winding map about* $c \in U$ if

a) *V is a disc about $f(c)$ and there exist a biholomorphic map $u : U \xrightarrow{\sim} \mathbb{E}$, $u(c) = 0$, and a linear map $v : \mathbb{E} \to V$, $v(0) = f(c)$ such that*

b) *f has the factorization $U \xrightarrow{u} \mathbb{E} \xrightarrow{z \mapsto z^n} \mathbb{E} \xrightarrow{v} V$, where $n := \nu(f, c)$.*[1]

Such maps are finite and locally biholomorphic in $U \backslash \{c\}$; the number n is called the *mapping degree* of f. In the small, holomorphic maps are always winding maps; in fact (cf. I.9.4.4):

(1) *If $f \in \mathcal{O}(G)$ is nonconstant, then for every point $c \in G$ there exists a neighborhood $U \subset G$ such that the induced map $f|U : U \to f(U)$ is a winding map of mapping degree $\nu(f, c)$ about c.*

Winding maps have the following "shrinkage property":

(2) *If $f : U \to V$ is a winding map of degree n about c, then for every disc $V_1 \subset V$ about $f(c)$ the induced map $f|U_1 : U_1 \to V_1$, where $U_1 := f^{-1}(V_1)$, is a winding map of degree n about c.*

Proof. Let $v \circ p \circ u$, with $p(z) := z^n$, be a factorization of f as in a) and b). Since v is linear, $B' := v^{-1}(V_1)$ is a disc about 0. If r is its radius, then $B := p^{-1}(B')$ is the disc about 0 of radius $s := \sqrt[n]{r}$. If we now set $u_1(z) := s^{-1}u(z)$, $z \in U_1$, and $v_1(z) := v(rz)$, $z \in \mathbb{E}$, then $v_1 \circ p \circ u_1$ gives a factorization of $f|U_1$ as a winding map. □

---

[1]The *multiplicity* $\nu(f, c)$ of f at $c \in G$ is the *order of the zero* of $f - f(c)$ at c. For nonconstant f, the inequality $1 \leq \nu(f, c) < \infty$ always holds; $\nu(f, c) = 1$ if and only if f is biholomorphic in a neighborhood of c; that is, if $f'(c) \neq 0$. For more on this, see I.8.1.4 and I.9.4.2.

We now generalize (1). Let $f \in \mathcal{O}(G)$ have finite fibers, and let $c_1, \dots, c_m$ be the distinct points of such a fiber $f^{-1}(p)$. Then f has the following representation about $f^{-1}(p)$:

Proposition. *There exist an open disc V about p and open pairwise disjoint neighborhoods $U_1, \dots, U_m \subset G$ of $c_1, \dots, c_m$ such that $V = f(U_\mu)$ and every induced map $f|U_\mu : U_\mu \to V$ is a winding map about c of degree $\nu(f, c_\mu)$, $1 \le \mu \le m$.*

If f is finite, the sets U_μ and V can be chosen so that $f^{-1}(V) = U_1 \cup U_2 \cup \dots \cup U_m$.

Proof. Using (1), we choose pairwise disjoint $\widetilde{U}_1, \dots, \widetilde{U}_m \subset G$ about $c_1, \dots, c_m$ such that $f|\widetilde{U}_\mu : \widetilde{U}_\mu \to f(\widetilde{U}_\mu)$ is a winding map about c of degree $\nu(f, c_\mu)$, $1 \le \mu \le m$. There exists a disc $V \subset \bigcap_{\mu=1}^{m} f(\widetilde{U}_\mu)$ about p. If we set $U_\mu := f^{-1}(V) \cap \widetilde{U}_\mu$, then, by (2), $f|U_\mu \to V$ is also a winding map about c of degree $\nu(f, c_\mu)$, $1 \le \mu \le m$.

If f is finite, then f is closed; hence, by 1(1) and the shrinkage property (2), V can be chosen so that $f^{-1}(V) = U_1 \cup \dots \cup U_m$. □

Corollary. *If $f : G \to G'$ is finite and locally biholomorphic, then every point $p \in G'$ has a neighborhood V for which $f^{-1}(V)$ can be decomposed into finitely many domains U_j, $j \in J$, in such a way that every induced map $f : U_j \to V$ is biholomorphic.*

Such maps are also called *finite-sheeted (unbranched) coverings.*

3. Radó's theorem. If $f \in \mathcal{O}(G)$ is nonconstant, then for every point $w \in \mathbb{C}$ the *number of points in the fiber* $f^{-1}(w)$ is measured by

$$\deg_w f := \sum_{c \in f^{-1}(w)} \nu(f, c) \quad \text{if } w \in f(G), \quad \deg_w f := 0 \text{ otherwise.}$$

We have the equivalence

$$1 \le \deg_w f < \infty \iff \textit{the fiber } f^{-1}(w) \textit{ is nonempty and finite.}$$

For polynomials q of degree $d \ge 1$,

$$\deg_w q = d \textit{ for all } w \in \mathbb{C} \textit{ (fundamental theorem of algebra).}$$

For winding maps $f : G \to G'$ about $c \in G$, the degree function is also constant: $\deg_w f = \nu(f, c)$ for all $w \in G'$. We prove the following general result.

Theorem (Radó). *A holomorphic map $f : G \to G'$ is finite if and only if its degree function* $\deg_w f$ *is finite and constant in G.*

Proof. For every point $p \in G'$ with $1 \le \deg_p f < \infty$, we choose V, $U_1, \dots, U_m$ as in Proposition 2. We set $U := U_1 \cup \dots \cup U_m$. Since $U_1, \dots, U_m$ are pairwise disjoint, it follows from Proposition 2 that =

$$(1) \quad \deg_w(f|U) = \sum_{\mu=1}^{m} \deg_w(f|U_\mu) = \sum_{\mu=1}^{m} \nu(f, c_\mu) = \deg_p f, \quad w \in V.$$

We can moreover arrange that $\overline{U}$ is compact and lies in G.

1. *Let f be finite.* Let $p \in G'$ be arbitrary. By Proposition 2, we may assume that $U = f^{-1}(V)$. Then $\deg_w f = \deg_w(f|U)$ for all $w \in V$; hence $\deg_w f = \deg_p f$ for all $w \in V$ by (1). Thus the degree function $\deg_w f$ is locally constant, and hence constant in G'.

2. *Let $\deg_w f$ be finite and constant in G'.* If f were not finite, there would exist a boundary sequence z_n in G whose image $f(z_n)$ had a limit $p \in G'$. By (1), it would follow that $\deg_w f = \deg_w(f|U)$ for all $w \in V$. Hence $f^{-1}(w) \in U$ for all $w \in V$. Since $p = \lim f(z_n)$, it thus follows that $z_n \in U$ for almost all n. But this is impossible because $\overline{U} \subset G$ is compact and z_n is a boundary sequence in G. □

Radó proved this theorem in 1922 ([R], pp. 57–58). The theorem and proof hold verbatim for *arbitrary* Riemann surfaces G, G'.

4. Mapping degree. For every finite map $f : G \to G'$,

$$\deg f := \deg_w f = \sum_{c \in f^{-1}(w)} \nu(f, c), \quad w \in G',$$

is a positive integer by Theorem 3; it is called the *mapping degree* of f. Polynomials of degree d define finite maps $\mathbb{C} \to \mathbb{C}$ of mapping degree d. The integers $d \ge 1$ that appeared in the theorems of Subsection 3 are always the degree of the corresponding finite maps. We emphasize:

(1) *The finite maps $f : G \to G'$ of degree 1 are exactly the biholomorphic maps.*

For every function $f \in \mathcal{O}(G)$, the set $S := \{z \in G : f'(z) = 0\}$ is called the *branch locus* of f. If f is nonconstant, then S is locally finite in G. For finite maps $f : G \to G'$, $f(S)$ is therefore always locally finite in G'. The following is immediate from Radó's theorem.

(2) *If $f : G \to G'$ is finite with degree d, then every fiber has at most d distinct points. The fibers over $G' \backslash f(S)$ are exactly those that have d distinct points. (The induced map $G \backslash f^{-1}(f(S)) \to G \backslash f(S)$ is a d-sheeted covering.)*

The following is another immediate consequence of Radó's theorem.

Degree theorem. *If $f : G \to G'$ and $g : G' \to G''$ are holomorphic and finite, then $g \circ f : G \to G''$ is also finite, and $\deg(g \circ f) = (\deg g) \cdot (\deg f)$.*

This statement is to be viewed as an analogue of the degree theorem $[M : K] = [M : L][L : K]$ of field theory (M is an extension field of L and L is an extension field of K). For lack of space we cannot pursue the interesting connections further; see the exercise.

Historical note. Radó introduced the general concept of finite holomorphic maps in 1922 [R]; he called them $(1, m)$ conformal maps, where m is the degree.

Exercise. Let $f : G \to G'$ be holomorphic and finite. Regard $\mathcal{O}(G)$ as a ring containing $\mathcal{O}(G')$ (with respect to the lifting $f^* : \mathcal{O}(G') \to \mathcal{O}(G)$, $h \mapsto h \circ f$) and prove:

a) For every $g \in \mathcal{O}(G)$, there exists a polynomial $\omega(Z) = Z^n + a_1 Z^{n-1} + \cdots + a_n \in \mathcal{O}(G')[Z]$, with $n = \deg f$, such that $\omega(g) = 0$.

b) If g is bounded, then $a_1, \ldots, a_n$ are also bounded.

c) If G is bounded, then $G' \neq \mathbb{C}$.

5. Glimpses. That f is holomorphic is used only superficially in the proof of the equivalences in 1. (To prove the implication iii) $\Rightarrow$ i), one needs only that f, as a nonconstant function, is nowhere locally constant.) The theorem can therefore be generalized. We sketch a more general situation. Let X and Y be metrizable locally compact spaces whose topologies have countable bases. A continuous map $f : X \to Y$ is called *proper* if every compact set in Y has a compact preimage under f. One can prove the following:

a) *A continuous map $X \to Y$ is proper if and only if it maps boundary sequences in X to boundary sequences in Y.*

b) *Every proper map is closed.*

Finite maps are now defined as proper maps whose fibers are all discrete sets; this definition is equivalent to ours in the holomorphic case.

Finite holomorphic maps play an important role in the function theory of several variables. With their aid, the n-dimensional local theory can be developed in a particularly elegant way; the reader can find this carried out systematically in [GR], Chapters 2–3.

All proper holomorphic maps between domains in $\mathbb{C}^n$, $1 \leq n < \infty$, are automatically finite. The situation changes if one studies maps between *arbitrary complex spaces*: since holomorphic maps can now be dimension-lowering (without being constant), there exist many proper (but not finite) holomorphic maps. For all such maps, Grauert's famous coherence theorem for image sheaves is valid; cf. [GR], Chapter 10, especially p. 207.

Bibliography

[A] Arens, R.: Topologies for homeomorphism groups, *Amer. Journ. Math.* 68, 593–610 (1946).

[B] Bieberbach, L.: Über einen Satz des Herrn Carathéodory, *Nachr. Königl. Gesellschaft Wiss. Göttingen, Math.-phys. Kl.*, 552–560 (1913).

[C] Cartan, H.: *Œuvres* 1, Springer, 1979.

[F$_1$] FATOU, P.: Sur les équations fonctionelles, *Bull. Soc. Math. France* 48, 208–314 (1920).

[F$_2$] FATOU, P.: Sur les fonctions holomorphes et bornées à l'intérieur d'un cercle, *Bull. Soc. Math. France* 51, 191–202 (1923).

[GR] GRAUERT, H. and R. REMMERT: *Coherent Analytic Sheaves*, Grdl. math. Wiss. 265, Springer, 1984.

[Hei$_1$] HEINS, M.H.: A note on a theorem of Radó concerning the $(1, m)$ conformal maps of a multiply-connected region into itself, *Bull. Amer. Math. Soc.* 47, 128–130 (1941).

[Hei$_2$] HEINS, M.H.: On the number of 1-1 directly conformal maps which a multiply-connected plane region of finite connectivity $p(> 2)$ admits onto itself, *Bull. Amer. Math. Soc.* 52, 454–457 (1946).

[Hu] HUBER, H.: Über analytische Abbildungen von Ringgebieten in Ringgebiete, *Comp. Math.* 9, 161–168 (1951).

[K] KOEBE, P.: Abhandlungen zur Theorie der konformen Abbildung, I. Die Kreisabbildung des allgemeinsten einfach und zweifach zusammenhängenden schlichten Bereichs und die Ränderzuordnung bei konformer Abbildung, *Journ. reine angew. Math.* 145, 177–223 (1915).

[R] RADÓ, T.: Zur Theorie der mehrdeutigen konformen Abbildungen, *Acta Litt. Sci. Szeged* 1, 55–64 (1922–23).

Part C

Selecta

10

The Theorems of Bloch, Picard, and Schottky

Une fonction entière, qui ne devient jamais ni à a ni à b est nécessairement une constante. (An entire function which is never equal to either a or b must be constant.) — E. Picard, 1879

The sine function assumes every complex number as a value; the exponential function omits only the value 0. These examples are significant for the value behavior of entire functions. A famous theorem of E. Picard says that *every* nonconstant entire function omits at most one value. This so-called little Picard theorem is an astonishing generalization of the theorems of Liouville and Casorati-Weierstrass.

The starting point of this chapter is a theorem of A. Bloch, which deals with the "size of image domains" $f(\mathbb{E})$ under holomorphic maps; this theorem is discussed in detail in Section 1. In Section 2 we obtain Picard's little theorem with the aid of Bloch's theorem and a lemma of Landau.

In Section 3 we introduce a classical theorem of Schottky, which leads to significantly sharpened forms of Picard's little theorem and the theorems of Montel and Vitali. The short proof of Picard's great theorem by means of Montel's theorem is given in Section 4.

§1. Bloch's Theorem

One of the queerest things in mathematics, ... the proof itself is crazy. — J. E. Littlewood

For every region $D \subset \mathbb{C}$, let $\mathcal{O}(\overline{D})$ denote the set of all functions that are holomorphic in an open neighborhood of $\overline{D} = D \cup \partial D$.

Bloch's theorem. *If $f \in \mathcal{O}(\overline{\mathbb{E}})$ and $f'(0) = 1$, then the image domain $f(\mathbb{E})$ contains discs of radius $\frac{3}{2} - \sqrt{2} > \frac{1}{12}$.*

The queer thing about this statement is that for a "large family" of functions a *universal* statement is made about the "size of the image domains." In our proof of this in Subsection 2, we give a center for a disc of the asserted radius. (That the point $f(0)$ is in general not such a center is shown by the function $f_n(z) := (e^{nz} - 1)/n = z + \cdots$, which omits the value $-1/n$.)

Corollary. *If f is holomorphic in the domain $G \subset \mathbb{C}$ and $f'(c) \neq 0$ at a point $c \in G$, then $f(G)$ contains discs of every radius $\frac{1}{12}s|f'(c)|$, $0 < s < d(c, \partial G)$.*

Proof. We may assume $c = 0$. For $0 < s < d(c, \partial G)$, we have $\overline{B_s(0)} \subset G$; hence $g(z) := f(sz)/sf'(0) \in \mathcal{O}(\overline{\mathbb{E}})$. Since $g'(0) = 1$, Bloch's theorem implies that $g(\mathbb{E})$ contains discs of radius $1/12$. Since $f(B_s(0)) = s|f'(0)|g(\mathbb{E})$, the assertion follows. □

In particular, the corollary contains the following:

If $f \in \mathcal{O}(\mathbb{C})$ is nonconstant, then $f(\mathbb{C})$ contains discs of every radius.

For connoisseurs of estimates, the bound $\frac{3}{2} - \sqrt{2} \approx 0.0858$ will be improved in Subsection 3 to $\frac{3}{2}\sqrt{2} - 2 \approx 0.1213$ and even, in Subsection 4, to $\sqrt{3}/4 \approx 0.43301$. The optimal value is unknown; see Subsection 5. — For the applications of Bloch's theorem in Sections 2, 3, and 4, any (poor) bound > 0 suffices.

1. Preparation for the proof. If $G \subset \mathbb{C}$ is a domain and $f \in \mathcal{O}(G)$ is nonconstant, then by the *open mapping theorem* $f(G)$ is again a domain. There is an obvious criterion for the size of discs in the image domain.

(1) *Let G be bounded, $f : \overline{G} \to \mathbb{C}$ continuous, and $f|G : G \to \mathbb{C}$ open. Let $a \in G$ be a point such that $s := \min_{z \in \partial G}|f(z) - f(a)| > 0$. Then $f(G)$ contains the disc $B_s(f(a))$.*

Proof. Since $\partial f(G)$ is compact, there exists a point $w_* \in \partial f(G)$ such that $d(\partial f(G), f(a)) = |w_* - f(a)|$. Since $\overline{G}$ is compact, there exists a sequence $z_\nu \in G$ with $\lim f(z_\nu) = w_*$ and $z_* := \lim z_\nu \in \overline{G}$. It follows that

$f(z_*) = w_* \in \partial f(G)$. Since $f|G$ is open, z_* cannot lie in G. Hence $z_* \in \partial G$ and therefore $|w_* - f(a)| \geq s$. It follows that $B_s(f(a)) \subset f(G)$. □

We apply (1) to holomorphic functions f. The number s certainly depends on $f'(a)$ and $|f|_G$ (example: $f(z) = \varepsilon z$ in $\mathbb{E}$). For discs $V := B_r(a)$, $r > 0$, there are good estimates from below for s.

Lemma. *Let $f \in \mathcal{O}(\overline{V})$ be nonconstant and satisfy $|f'|_V \leq 2|f'(a)|$. Then*

$$B_R(f(a)) \subset f(V), \quad \text{with} \quad R := (3 - 2\sqrt{2})r|f'(a)|. \quad (3 - 2\sqrt{2} > \tfrac{1}{6})$$

Proof. We may assume that $a = f(a) = 0$. Set $A(z) := f(z) - f'(0)z$; then

$$A(z) = \int_{[0,z]} [f'(\zeta) - f'(0)]d\zeta, \quad \text{whence} \quad |A(z)| \leq \int_0^1 |f'(zt) - f'(0)|\,|z|dt.$$

For $v \in V$, Cauchy's integral formula and standard estimates give

$$f'(v) - f'(0) = \frac{v}{2\pi i}\int_{\partial V} \frac{f'(\zeta)d\zeta}{\zeta(\zeta - v)}, \quad |f'(v) - f'(0)| \leq \frac{|v|}{r - |v|}|f'|_V.$$

It follows that

$$(*) \qquad |A(z)| \leq \int_0^1 \frac{|zt|\;|f'|_V}{r - |zt|}|z|dt \leq \frac{1}{2}\frac{|z|^2}{r - |z|}|f'|_V.$$

Now let $\rho \in (0, r)$. The inequality $|f(z) - f'(0)z| \geq |f'(0)|\rho - |f(z)|$ holds for z such that $|z| = \rho$. Since $|f'|_V \leq 2|f'(0)|$, it follows from $(*)$ that

$$|f(z)| \geq \left(\rho - \frac{\rho^2}{r - \rho}\right)|f'(0)|.$$

Now $\rho - \rho^2/(r - \rho)$ assumes its maximum value, $(3 - 2\sqrt{2})r$, at $\rho^* := (1 - \frac{1}{2}\sqrt{2})r \in (0, r)$. It follows that $|f(z)| \geq (3 - 2\sqrt{2})r|f'(0)| = R$ for all $|z| = \rho^*$. Setting $G := B_{\rho^*}(0)$ in (1) shows that $B_R(0) \subset f(G) \subset f(V)$. □

Exercise. Use (1) to show that, for all $f \in \mathcal{O}(\mathbb{E})$ with $f(0) = 0$ and $f'(0) = 1$, the inclusion $f(\mathbb{E}) \supset B_r(0)$ holds with $r := 1/6\;|f|_{\mathbb{E}}$.

2. Proof of Bloch's theorem. To every function $f \in \mathcal{O}(\overline{\mathbb{E}})$, let us assign the function $|f'(z)|(1 - |z|)$, which is continuous on $\overline{\mathbb{E}}$. It assumes its *maximum* M at a point $p \in \mathbb{E}$. Bloch's theorem is contained in the following

Theorem. *If $f \in \mathcal{O}(\overline{\mathbb{E}})$ is nonconstant, then $f(\mathbb{E})$ contains the disc about $f(p)$ of radius $(\frac{3}{2} - \sqrt{2})M > \frac{1}{12}|f'(0)|$.*

Proof. With $t := \frac{1}{2}(1 - |p|)$, we have

$$M = 2t|f'(p)|, \quad B_t(p) \subset \mathbb{E}, \quad 1 - |z| \geq t \quad \text{for} \quad z \in B_t(p).$$

From $|f'(z)|(1 - |z|) \leq 2t|f'(p)|$, it follows that $|f'(z)| \leq 2|f'(p)|$ for all $z \in B_t(p)$. Hence, by Lemma 1, $B_R(f(p)) \subset f(\mathbb{E})$ for $R := (3-2\sqrt{2})t|f'(p)|$. □

Historical note. A. Bloch discovered this theorem in 1924 (in fact in a sharper form; cf. Subsection 4); see [Bl$_1$], p. 2051, and [Bl$_2$]. G. Valiron and E. Landau simplified Bloch's arguments considerably; see [L$_2$], which contains a "three-line proof in telegraphic style". Landau reports on the early history of the theorem in [L$_3$].

The proof given above is due to T. Estermann ([E], 1971). It is more natural than Landau's proof ([L$_4$], pp. 99–101) and, for those who like bounds, yields $\frac{3}{2} - \sqrt{2} > \frac{1}{12}$, which is better than Landau's $\frac{1}{16}$; for more on this, see Subsection 5.

A theorem of the Bloch type was first proved in 1904 by Hurwitz. He used methods from the theory of elliptic modular functions to prove (*Math. Werke* 1, Satz IV, p. 602):

If $f \in \mathcal{O}(\mathbb{E})$ *satisfies* $f(0) = 0$, $f'(0) = 1$, *and* $f(\mathbb{E}^\times) \subset \mathbb{C}^\times$, *then*

$$f(\mathbb{E}) \supset \overline{B_s(0)} \quad \textit{for} \quad s \geq \frac{1}{58} = 0.01724.$$

Carathéodory showed in 1907 (*Ges. Math. Schriften* 3, pp. 6–9) that in Hurwitz's situation 1/16 (rather than 1/58) is the *best possible* bound.

3*. Improvement of the bound by the solution of an extremal problem. In Theorem 2, the auxiliary function $|f'(z)|(1 - |z|)$ was introduced without motivation. Here we discuss a variant whose proof is more transparent: the other auxiliary function $|f'(z)|(1 - |z|^2)$ and the better bound $\frac{3}{2}\sqrt{2} - 2 > \frac{1}{12}\sqrt{2}$ appear *automatically.*

Let $f \in \mathcal{O}(\overline{\mathbb{E}})$ be *nonconstant.* The hope that $f(\mathbb{E})$ contains larger discs as $|f'(0)|$ increases leads to an

Extremal problem. Find a function $F \in \mathcal{O}(\overline{\mathbb{E}})$ *such that* $F(\mathbb{E}) = f(\mathbb{E})$ *and* F *has the greatest possible derivative at* 0 *(extremal function).*

To make this precise we consider, for a given f, the family

$$(1) \qquad \mathcal{F} := \{h = f \circ j,\ j \in \operatorname{Aut} \mathbb{E}\}, \quad \text{where} \quad j(z) := \frac{\varepsilon z - w}{\overline{w}\varepsilon z - 1}, \ \varepsilon \in S^1, \ w \in \mathbb{E}.$$

Since $j \in \mathcal{O}(\overline{\mathbb{E}})$ and $j'(0) = \varepsilon(|w|^2 - 1)$, it is clear that

$$(2) \qquad h \in \mathcal{O}(\overline{\mathbb{E}}), \quad h(\mathbb{E}) = f(\mathbb{E}), \quad |h'(0)| = |f'(w)|(1 - |w|^2)$$

for every $h = f \circ j \in \mathcal{F}$. Since $f' \in \mathcal{O}(\overline{\mathbb{E}})$, the (auxiliary) function on the right-hand side assumes its maximum $N > 0$ at a point $q \in \mathbb{E}$. Thus one solution of the extremal problem is the function

$$(3)\qquad \begin{aligned} &F(z) := f\left(\frac{z-q}{\bar{q}z-1}\right) \in \mathcal{O}(\overline{\mathbb{E}}), \quad \text{where} \quad F(0) = f(q) \quad \text{and} \\ &|F'(0)| = \max_{|w|\le 1} |f'(w)|(1-|w|^2). \end{aligned}$$

The following estimate for the derivative of F is now crucial:

$$(4)\qquad \begin{aligned} &|F'(z)| \le \frac{N}{1-|z|^2} \quad \text{for} \quad z \in \mathbb{E}; \quad \text{in particular,} \\ &\max_{|z|\le r} |F'(z)| \le \frac{N}{1-r^2} \quad \text{for} \quad 0<r<1. \end{aligned}$$

Proof. Since $\mathcal{F} = \{F \circ j,\ j \in \operatorname{Aut}\mathbb{E}\}$ (group property) and *every* $j \in \operatorname{Aut}\mathbb{E}$ has the form (1), the inequality $N \ge |(F\circ j)'(0)| = |F'(w)|(1-|w|^2)$ holds for all $w \in \mathbb{E}$. □

The hopes placed in F are now fully justified.

Bloch's theorem (variant). *Let $f \in \mathcal{O}(\overline{\mathbb{E}})$. Suppose that the function $|f'(z)|(1-|z|^2)$ assumes its maximum $N > 0$ at $q \in \mathbb{E}$. Then $f(\mathbb{E})$ contains the disc about $f(q)$ with radius $(\frac{3}{2}\sqrt{2}-2)N$. — If $f'(0) = 1$, then $f(\mathbb{E})$ contains discs of radius $\frac{3}{2}\sqrt{2}-2 > \frac{1}{12}\sqrt{2}$.*

Proof. Choose F as in (3). Since $|F'(0)| = N$ and, by (4), $|F'(z)| \le N/(1-|z|^2)$, it follows that $|F'(z)| \le 2|F'(0)|$ for all $|z| \le \frac{1}{2}\sqrt{2}$. By Lemma 1, $f(\mathbb{E}) = F(\mathbb{E})$ therefore contains the disc about $f(q) = F(0)$ of radius $(\frac{3}{2}\sqrt{2}-2)N$. □

The extremal problem has led us to the auxiliary function $|f'(z)|(1-|z|^2)$. It is clear that $M \le N$; thus the new lower bound is obviously better than that in Theorem 2.

Remark. The auxiliary function $|f'(z)|(1-|z|^2)$ and its maximum N in $\mathbb{E}$ were introduced in 1929 by Landau ([L3], p. 83). All functions in the set

$$\mathcal{B} := \left\{ f \in \mathcal{O}(\mathbb{E}) : \sup_{z\in\mathbb{E}} |f'(z)|(1-|z|^2) < \infty \right\}$$

have come to be called *Bloch functions*. It can be shown that

$\mathcal{B}$ is a $\mathbb{C}$-vector space. In fact, $\mathcal{B}$ is a Banach space when it is equipped with the norm $\|f\| := |f(0)| + \sup_{z\in\mathbb{E}} |f'(z)|(1-|z|^2)$, which satisfies the inequality $\|f\| \le 2\sup_{z\in\mathbb{E}} |f(z)|$. □

A still sharper version of Bloch's theorem is given in the next subsection.

4*. Ahlfors's theorem. If $f : G \to \mathbb{C}$ is holomorphic, a disc $B \subset f(G)$ is called *schlicht (with respect to f)* if there exists a domain $G^* \subset G$ that is mapped biholomorphically onto B by f.

Ahlfors's theorem. *Let* $f \in \mathcal{O}(\overline{\mathbb{E}})$; *set* $N := \max_{|z|\leq 1}|f'(z)|(1-|z|^2) > 0$. *Then* $f(\mathbb{E})$ *contains schlicht discs of radius* $\frac{1}{4}\sqrt{3}N$.

This theorem makes Bloch's theorem look weak: not only do we now have schlicht discs instead of discs, but the new bound $\sqrt{3}/4 \approx 0.433$ is more than three times the old bound $\frac{3}{2}\sqrt{2} - 2 \approx 0.121$.

Ahlfors obtains the theorem from his differential-geometric version of Schwarz's lemma ([A], p. 364). In what follows, we give what is probably the simplest proof at present, that of M. Bonk ([Bon], 1990). The next lemma is crucial.

Lemma. *Let* $F \in \mathcal{O}(\mathbb{E})$ *satisfy* $|F'(z)| \leq 1/(1-|z|^2)$ *and* $F'(0) = 1$. *Then*

$$(*) \qquad \operatorname{Re} F'(z) \geq \frac{1-\sqrt{3}|z|}{(1-\sqrt{3}|z|)^3} \quad \text{for all } z \text{ with } |z| \leq 1/\sqrt{3}.$$

In order to obtain the theorem from this, we need a biholomorphy criterion.

a) *Let* $G \subset \mathbb{C}$ *be convex and* $h \in \mathcal{O}(G)$, *and suppose that* $\operatorname{Re} h'(z) > 0$ *for all* $z \in G$. *Then* h *maps* G *biholomorphically onto* $h(G)$.

Proof. For u, $v \in G$ the line segment $\gamma(t) = u + (v-u)t$, $0 \leq t \leq 1$, lies in G. Hence

$$h(v) - h(u) = (v-u)\left[\int_0^1 \operatorname{Re} h'(\gamma(t))dt + i\int_0^1 \operatorname{Im} h'(\gamma(t))dt\right] \neq 0$$

if $u \neq v$, since the first integral on the right-hand side is positive because $\operatorname{Re} h'(z) > 0$. □

We now prove the theorem. We may assume that $N = 1$. Choose F as in 3*(3); then $F'(0) = \eta \in S^1$. First let $\eta = 1$. By the lemma, $\operatorname{Re} F'(z) > 0$ in $B := B_\rho(0)$, $\rho := 1/\sqrt{3}$; hence $F|B : B \to F(B)$ is *biholomorphic* by a). For all $\zeta = \rho e^{i\varphi} \in \partial B$, the lemma implies that

$$|F(\zeta) - F(0)| = \left|\int_0^\rho F'(te^{i\varphi})dt\right| \geq \int_0^\rho \operatorname{Re} F'(te^{i\varphi})dt$$

$$\geq \int_0^{1/\sqrt{3}} \frac{1-\sqrt{3}t}{(1-1/\sqrt{3}t)^3}dt = \frac{1}{4}\sqrt{3}.$$

Hence, by 1(1), $F(B)$ contains discs of radius $\sqrt{3}/4$. Since $f = F \circ g$ with $g \in \operatorname{Aut} \mathbb{E}$, f maps the domain $G := g^{-1}(B) \subset \mathbb{E}$ *biholomorphically* onto a domain G^* that contains discs of radius $\sqrt{3}/4$.

For arbitrary $\eta \in S^1$, one works with $\eta^{-1}F$. Then $f : G \to \eta G^*$ is biholomorphic and ηG^*, like G^*, contains schlicht discs of radius $\sqrt{3}/4$. □

We now come to the proof of the lemma. We first observe:

It suffices to prove the estimate $(*)$ *for all real* $z \in [0, 1/\sqrt{3}]$.

Indeed, $F_\varphi(z) := e^{-i\varphi}F(e^{i\varphi}z)$ satisfies the hypotheses of the lemma for every $\varphi \in \mathbb{R}$. Thus $(*)$ holds for F and $z = re^{i\varphi}$ if it holds for F_φ and $z = r$.

The proof itself is rather technical and requires:

b) *For* $z := p(w) := \sqrt{3}(1-w)/(3-w)$ *and* $q(w) := \frac{9}{4}w(1-\frac{1}{3}w)^2$, *the following statements hold:*

$$p(\overline{\mathbb{E}}) \subset \mathbb{E}; \quad p \textit{ maps } [0,1] \textit{ onto } [0, \tfrac{1}{\sqrt{3}}];$$

$$q(p^{-1}(z)) = \frac{1-\sqrt{3}z}{\left(1-\frac{1}{\sqrt{3}}z\right)^3}; \quad \textit{and} \quad |q(w)|(1-|p(w)|^2) = 1 \textit{ for all } w \in \partial\mathbb{E}.$$

c) *If* $h \in \mathcal{O}(\mathbb{E})$ *with* $h'(0) = 1$ *and* $|h'(z)| \le 1/(1-|z|^2)$, *then* $h''(0) = 0$.

The proof of b) is a routine calculation. — For the proof of c), consider the antiderivative

$$\int_0^z h'(\zeta)d\zeta = z + az^2 + \cdots \in \mathcal{O}(\mathbb{E})$$

of h'. For all $z = re^{i\varphi} \in \mathbb{E}$,

$$|z + az^2 + \cdots| \le \int_0^r |h'(e^{i\varphi}t)|dt \le \int_0^r \frac{dt}{1-|t|^2} = |z| + \frac{1}{3}|z|^3 + \cdots.$$

Since the quadratic term in $|z|$ is missing on the right-hand side, considering small $|z|$ shows that $a = 0$ and hence that $h''(0) = 2a = 0$. □

After these preliminaries, the proof of $(*)$ in the lemma goes as follows. Consider the auxiliary function

$$H(w) := \left(\frac{F'(p(w))}{q(w)} - 1\right)\frac{w}{(1-w)^2}$$

($w/(1-w)^2$ is Koebe's extremal function). Since $p(\overline{\mathbb{E}}) \subset \mathbb{E}$, H is holomorphic everywhere in $\overline{\mathbb{E}}\backslash\{1\}$. Since $F''(0) = 0$ by c), H is also holomorphic at $1 \in \mathbb{C}$. It follows that $H \in \mathcal{O}(\overline{\mathbb{E}})$. Now $w/(1-w)^2$ is *real and negative* ($\le -\frac{1}{4}$) for all $w \in \partial\mathbb{E}\backslash\{1\}$. Hence

$$\operatorname{Re} H(w) = \frac{w}{(1-w)^2} \operatorname{Re}\left(\frac{F'(p(w))}{q(w)} - 1\right) \quad \text{for all } w \in \partial\mathbb{E}.$$

The inequality $|F'(z)|(1-|z|^2) \leq 1$ and the last equation in b) imply that $|F'(p(w))| \leq |q(w)|$ on $\partial\mathbb{E}$. Since $\text{Re}(a-1) \leq 0$ for every $a \in \overline{\mathbb{E}}$, it follows that $\text{Re } H(w) \geq 0$ for all $w \in \partial\mathbb{E}$. Applying the maximum principle to $e^{-H(w)}$ now gives $\text{Re } H(w) \geq 0$ for all $w \in \overline{\mathbb{E}}$. Since $q(w) \geq 0$ and $w/(1-w)^2 \geq 0$ for $w \in [0,1)$, it also follows that

$$\text{Re } F'(p(w)) \geq q(w) \quad \text{for all } w \in [0,1].$$

By the statements in b), this is (*) for all $z \in [0, 1/\sqrt{3}]$. □

The next result follows immediately from Ahlfors's theorem (see the introduction):

If $f \in \mathcal{O}(\mathbb{C})$ is nonconstant, then $f(\mathbb{C})$ contains arbitrarily large schlicht discs.

5*. Landau's universal constant. Bloch's theorem and its variation prompted Landau to introduce "universal constants"; cf. [L_3], pp. 609–615, and [L_4], p. 149. For every $h \in \mathcal{F} := \{f \in \mathcal{O}(\overline{\mathbb{E}}) : f'(0) = 1\}$, let L_h denote the radius of the largest disc that lies in $h(\mathbb{E})$ and let B_h denote the radius of the largest disc that is the biholomorphic image under h of a subdomain of $\mathbb{E}$. Then

$$L := \inf\{L_h : h \in \mathcal{F}\} \quad \text{and} \quad B := \inf\{B_h : h \in \mathcal{F}\}$$

are called *Landau's* and *Bloch's constants*, respectively. Landau correspondingly defines the numbers A_h and A for the family $\mathcal{F}^* := \{h \in \mathcal{F} : h \text{ injective}\}$. It is trivial that $B \leq L \leq A$. Only bounds are known for B, L, and A; thus, in the preceding subsection, we showed first that $L \geq \frac{3}{2} - \sqrt{2} \approx 0.0858$ and then that $L \geq \frac{3}{2}\sqrt{2} - 2 \approx 0.1213$. Ahlfors's theorem even says that $B \geq \frac{1}{4}\sqrt{3} \approx 0.4330$. In [Bon], Bonk shows more: $B > \frac{1}{4}\sqrt{3} + 10^{-14}$. Since $\frac{1}{2}\log\frac{1+z}{1-z} \in \mathcal{F}^*$, certainly $A \leq \frac{1}{4}\pi \approx 0.7853$. Thus

$$0.4330 + 10^{-14} < B \leq L \leq A \leq 0.7853.$$

Such estimates and refinements continue to fascinate function theorists; it has been proved (cf. [L_4], [M], and [Bon]) that

$$0.5 < L < 0.544, \quad 0.433 + 10^{-14} < B < 0.472, \quad 0.5 \leq A.$$

More recently, Yanagihara [Y] proved that $0.5 + 10^{-335} < L$. It actually holds that $B < L < A$. The

Ahlfors-Grunsky conjecture:

$$B = \sqrt{\frac{\sqrt{3}-1}{2}} \cdot \frac{\Gamma(\frac{1}{3})\Gamma(\frac{11}{12})}{\Gamma(\frac{1}{4})} = 0.4719\ldots$$

has been unsolved since 1936.

The theorem of Hurwitz and Carathéodory also has a corresponding universal constant. "Diese möchte ich aber nicht die Carathéodory Konstant C nennen, da Herr Carathéodory festgestellt hat, daß sie schon einen anderen Namen, nämlich 1/16, hatte." (But I would not like to call it the Carathéodory constant C, as Mr. Carathéodory determined that it already had another name, namely 1/16.) ([L3], p. 78)

§2. Picard's Little Theorem

Nonconstant polynomials assume *all* complex numbers as values. In contrast, nonconstant entire functions can omit values, as the *nonvanishing* exponential function shows. With the aid of Bloch's theorem, we will prove:

Theorem (Picard's little theorem). *Every nonconstant entire function omits at most one complex number as value.*

This statement can also be formulated as follows:

Let $f \in \mathcal{O}(\mathbb{C})$ *and suppose* $0 \notin f(\mathbb{C})$ *and* $1 \notin f(\mathbb{C})$. *Then* f *is constant.*

The theorem follows immediately from this: Indeed, suppose $h \in \mathcal{O}(\mathbb{C})$ omits the values a, b, where $a \neq b$; then $[h(z) - a]/(b - a) \in \mathcal{O}(\mathbb{C})$ omits the values 0 and 1 and is therefore constant. Hence h is also constant. □

In $\mathbb{C}$, meromorphic functions can omit *two* values; for instance, $1/(1+e^z)$ is never 0 or 1. This example is significant in view of

Picard's little theorem for meromorphic functions. *Every function* $h \in \mathcal{M}(\mathbb{C})$ *that omits three distinct values* $a, b, c \in \mathbb{C}$ *is constant.*

The function $1/(h-a)$ is then entire and omits $1/(b-a)$ and $1/(c-a)$. □

In Subsections 1 and 2, we give the Landau-König proof of Picard's little theorem (cf. [L4], pp. 100–102, and [K]).

1. Representation of functions that omit two values. If $G \subset \mathbb{C}$ is simply connected, then the units of $\mathcal{O}(G)$ have logarithms and square roots in $\mathcal{O}(G)$; cf. 8.2.6. Our next result follows just from this, by elementary manipulations.

Lemma. *Let* $G \subset \mathbb{C}$ *be simply connected, and let* $f \in \mathcal{O}(G)$ *be such that* $1 \notin f(G)$ *and* $-1 \notin f(G)$. *Then there exists an* $F \in \mathcal{O}(G)$ *such that*

$$f = \cos F.$$

Proof. Since $1 - f^2$ has no zeros in G, there exists a function $g \in \mathcal{O}(G)$ such that $(f + ig)(f - ig) = f^2 + g^2 = 1$. Then $f + ig$ has no zeros in G, and hence $f+ig=e^{iF}$ with $F\in\mathcal{O}(G)$. It follows that $f-ig=e^{-iF}$; thus $f=\frac{1}{2}(e^{iF}+e^{-iF})=\cos F$. □

With the aid of the lemma, we now prove:

Theorem. *Let $G\subset\mathbb{C}$ be simply connected, and let $f\in\mathcal{O}(G)$ be such that $0\notin f(G)$ and $1\notin f(G)$. Then there exists $g\in\mathcal{O}(G)$ such that*

$$(1)\qquad f=\frac{1}{2}[1+\cos\pi(\cos\pi g)].$$

If $g\in\mathcal{O}(G)$ is any function for which (1) *holds, then $g(G)$ contains no disc of radius* 1.

Proof. a) The function $2f-1$ omits the values ±1 in G. Thus, by the lemma, we have $2f-1=\cos\pi F$. The function $F\in\mathcal{O}(G)$ must omit all integer values. Hence there exists a $g\in\mathcal{O}(G)$ with $F=\cos\pi g$.

b) We set $A:=\{m\pm i\pi^{-1}\log(n+\sqrt{n^2-1}),\ m\in\mathbb{Z},\ n\in\mathbb{N}\backslash\{0\}\}$ and prove first that $A\cap g(G)=\emptyset$. Let $a:=p\pm i\pi^{-1}\log(q+\sqrt{q^2-1})\in A$; then $\cos\pi a=\frac{1}{2}(e^{i\pi a}+e^{-i\pi a})=\frac{1}{2}(-1)^p[(q+\sqrt{q^2-1})^{-1}+(q+\sqrt{q^2-1})]=(-1)^p q$. Hence $\cos\pi(\cos\pi a)=\pm1$ in the case $p,q\in\mathbb{Z}$. Since $0,1\notin f(G)$, the set $g(G)\cap A$ is empty.

The points of A are the vertices of a "rectangular grid" in $\mathbb{C}$. The "length" of every rectangle is 1. Since

$$\log(n+1+\sqrt{(n+1)^2-1})-\log(n+\sqrt{n^2-1})=\log\frac{1+\frac{1}{n}+\sqrt{1+\frac{2}{n}}}{1+\sqrt{1-\frac{1}{n^2}}}$$

$$\le\log\left(1+\tfrac{1}{n}+\sqrt{1+\tfrac{2}{n}}\right)\le\log(2+\sqrt{3})<\pi$$

by the monotonicity of $\log x$, the "height" of each rectangle is less than 1. Thus for every $w\in\mathbb{C}$ there exists an $a\in A$ such that $|\operatorname{Re}a-\operatorname{Re}w|\le1/2$ and $|\operatorname{Im}a-\operatorname{Im}w|<1/2$, i.e. $|a-w|<1$. Every disc of radius 1 therefore intersects A. But $g(G)\cap A=\emptyset$; hence $g(G)$ contains no disc of radius 1. □

In the first edition of this book, Landau's equation $f=-\exp[\pi i\cosh(2g)]$ was used instead of (1). The presentation with the iterated cosine seems easier and more natural; it was given in 1957 by Heinz König [K], who at that time was not familiar with the second edition of [L4] and used Schottky's theorem in the proof.

2. Proof of Picard's little theorem. We have $f=\frac{1}{2}(1+\cos\pi(\cos\pi g))$ by Theorem 1, where $g\in\mathcal{O}(\mathbb{C})$ and $g(\mathbb{C})$ contains no disc of radius 1. By the corollary to Bloch's theorem (see the introduction to §1), g is constant. Hence f is also constant. □

Remark. Picard's little theorem can also be formulated as follows:

Suppose that $f, g \in \mathcal{O}(\mathbb{C})$ *and* $1 = e^f + e^g$. *Then* f *and* g *are constant.*

This statement is equivalent to Picard's statement. (Prove this.)

Exercises. 1) Let $f, g, h \in \mathcal{O}(\mathbb{C})$. Then the following assertions are true.
a) If $h = e^f + e^g$, then h has either no zeros in $\mathbb{C}$ or infinitely many zeros in $\mathbb{C}$.
b) If h is a nonconstant polynomial, then he^f assumes every value.

Hint for a). Transform the problem and apply Picard's little theorem to $g - f$.

2) Construct a function $f \in \mathcal{O}(\mathbb{E})$ that maps $\mathbb{E}$ *locally biholomorphically onto* $\mathbb{C}$.

Hint. Set $h(z) := ze^z$, $k(z) := 4z/(1-z)^2$ (the Koebe function), and $f := h \circ k$. Show that $k(\mathbb{E}) = \mathbb{C} \backslash (-\infty, -1]$ and $h((-\infty, -1)) = h((-1, 0))$.

Glimpse. A "three-line proof" of Picard's little theorem is possible if one knows that there exists a holomorphic covering $u : \mathbb{E} \to \mathbb{C} \backslash \{0, 1\}$ (uniformization). For one can then, by a general principle from topology, lift every holomorphic map $f : \mathbb{C} \to \mathbb{C} \backslash \{0, 1\}$ to a holomorphic map $\widetilde{f} : \mathbb{C} \to \mathbb{E}$ with $f = u \circ \widetilde{f}$. Since $\widetilde{f}$ is constant by Liouville, f is constant. — There is also a proof of Picard's little theorem by means of the *theory of Brownian motion*; cf. R. Durrett: *Brownian Motion and Martingales in Analysis*, Wadsworth, Inc., 1984, 139–143.

3. Two amusing applications. In general, holomorphic maps $f : \mathbb{C} \to \mathbb{C}$ have no fixed points; $f(z) := z + e^z$, for example, has none. But the following does hold.

Fixed-point theorem. *Let* $f : \mathbb{C} \to \mathbb{C}$ *be holomorphic. Then* $f \circ f : \mathbb{C} \to \mathbb{C}$ *always has a fixed point unless* f *is a translation* $z \mapsto z + b$, $b \neq 0$.

Proof. Suppose $f \circ f$ has no fixed points. Then f also has no fixed points, and it follows that $g(z) := [f(f(z)) - z]/[f(z) - z] \in \mathcal{O}(\mathbb{C})$. This function omits the values 0 and 1 (!); hence, by Picard, there exists a $c \in \mathbb{C} \backslash \{0, 1\}$ with

$$f(f(z)) - z = c(f(z) - z), \quad z \in \mathbb{C}.$$

Differentiation gives $f'(z)[f'(f(z)) - c] = 1 - c$. Since $c \neq 1$, f' has no zeros and $f'(f(z))$ is never equal to c. Thus $f' \circ f$ omits the values 0 and $c \neq 0$; by Picard, $f' \circ f$ is therefore constant. It follows that $f' =$ constant, hence that $f(z) = az + b$. Since f has no fixed points, $a = 1$ and $b \neq 0$. □

Exercise. Show that every entire periodic function has a fixed point.

The entire functions $f := \cos \circ h$ and $g := \sin \circ h$, where $h \in \mathcal{O}(\mathbb{C})$, satisfy the equation $f^2 + g^2 = 1$. (It is easy to show that these are all the solutions by entire functions of the equation $(f + ig)(f - ig) = 1$.) We investigate the solvability of the Fermat equation $X^n + Y^n = 1$, $n \geq 3$, by functions that are meromorphic in $\mathbb{C}$.

Proposition. *If $f, g \in \mathcal{M}(\mathbb{C})$ and $f^n + g^n = 1$ with $n \in \mathbb{N}$, $n \geq 3$, then either f and g are constant or they have common poles.*

Proof. Let $P(f) \cap P(g) = \emptyset$. It follows from the equation $f^n + g^n = 1$ that $P(f) = P(g)$; hence $f, g \in \mathcal{O}(\mathbb{C})$. Suppose that $g \not\equiv 0$. Since $Z(f) \cap Z(g) = \emptyset$, $f/g \in \mathcal{M}(\mathbb{C})$ assumes the value $a \in \mathbb{C}$ at $w \in \mathbb{C}$ if and only if $f(w) = ag(w)$. It now follows from the factorization

$$1 = \prod_{1}^{n}(f - \zeta_\nu g), \quad \zeta_1, \ldots, \zeta_n \text{ the } n \text{ roots of } X^n + 1,$$

that f/g assumes none of the n *distinct* values $\zeta_1, \ldots, \zeta_n$. Since $n \geq 3$, Picard implies that $f = cg$ with a constant $c \neq \zeta_1, \ldots, \zeta_n$. It follows that $(c^n + 1)g^n = 1$. Hence g, and therefore f, is constant. □

Remark 1. The statement just proved is representative of theorems of the following type. One considers polynomials $F(z_1, z_2)$ of two complex variables, for instance $z_1^n - z_2^n - 1$, and their zero sets X in $\mathbb{C}^2$.

If the unit disc $\mathbb{E}$ is the universal cover of the "projective curve" $\overline{X}$, there exist no nonconstant functions $f, g \in \mathcal{O}(\mathbb{C})$ such that $F(f, g) = 0$.

The hypothesis is satisfied if and only if the curve $\overline{X}$ has genus $g > 1$; the genus of the Fermat curves $z_1^n + z_2^n - 1$ is $g = \frac{1}{2}(n-1)(n-2)$.

Remark 2. There exist *nonconstant functions $f, g \in \mathcal{M}(\mathbb{C})$ with common poles* such that $f^3 + g^3 = 1$. Indeed, the equation $X^3 + Y^3 = 1$ describes an *affine elliptic* curve in $\mathbb{C}^2$, the universal cover of the *projective* curve is $\mathbb{C}$, and the projection of $\mathbb{C}$ onto the curve determines such functions f, g. Anyone familiar with the Weierstrass $\wp$-function can give such functions explicitly. Starting with

$$(a + b\wp')^3 + (a - b\wp')^3 = \wp^3 \quad \text{with constants} \quad a, b \in \mathbb{C}$$

leads immediately, because of the differential equation $\wp'^2 = 4\wp^3 - g_2\wp - g_3$, to

$$24ab^2 = 1, \quad g_2 = 0, \quad 8a^3 = g_3.$$

Now it is well known that the value $g_2 = 0$ corresponds to the "triangular lattice" $\{m + ne^{2\pi i/3} : m, n \in \mathbb{Z}\}$ (and $g_3 = \pm\Gamma(\frac{1}{3})^{18}/(2\pi)^6$). Thus, if we choose the $\wp$-function corresponding to this lattice, we have $f^3 + g^3 = 1$ for

$$f := \frac{a + b\wp'}{\wp}, \quad g := \frac{a - b\wp'}{\wp} \quad \text{with} \quad a := \frac{1}{2}\sqrt[3]{g_3}, \quad b := \frac{1}{\sqrt{24a}}.$$

§3. Schottky's Theorem and Consequences

The growth of holomorphic functions that omit 0 and 1 can be estimated by a *universal* bound. Let $S(r)$ denote the set of all functions $f \in \mathcal{O}(\overline{\mathbb{E}})$ with $|f(0)| \leq r$ that do not assume the values 0 and 1. We choose any constant $\beta > 0$ for which Bloch's theorem holds ($\beta = \frac{1}{12}$, for instance). In $(0,1) \times (0,\infty)$, we consider the positive function

$$L(\Theta, r) := \exp\left[\pi \exp \pi \left(3 + 2r + \frac{\Theta}{\beta(1-\Theta)}\right)\right].$$

Schottky's theorem. *For any function $f \in S(r)$,*

$$|f(z)| \le L(\Theta, r) \quad \textit{for all } z \in \mathbb{E} \textit{ with } |z| \le \Theta,\ 0 < \Theta < 1.$$

It may seem surprising at first glance, but this peculiar theorem is stronger than Picard's little theorem, as we see in Subsection 2. In Subsection 3 we use Schottky's theorem to obtain substantially sharpened versions of Montel's and Vitali's theorems. Of course, the explicit form of the bounding function $L(\Theta, r)$ is not significant; our $L(\Theta, r)$ can be considerably improved.

1. Proof of Schottky's theorem. The proof works, with the aid of Bloch's theorem, by a clever choice of the function g in Theorem 2.1. We first observe:

(*) *If $\cos \pi a = \cos \pi b$, then $b = \pm a + 2n$, $n \in \mathbb{Z}$. For every $w \in \mathbb{C}$ there exists a $v \in \mathbb{C}$ such that $\cos \pi v = w$ and $|v| \le 1 + |w|$.*

Proof. Since $\cos \pi a - \cos \pi b = -2 \sin \frac{\pi}{2}(a+b) \sin \frac{\pi}{2}(a-b)$, the first assertion is clear. To see the second, choose $v = \alpha + i\beta$ with $w = \cos \pi v$ such that $|\alpha| \le 1$. Since $|w|^2 = \cos^2 \pi\alpha + \sinh^2 \pi\beta$ and $\sinh^2 \pi\beta \ge \pi^2\beta^2$, it follows that

$$|v| = \sqrt{\alpha^2 + \beta^2} \le \sqrt{1 + |w|^2/\pi^2} \le 1 + |w|. \qquad \square$$

We quickly obtain a sharpened version of Theorem 2.1.

Theorem. *If $f \in \mathcal{O}(\overline{\mathbb{E}})$ omits the values 0 and 1, then there exists a function $g \in \mathcal{O}(\overline{\mathbb{E}})$ with the following properties:*

1) $f = \frac{1}{2}[1 + \cos \pi(\cos \pi g)]$ *and* $|g(0)| \le 3 + 2|f(0)|$.
2) $|g(z)| \le |g(0)| + \Theta/\beta(1-\Theta)$ *for all z such that* $|z| \le \Theta$, $0 < \Theta < 1$.

Proof. ad 1). First, the equation $2f - 1 = \cos \pi \widetilde{F}$ holds with $\widetilde{F} \in \mathcal{O}(\overline{\mathbb{E}})$. By (*), there exists a $b \in \mathbb{C}$ such that $\cos \pi b = 2f(0) - 1$ and $|b| \le 1 + |2f(0) - 1| \le 2 + 2|f(0)|$. It follows from (*) that $b = \pm\widetilde{F}(0) + 2k$, $k \in \mathbb{Z}$. For $F := \pm\widetilde{F} + 2k \in \mathcal{O}(\overline{\mathbb{E}})$, we now have $2f - 1 = \cos \pi F$ with $F(0) = b$. Since F omits all integer values, there exists a $\widetilde{g} \in \mathcal{O}(\overline{\mathbb{E}})$ with $F = \cos \pi \widetilde{g}$. By (*) there exists an $a \in \mathbb{C}$ such that $\cos \pi a = b$ and $|a| \le 1 + |b| \le 3 + 2|f(0)|$. Since $\cos \pi a = \cos \pi \widetilde{g}(0)$, we can pass — just as for $\widetilde{F}$ — to a function

$g = \pm\widetilde{g} + 2m$, with $g(0) = a$ and $F = \cos\pi g$. Property 1) holds for these functions g.

ad 2). By Theorem 2.1, $g(\mathbb{E})$ contains no disc of radius 1. Since $d(z, \partial E) \geq 1 - \Theta$ when $|z| \leq \Theta$, the corollary to Bloch's theorem (see the introduction to §1) implies that $\beta(1-\Theta)|g'(z)| \leq 1$, i.e. $|g'(z)| \leq 1/\beta(1-\Theta)$. Hence, for all z with $|z| \leq \Theta$,

$$g(z) - g(0) = \int_0^z g'(\zeta)d\zeta \quad \text{and} \quad \left|\int_0^z g'(\zeta)d\zeta\right| \leq \Theta/\beta(1-\Theta). \qquad \square$$

This theorem immediately yields Schottky's theorem. For all w, $|\cos w| \leq e^{|w|}$ and $\frac{1}{2}|1 + \cos w| \leq e^{|w|}$. Hence 1) and 2) imply that, for all z such that $|z| \leq \Theta$,

$$|f(z)| \leq \exp[\pi \exp(\pi|g(z)|)] \leq \exp[\pi \exp \pi(3 + 2|f(0)| + \Theta/\beta(1-\Theta))].$$

The assertion follows since $|f(0)| \leq r$. $\square$

2. Landau's sharpened form of Picard's little theorem. *There exists a positive function $R(a)$, defined on $\mathbb{C}\backslash\{0,1\}$, for which there is no function $f \in \mathcal{O}(\overline{B}_{R(a)}(0))$ such that $f(0) = a$, $f'(0) = 1$, and f omits the values 0 and 1.*

Proof. Set $R(a) := 3L(\frac{1}{2}, |a|)$. If $f(z) = a + z + \cdots \in \mathcal{O}(\overline{B}_{R(a)}(0))$ omitted the values 0 and 1, then $g(z) := f(Rz) = a + Rz + \cdots \in \mathcal{O}(\overline{\mathbb{E}})$, where $R := R(a)$, would also omit these values. By Schottky, we would have $\max\{|g(z)| : |z| \leq \frac{1}{2}\} \leq \frac{1}{3}R$. But $R \leq 2\max\{|g(z)| : |z| \leq \frac{1}{2}\}$ by Cauchy's inequalities, giving a contradiction. $\square$

Landau's theorem contains Picard's little theorem: If $f \in \mathcal{O}(\mathbb{C})$ is nonconstant, choose ζ such that $a := f(\zeta)$, $f'(\zeta) \neq 0$. Then, for $a \in \mathbb{C}\backslash\{0,1\}$,

$$h(z) := f(\zeta + z/f'(\zeta)) = a + z + \cdots \in \mathcal{O}(\mathbb{C})$$

is not always different from 0 and 1 in $\overline{B}_{R(a)}(0)$. $\square$

Landau's theorem can be proved quite easily with uniformization theory, and the best possible bounding function $R(a)$ can be given explicitly by means of the modular function $\lambda(\tau)$.

Historical note. In 1904 Landau "appended" his theorem, as an "unexpected fact," to Picard's theorem ([L$_1$], p. 130 ff). He "hesitated for a long time to publish it, as the proof seemed correct but the theorem too improbable" (*Coll. Works* 4, p. 375). His situation was similar to that of Stieltjes ten years earlier; cf. Chapter 7.3.4. — The classical form of the theorem can be found in [L$_4$] (p. 102). The precise value of the Landau radius $R(a)$ was given in 1905 by Carathéodory (*Ges. Math. Schriften* 3, pp. 6–9).

3. Sharpened forms of Montel's and Vitali's theorems. Let G be a domain in $\mathbb{C}$ and set $\mathcal{F} := \{f \in \mathcal{O}(G) : f \text{ omits the values } 0 \text{ and } 1\}$. For $w \in G$ and $r \in (0, \infty)$, let $\mathcal{F}_*$ be a subfamily of $\mathcal{F}$ such that $|g(w)| \le r$ for all $g \in \mathcal{F}_*$.

(1) *There exists a neighborhood B of w such that $\mathcal{F}_*$ is bounded in B.*

Proof. Let $\overline{B}_{2t}(w) \subset G$, $t > 0$. We may assume that $w = 0$ and $2t = 1$. By Schottky, $\sup\{|g|_{B_t(w)} : f \in \mathcal{F}_*\} \le L(\frac{1}{2}, r) < \infty$. □

We now fix a point $p \in G$ and set $\mathcal{F}_1 := \{f \in \mathcal{F} : |f(p)| \le 1\}$.

(2) *The family $\mathcal{F}_1$ is locally bounded in G.*

Proof. The set $U := \{w \in G : \mathcal{F}_1 \text{ is bounded in a neighborhood of } w\}$ is open in G; by (1), $p \in U$. If U were not equal to G, then by (1) there would exist a point $w \in \partial U \cap G$ and a sequence $f_n \in \mathcal{F}_1$ with $\lim f_n(w) = \infty$. Set $g_n := 1/f_n$; then $g_n \in \mathcal{F}$. Since $\lim g_n(w) = 0$, the family $\{g_n\}$ is bounded in a neighborhood of w by (1). By Montel (Theorem 7.1.1), there is a subsequence g_{n_k} that converges uniformly in a disc B about w to some $g \in \mathcal{O}(B)$. Since all the g_n are nonvanishing and $g(w) = 0$, it follows from Hurwitz (Corollary 7.5.1) that $g \equiv 0$. But then $\lim f_{n_k}(z) = \infty$ at points of U as well. Contradiction. □

We now generalize the concept of normal families so as to admit sequences that converge compactly to ∞ in G. The next theorem then follows from (2).

Sharpened version of Montel's theorem. *The family $\mathcal{F}$ is normal in G.*

Proof. Let f_n be a sequence in $\mathcal{F}$. If f_n has subsequences in $\mathcal{F}_1$, then the assertion is clear by (2). If only finitely many f_n lie in $\mathcal{F}_1$, then almost all $1/f_n$ lie in $\mathcal{F}_1$. Choose from these a subsequence g_n that converges compactly in G. If its limit g is nonvanishing, then the subsequence $1/g_n$ of the sequence f_n converges compactly in G to $1/g$. If g has zeros, then $g \equiv 0$ (Hurwitz) and $1/g_n$ converges compactly to ∞. □

The sharpened version of Vitali's theorem mentioned in 7.3.4 now follows immediately.

Carathéodory-Landau theorem (1911). *Let $a, b \in \mathbb{C}$, $a \neq b$, and let $f_1, f_2, \ldots$ be a sequence of holomorphic maps $G \to \mathbb{C}\backslash\{a, b\}$. Suppose that $\lim f_n(w) \in \mathbb{C}$ exists for a set of points in G that has at least one accumulation point in G. Then the sequence f_n converges compactly in G.*

Proof. We may assume that $a = 0$ and $b = 1$. The sequence $f_n \in \mathcal{F}$ must then be locally bounded in G. □

Exercises. 1) Let A and B be disjoint bounded sets in $\mathbb{C}$ with a positive distance $d(A,B)$. Then $\{f \in \mathcal{O}(G) : A \not\subset f(G) \text{ and } B \not\subset f(G)\}$ is a normal family in G. Formulate a corresponding Vitali theorem.

2) Let $m \in \mathbb{N}$. The family $\{f \in \mathcal{O}(G)$: *f never equals* 0 *and equals* 1 *at most m times in G*$\}$ is normal in G.

§4. Picard's Great Theorem

Theorem (Picard's great theorem). *Let $c \in \mathbb{C}$ be an isolated essential singularity of f. Then, in every neighborhood of c, f assumes every complex number as a value infinitely many times, with at most one exception.*

This contains a sharpened form of Picard's little theorem:

Every entire transcendental function f assumes every complex number as a value infinitely many times, with at most one exception.

Apply the theorem to $g(z) := f(1/z) \in \mathcal{O}(\mathbb{C}^\times)$.

1. Proof of Picard's great theorem. It suffices to prove:

If $f \in \mathcal{O}(\mathbb{E}^\times)$ and $0, 1 \notin f(\mathbb{E}^\times)$, then f or $1/f$ is bounded in a neighborhood of 0.

Proof. By Montel, there exists a subsequence (f_{n_k}) of the sequence $f_n(z) := f(z/n) \in \mathcal{O}(\mathbb{E}^\times)$ such that the sequence (f_{n_k}) or $(1/f_{n_k})$ is bounded on $\partial B_{\frac{1}{2}}(0)$. In the first case, $|f(z/n_k)| \le M$ for $|z| = \frac{1}{2}$ and $n_k \ge 1$ with $M \in (0,\infty)$. Hence $|f(z)| \le M$ on every circle about 0 with radius $1/(2n_k)$. By the maximum principle, $|f(z)| \le M$ on every annulus $1/(2n_{k+1}) \le |z| \le 1/(2n_k)$. Hence f is bounded in a neighborhood of 0. In the second case, it follows similarly that $1/f$ is bounded in a neighborhood of 0. □

The proof shows that the family $\{f(z/n)\}$ is not normal in $\mathbb{E}^\times$ if f has an essential singularity at 0. Then there exists (!) a $c \in \mathbb{E}^\times$ such that this family is not normal in any neighborhood of c. The next theorem follows easily from this.

Sharpened version of Picard's great theorem. *If $f \in \mathcal{O}(\mathbb{E}^\times)$ has an essential singularity at* 0, *then there exist $c \in \mathbb{E}^\times$ and $a \in \mathbb{C}$ such that, in every disc $B_{\varepsilon/n}(c/n)$, $0 < \varepsilon < |c|$, f assumes every value in $\mathbb{C}\backslash\{a\}$.*

The proof is left to the reader.

2. On the history of the theorems of this chapter. E. Picard proved his theorems in 1879 with the aid of elliptic modular functions ([P], p. 19 and p. 27); his results mark the beginning of a development which eventually culminated in the value distribution theory of R. Nevanlinna. In 1896, E. Borel derived Picard's little theorem with elementary function-theoretic tools ([Bor], p. 571). In 1904 Landau, in [L_1], through a modification of Borel's train of thought, proved among other things the existence of the "radius function" $R(a)$. In the same year, F. Schottky was able to generalize Landau's result ([Sch], p. 1258).

The theory took a surprising turn in 1924, when A. Bloch discovered the theorem named after him. As we have seen, everything now follows from this theorem.

In 1971 T. Estermann, in [E], managed to prove Picard's great theorem without resorting to Schottky's theorem.

Bibliography

[A] Ahlfors, L. V.: An extension of Schwarz's lemma, *Trans. Amer. Math. Soc.* 43, 359–364 (1938); *Coll. Papers* 1, 350–364.

[Bl_1] Bloch, A.: Les théorèmes de M. Valiron sur les fonctions entières et la théorie de l'uniformisation, *C.R. Acad. Sci. Paris* 178, 2051–2052 (1924).

[Bl_2] Bloch, A.: Les théorèmes de M. Valiron sur les fonctions entières et la théorie de l'uniformisation, *Ann. Sci. Univ. Toulouse* 17, Sér. 3, 1–22 (1925).

[Bon] Bonk, M.: On Bloch's constant, *Proc. Amer. Math. Soc.* 110, 2, 889–894 (1990).

[Bor] Borel, É.: Démonstration élémentaire d'un théorème de M. Picard sur les fonctions entières, *C.R. Acad. Sci. Paris* 122, 1045–1048 (1896); *Œuvres* 1, 571–574.

[E] Estermann, T.: Notes on Landau's proof of Picard's 'Great' Theorem, 101–106; *Studies in Pure Mathematics* presented to R. Rado, ed. L. Mirsky, Acad. Press London, New York, 1971.

[K] König, H.: Über die Landausche Verschärfung des Schottkyschen Satzes, *Arch. Math.* 8, 112–114 (1957).

[L_1] Landau, E.: Über eine Verallgemeinerung des Picardschen Satzes, *Sitz.-Ber. Königl. Preuss. Akad. Wiss. Berlin*, Jhrg. 1904, 1118–1133; *Coll. Works* 2, 129–144.

[L_2] Landau, E.: Der Picard-Schottkysche Satz und die Blochsche Konstante, *Sitz.-Ber. Königl. Preuss. Akad. Wiss. Berlin*, Jhrg. 1926, 467–474; *Coll. Works* 9, 17–24.

[L_3] Landau, E.: Über die Blochsche Konstante und zwei verwandte Weltkonstanten, *Math. Zeitschr.* 30, 608–634 (1929); *Coll. Works* 9, 75–101.

[L_4] Landau, E.: *Darstellung und Begründung einiger neuerer Ergebnisse der Funktionentheorie*, 1st ed. 1916; 2nd ed. 1929; 3rd ed. with supplements by D. Gaier, 1986, Springer, Heidelberg.

[M] Minda, C. D.: Bloch constants, *Journ. Analyse Math.* 41, 54–84 (1982).

[P] Picard, É.: *Œuvres* 1.

[Sch] SCHOTTKY, F.: Über den Picard'schen Satz und die Borel'schen Ungleichungen, *Sitz.-Ber. Königl. Preuss. Akad. Wiss. Berlin*, Jhrg. 1904, 1244–1262.

[Y] YANAGIHARA, H.: On the locally univalent Bloch constant, *Journ. Anal. Math.* 65, 1–17 (1995).

11

Boundary Behavior of Power Series

A function holomorphic in a domain is completely determined as soon as one of its Taylor series $\sum a_\nu(z-c)^\nu$ is known. Thus all the properties of the function are, in principle, stored in the sequence of coefficients a_ν. As early as 1892, J. Hadamard, in his thesis [H], considered the following problem:

What relationships are there between the coefficients of a power series and the singularities of the function it represents?

Hadamard says in this regard (loc. cit. p. 8): "Le développement de Taylor, en effet, ne met pas en évidence les propriétés de la fonction représentée et semble même les masquer complètement." (Indeed, the Taylor expansion does not reveal the properties of the function represented, and even seems to mask them completely.) Hadamard's question has led to many beautiful results; this chapter contains a selection. In presenting them, we formulate the question more narrowly:

What relationships are there between the coefficients and partial sums of a power series and the possibility that the corresponding function can be extended holomorphically or meromorphically to certain boundary points of the disc of convergence?

We discuss theorems of Fatou, Hadamard, Hurwitz, Ostrowski, Pólya, Porter, M. Riesz, and Szegö. The four sections of this chapter can be read independently of each other; the bibliography is given section by section.

§1. Convergence on the Boundary

Even if a function $f \in \mathcal{O}(\mathbb{E})$ can be extended holomorphically to a boundary point $c \in \mathbb{E}$, its Taylor series about 0 may diverge at c. The examples

$$\sum z^\nu \text{ with } c := \pm 1, \quad \sum z^\nu/\nu^2 \text{ with } c := 1, \quad \sum z^\nu/\nu \text{ with } c := -1$$

show that, in general, convergence or divergence at boundary points has nothing to do with the possibility of holomorphic extension to these points. But it was discovered in the early twenties that there are transparent relationships for special series. In Subsection 1 we present three classical theorems of Fatou, M. Riesz, and Ostrowski, which link the extension problem for a power series with the boundedness or convergence of its series of partial sums. These theorems are proved in Subsections 2 and 3; Vitali's theorem again proves helpful. In Subsection 4 we discuss Ostrowski's theorem.

If B is the disc of convergence of $f = \sum a_\nu z^\nu$, a *closed* circular arc L in ∂B is called an *arc of holomorphy* of f if f can be holomorphically extended to every point of L. We have $L \neq \partial B$, since at least one singular point of f must lie on ∂B (cf. I.8.1.5).

1. Theorems of Fatou, M. Riesz, and Ostrowski. The sequence of partial sums $s_n(z) = (1 - z^{n+1})/(1 - z)$ of the geometric series $\sum z^\nu$ is uniformly bounded on every arc of holomorphy $L \subset \partial\mathbb{E}\backslash\{1\}$. In contrast, the sequence of partial sums $t_n(z)$ of the derivative $\sum \nu z^{\nu-1}$ no longer has this property; for example, $t_{2m+1}(-1) = m+1$, $m \in \mathbb{N}$. The reason for this different behavior is that the coefficients of the series are bounded in the first case, but not in the second.

M. Riesz's boundedness theorem. *If $f = \sum a_\nu z^\nu$ is a power series with bounded sequence of coefficients, then the sequence of its partial sums $s_n := \sum_0^n a_\nu z^\nu$ is (uniformly) bounded on every arc of holomorphy L of f.*

The boundedness of the sequence of coefficients alone is not enough to guarantee the convergence of the sequence s_n on L (geometric series). But we do have the

Convergence theorem of Fatou and M. Riesz. *If $f = \sum a_\nu z^\nu$ is a power series with* $\lim a_\nu = 0$, *then its sequence of partial sums s_n converges uniformly on every arc of holomorphy L of f (to the holomorphic extension of f to L).*

This theorem has an amusing consequence:

If $\sum a_\nu z^\nu$ can be continued holomorphically to 1, *then (as in p-adic analysis)*

$$\sum a_\nu \textit{ is convergent} \Leftrightarrow \lim a_\nu = 0.$$

A power series $\sum a_\nu z^\nu$ is called a *lacunary series* if there exists a sequence $m_\nu \in \mathbb{N}$ such that

$$(*) \quad a_j = 0 \quad \text{when} \quad m_\nu < j < m_{\nu+1},\ \nu \in \mathbb{N}; \quad \lim(m_{\nu+1} - m_\nu) = \infty.$$

The method used to prove the Fatou-Riesz theorem also yields

Ostrowski's convergence theorem. *If $f = \sum a_{m_\nu} z^{m_\nu}$ is a lacunary series with bounded sequence of coefficients, then its sequence of partial sums s_{m_ν} converges uniformly on every arc of holomorphy L of f.*

The theorems just stated are proved in the next two subsections. We may assume in all cases that the power series has radius of convergence 1.

2. A lemma of M. Riesz. For every arc of holomorphy $L \subset \partial\mathbb{E}$ of a power series $f = \sum a_\nu z^\nu$ with radius of convergence 1, there exists a *compact circular sector* S with vertex at 0 such that L lies in the *interior* $\mathring{S}$ of S and f has a holomorphic extension $\widehat{f}$ to S.[1] Let z_1 and z_2 be the corners $\neq 0$ of S, and let w_1 and w_2 be the points of intersection of $\partial\mathbb{E}$ with $[0, z_1]$ and $[0, z_2]$, respectively. (See Figure 11.1.) Then $|w_1| = |w_2| = 1$ and $s := |z_1| = |z_2| > 1$.

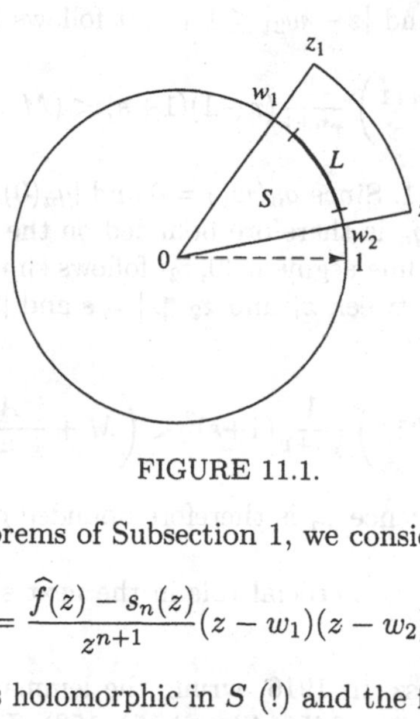

FIGURE 11.1.

To prove the theorems of Subsection 1, we consider the functions

$$g_n(z) := \frac{\widehat{f}(z) - s_n(z)}{z^{n+1}}(z - w_1)(z - w_2), \quad n \in \mathbb{N}.$$

Every function g_n is holomorphic in S (!) and the following holds.

[1]By definition, f can be extended holomorphically to *every* point of L. The reader should convince himself that this extension determines a holomorphic function in a neighborhood of L that agrees with f in $\mathbb{E}$ (one argues as in the proof of the existence of singular points on the boundary of the disc of convergence; cf. I.8.1.5).

M. Riesz's lemma. *Suppose that the power series $f = \sum a_\nu z^\nu$ (with radius of convergence 1) has a bounded sequence of coefficients and that $\widehat{f}$ is a holomorphic extension of f to S. Then the sequence g_n is bounded in S.*

Proof. Because of the maximum principle, it suffices to show that the sequence $g_n|\partial S$ is bounded. We verify this directly. Let $A := \sup|a_\nu| < \infty$, $M = |\widehat{f}|_S < \infty$. If $z = rw_1$ with $0 < r < 1$, then, first,

$$|\widehat{f}(z) - s_n(z)| = \left|\sum_{n+1}^{\infty} a_\nu z^\nu\right| \leq A(r^{n+1} + r^{n+2} + \cdots) = \frac{A}{1-r}r^{n+1}.$$

Since $|z - w_1| = 1 - r$ and $|z - w_2| < 2$, it follows that

$$|g_n(z)| \leq \frac{A}{1-r}r^{n+1}\frac{1}{r^{n+1}}(1-r)2 = 2A, \quad n \in \mathbb{N}.$$

If $z = rw_1$ with $1 < r \leq s$, then, first,

$$|\widehat{f}(z) - s_n(z)| \leq M + A(1 + r + \cdots + r^n) < M + Ar^{n+1}/(r-1).$$

Since $|z - w_1| = r - 1$ and $|z - w_2| \leq 1 + s$, it follows that

$$|g_n(z)| \leq \left(M + \frac{A}{r-1}r^{n+1}\right)\frac{1}{r^{n+1}}(r-1)(1+s) < (M+A)(1+s), \quad n \in \mathbb{N},$$

because $(r-1)/r^{n+1} < 1$. Since $g_n(w_1) = 0$ and $|g_n(0)| = |a_{n+1}w_1w_2| \leq A$ for all n, the sequence g_n is therefore bounded on the line segment $[0, z_1]$. Its boundedness on the line segment $[0, z_2]$ follows similarly.

On the circular arc between z_1 and z_2, $|z| = s$ and $|(z - w_1)(z - w_2)| \leq (1+s)^2$. Hence

$$|g_n(z)| \leq \left(M + \frac{A}{s-1}s^{n+1}\right)\frac{1}{s^{n+1}}(1+s)^2 < \left(M + \frac{A}{s-1}\right)(1+s)^2, \quad n \in \mathbb{N},$$

because $s > 1$. The sequence g_n is therefore bounded on ∂S. □

M. Riesz's lemma plays a crucial role in the next subsection as well as in 4.1.

Historical note. M. Riesz, in 1916, wrung the lemma out of older proofs of Fatou's theorem ([R], pp. 145–148 and 151–153). The trick of considering the "auxiliary sequence" g_n is based on an old idea of Riemann: the convergence (here, for the present, only boundedness) of a series is to be investigated in certain sets; the behavior of the series is improved at two auxiliary points by multiplying it by an appropriate function; cf. [R], p. 146.

3. Proof of the theorems in 1. In all three cases, S and $\widehat{f}$ can be chosen as in Subsection 2. Since $|z| = 1$ for $z \in L$, the inequality

$$(1) \quad |\widehat{f} - s_n|_L \le a^{-1}|g_n|_S, \quad \text{where} \quad a := \min_{z\in L}\{|(z-w_1)(z-w_2)|\} > 0,$$

holds for all $n \in \mathbb{N}$.

Proof of M. Riesz's boundedness theorem. By Lemma 2 there exists a $B > 0$ such that $|g_n|_S \le B$, $n \ge \mathbb{N}$. Hence it follows from (1) that $|s_n|_L \le |\widehat{f}|_L + a^{-1}B$ for all $n \in \mathbb{N}$. □

Proof of the convergence theorem of Fatou and M. Riesz. Since $L \subset \mathring{S}$, it suffices by (1) to show that the sequence g_n converges compactly to zero in $\mathring{S}$. We fix a $q \in (0,1)$. The sequence g_n is bounded in S by Lemma 2; hence, by Vitali's theorem, all we need to prove is that $\lim g_n(z) = 0$ for all z with $|z| = q$. Setting $\varepsilon_n := \sup_{\nu \ge n} |a_\nu|$, we have

$$|\widehat{f}(z) - s_n(z)| \le \sum_{n+1}^{\infty} |a_\nu| q^\nu \le \varepsilon_n q^{n+1}/(1-q), \quad |z| = q, \quad n \in \mathbb{N}.$$

Since $|(z-w_1)(z-w_2)| \le (1+q)^2$, it is clear that

$$|g_n(z)| \le \varepsilon_n \frac{q^{n+1}}{1-q}\frac{1}{q^{n+1}}(1+q)^2 = \varepsilon_n \frac{(1+q)^2}{1-q}, \quad |z| = q, \quad n \in \mathbb{N}.$$

But $\lim \varepsilon_n = 0$ since $\lim a_\nu = 0$; hence $\lim g_n(z) = 0$ if $|z| = q$. □

Proof of Ostrowski's convergence theorem. Now f is a lacunary series, and it must be shown that $\lim |\widehat{f} - s_{m_\nu}|_L = 0$. By (1), it suffices to show that the sequence g_{m_ν} tends compactly to zero in $\mathring{S}$. Again let $q \in (0,1)$. By Vitali, it suffices to show that $\lim g_{m_\nu}(z) = 0$ for $|z| = q$. Setting $A := \sup |a_{m_\nu}|$, we have

$$|\widehat{f}(z) - s_{m_\nu}(z)| \le \sum_{m_\nu+1}^{\infty} Aq^\mu = Aq^{m_{\nu+1}}/(1-q),$$

$$|g_{m_\nu}(z)| \le A\frac{q^{m_{\nu+1}}}{1-q}\frac{1}{q^{m_\nu+1}}(1+q)^2 = A\frac{(1+q)^2}{(1-q)q} q^{m_{\nu+1}-m_\nu}.$$

But $\lim q^{m_{\nu+1}-m_\nu} = 0$ since $\lim(m_{\nu+1} - m_\nu) = \infty$; hence $\lim g_{m_\nu}(z) = 0$ for all z such that $|z| = q$. □

Since the logarithm function $\log(1-z) = -\sum_1^\infty z^\nu/\nu$ can be extended holomorphically to every point $c \in \partial\mathbb{E}\backslash\{1\}$, the theorem of Fatou and F. and M. Riesz yields as a byproduct that

$\sum z^\nu/\nu$ is compactly convergent on $\partial\mathbb{E}\backslash\{1\}$.

Of course, this can also be seen in an elementary way, by means of *Abel summation*; see, for example, Exercise I.4.2.2.

Historical note. P. Fatou proved his theorem in 1906 for the case that L is a point and the sequence a_ν tends to zero like $1/\nu$ ([F], p. 389). The sharper form, for circular arcs and arbitrary sequences a_ν converging to zero, was given by M. Riesz in 1911 ([R], p. 77); he gave the elegant proof by means of Lemma 2, which also yields the boundedness theorem, in 1916 ([R], pp. 145–164). In 1921 A. Ostrowski used Vitali's theorem to prove the analogue of Fatou's theorem for lacunary series ([O], pp. 19–21).

4. A criterion for nonextendibility. *Let $\sum a_{m_\nu} z^{m_\nu}$ be a lacunary series that diverges at all points of $\partial\mathbb{E}$ and has a bounded sequence of coefficients. Then its domain of holomorphy is $\mathbb{E}$.*

This follows immediately from Ostrowski's theorem in 1.

Corollary. *Every infinite lacunary series $\sum a_{m_\nu} z^{m_\nu}$ has $\mathbb{E}$ as domain of holomorphy.*

The series $\sum z^{2^\nu}$ and $\sum z^{m_\nu}$ thus have $\partial\mathbb{E}$ as their natural boundary. We see moreover that the *theta series* $\theta(z) = 1 + 2\sum z^{\nu^2}$ has the disc $\mathbb{E}$ as domain of holomorphy. Kronecker pointed out this mathematically natural example (in contrast to the artificial examples of Weierstrass) as early as 1863; for more on this, see [Sch], p. 214, and [K], p. 118 and p. 182.

Kronecker obtains nonextendibility from classical transformation formulas for theta functions; see for instance his terse hints in [K], p. 118. One argues as follows: the "theta function"

$$\widetilde{\vartheta}(\tau) := \sum_{-\infty}^{\infty} e^{\pi i \nu^2 \tau}, \quad \tau \in \mathbb{H},$$

satisfies the (by no means obvious) transformation formulas

$$\widetilde{\vartheta}(\tau) = \frac{\zeta}{\sqrt{c\tau + d}}\widetilde{\vartheta}\left(\frac{a\tau + b}{c\tau + d}\right), \quad \text{where } \begin{pmatrix} a & b \\ c & d \end{pmatrix} \in SL(2,\mathbb{Z});\ ab,\ cd \text{ even},\ \zeta^8 = 1.$$

Let p and q be relatively prime integers with even product pq. One concludes from these formulas that as $\tau \in \mathbb{H}$ approaches the point p/q *vertically*, $\widetilde{\vartheta}$ tends to ∞ like $1/\sqrt{q\tau - p}$.

Since

$$\theta(z) = \widetilde{\vartheta}(\tau) \quad \text{with} \quad z = e^{\pi i \tau},$$

$\theta(z)$ becomes infinitely large in the course of a radial approach from $\mathbb{E}$ to any root of unity of the form $\exp(\pi i p/q)$, where p and q are as above. Since these roots of unity are dense in $\partial\mathbb{E}$, $\theta(z)$ cannot be extended holomorphically to any boundary point of $\mathbb{E}$.

Of course, the growth statement obtained here for $\theta(z)$ does not contradict the equation $\theta(z) = \prod_1^\infty (1 - z^\nu)(1 + z^\nu)(1 + z^{2\nu-1})^2$, $z \in \mathbb{E}$ (which follows immediately from 1.5*.1, (J), if one sets $z = 1$ there and writes z instead of q): the factors vanishing at the roots of unity suggest only to a cursory glance that $\theta(z)$ could tend to 0.

Bibliography for Section 1

[F] FATOU, P.: Séries trigonométriques et séries de Taylor, *Acta Math.* 30, 335–400 (1906).

[H] HADAMARD, J.: Essai sur l'étude des fonctions données par leur développement de Taylor, *Journ. Math. Pur. Appl.* 8, 4th series, 101–186 (1892); *Œuvres* 1, 7–92.

[K] KRONECKER, L.: *Theorie der einfachen und der vielfachen Integrale*, ed. E. NETTO, Teubner, Leipzig, 1894.

[O] OSTROWSKI, A.: Über eine Eigenschaft gewisser Potenzreihen mit unendlich vielen verschwindenden Koeffizienten, *Sitz. Ber. Preuss. Akad. Wiss., Math.-Phys. Kl.* 1921, 557–563; *Coll. Math. Pap.* 5, 13–21.

[R] RIESZ, M.: *Coll. Papers*, Springer, 1980.

[Sch] SCHWARZ, H. A.: Über diejenigen Fälle, in welchen die Gaussische hypergeometrische Reihe eine algebraische Function ihres vierten Elementes darstellt, *Journ. reine angew. Math.* 75, 292–335 (1873); *Ges. Math. Abh.* 2, 211–259.

§2. Theory of Overconvergence. Gap Theorem

We can get an analytic extension of our power series merely by inserting parentheses.

— M. B. Porter, 1906

If a power series $\sum a_\nu z^\nu$ has a finite radius of convergence $R > 0$, it is quite possible that sequences of sections of this series may converge compactly in domains that properly contain $B_R(0)$. This phenomenon, called *overconvergence*, is based on the fact that divergent series can become convergent through the insertion of parentheses. A simple example is given in Subsection 1.

There is a close relationship between overconvergence and gaps in the sequence of exponents of the power series. Ostrowski's overconvergence theorem, in Subsection 2, deals with this. A simple corollary is Hadamard's gap theorem in Subsection 3. In Subsection 4 we describe an elegant procedure for constructing overconvergent power series.

1. Overconvergent power series. A power series $\sum a_\nu z^\nu$ with finite radius of convergence $R > 0$ is called *overconvergent* if there exists a *sequence of sections*

$$s_{m_k}(z) := \sum_0^{m_k} a_\nu z^\nu \quad \text{with} \quad m_0 < m_1 < \cdots < m_k < \cdots$$

that converges compactly in a domain that *properly* contains $B_R(0)$. Probably the simplest example is due to Ostrowski ([O], 1926, p. 160). He starts with the polynomial series

$$(1) \qquad \sum_{\nu=0}^{\infty} d_\nu [z(1-z)]^{4^\nu}, \quad d_\nu^{-1} := \max_{0 \le j \le 4^\nu} \binom{4^\nu}{j};$$

clearly d_ν^{-1} is the coefficient of the polynomial $[z(1-z)]^{4^\nu}$ with greatest absolute value. Since $[z(1-z)]^{4^\nu}$ contains only those terms cz^j for which $4^\nu \le j \le 2 \cdot 4^\nu$, successive addition of these polynomials yields a *formal* power series

$$\sum a_\nu z^\nu \quad \text{with} \quad s_{2\cdot 4^k}(z) = \sum_0^{2\cdot 4^k} a_\nu z^\nu = \sum_0^k d_\nu [z(1-z)]^{4^\nu}, \quad k \in \mathbb{N}.$$

Proposition. *The series $\sum a_\nu z^\nu$ is overconvergent: Its radius of convergence is* 1, *but the sequence of sections $s_{2\cdot 4^k}(z)$ converges compactly in the Cassini domain*

$$W := \{z \in \mathbb{C} : |z(z-1)| < 2\} \supset (\overline{\mathbb{E}} \backslash \{-1\}) \cup (\overline{B_1(1)} \backslash \{2\})$$

(see the right-hand part of Figure 5.2, p. 122).

Proof. By the definition of d_ν in (1), all the coefficients in $d_\nu[z(1-z)]^{4^\nu}$ have absolute value ≤ 1 and equality holds at least once. Hence $|a_\nu| \le 1$ for all $\nu \in \mathbb{N}$, with equality occurring infinitely many times. It follows that $\overline{\lim} \sqrt[\nu]{|a_\nu|} = 1$.

Since $\lim \sqrt[4^\nu]{d_\nu} = \frac{1}{2}$,[2] the power series $\sum d_\nu w^{4^\nu}$ has radius of convergence 2. Hence the sequence $s_{2\cdot 4^k}(z)$ converges compactly in W. □

The only singular point of $\sum a_\nu z^\nu$ on $\partial\mathbb{E}$ is -1. The sequence of sections $s_{2\cdot 4^k}(z)$ thus converges compactly in a neighborhood of the set of all the *boundary points that are not singular.* We now turn to the general theorem hidden behind this insight.

2. Ostrowski's overconvergence theorem. A power series $f = \sum a_\nu z^\nu$ is called an *Ostrowski series* if there exist a $\delta > 0$ and two sequences $m_0, m_1, \ldots$ and $n_0, n_1, \ldots$ in $\mathbb{N}$ such that

---

[2] Since $d_\nu^{-1} = \max_{0 \le j \le 4^\nu} \binom{4^\nu}{j}$, we have $2^{-4^\nu} \le d_\nu \le (4^\nu + 1)2^{-4^\nu}$. Indeed, for the largest number $\binom{m}{j}$ among all binomial coefficients $\binom{m}{\mu}$, we have $\frac{1}{m+1} 2^m \le \binom{m}{j} \le 2^m$, as can immediately be deduced from the equation $(1+1)^m = \binom{m}{0} + \binom{m}{1} + \cdots + \binom{m}{m}$.

a) $0 \le m_0 < n_0 \le m_1 < n_1 \le \cdots \le m_\nu < n_\nu \le m_{\nu+1} < \cdots$; $n_\nu - m_\nu > \delta m_\nu$, $\nu \in \mathbb{N}$;

b) $a_j = 0$ if $m_\nu < j < n_\nu$, $\nu \in \mathbb{N}$.

Such series thus have *infinitely* many gaps (between m_ν and n_ν), which grow *uniformly;* between two successive gaps, however, there may be arbitrarily long (finite) sections without gaps (between n_ν and $m_{\nu+1}$). Thus Ostrowski series are not necessarily lacunary series in the sense of 1.1. The series in 1 is an Ostrowski series with $m_\nu = 2 \cdot 4^\nu$, $n_\nu = 4^{\nu+1}$, and, for example, $\delta = 0.9$.

Ostrowski's overconvergence theorem. *Let $f = \sum a_\nu z^\nu$ be an Ostrowski series with radius of convergence $R > 0$, and let $A \subset \partial B_R(0)$ denote the set of all the boundary points f that are not singular. Then the sequence of sections $s_{m_k}(z) = \sum_0^{m_k} a_\nu z^\nu$ converges compactly in a neighborhood of $B_R(0) \cup A$.*

Proof (following Estermann [E]). Let $R = 1$ and let $c \in \partial\mathbb{E}$. We introduce the polynomial

$$q(w) := \frac{1}{2}c(w^p + w^{p+1}), \quad \text{where} \quad p \in \mathbb{N} \quad \text{and} \quad p \ge \delta^{-1},$$

and consider the following function, which is holomorphic in $q^{-1}(\mathbb{E}) = \{w \in \mathbb{C} : |q(w)| < 1\}$:

$$g(w) := f(q(w)) = \sum a_\nu q(w)^\nu \quad \text{(Porter-Estermann trick).}$$

We denote by $\sum b_\nu w^\nu$ the Taylor series of g about $0 \in q^{-1}(\mathbb{E})$ and by $s_n(z)$ and $t_n(z)$ the nth partial sums of $\sum a_\nu z^\nu$ and $\sum b_\nu w^\nu$, respectively, and claim that

$$(*) \qquad t_{(p+1)m_k}(w) = s_{m_k}(q(w)) \quad \text{for all} \quad w \in \mathbb{C},\ k \in \mathbb{N}.$$

By the Weierstrass double series theorem I.8.4.2, $\sum b_\nu w^\nu$ comes from the series $\sum a_\nu q(w)^\nu$ by multiplying out the polynomials $q(w)^\nu$ and grouping the resulting series $\sum a_\nu(\ldots)$ according to powers of w. The polynomial $s_{m_k}(q(w))$ has degree $\le (p+1)m_k$. By b), each polynomial $a_\mu q(w)^\mu$, $\mu > m_k$, contains only monomials aw^j with $j \ge pn_k$. Since $pn_k > pm_k + p\delta m_k \ge (p+1)m_k$ by a) and since $p \ge \delta^{-1}$, no such polynomial contributes to the partial sum $t_{(p+1)m_k}(w)$ (which is a polynomial of degree $\le (p+1)m_k$). $(*)$ follows.

After these technical preliminaries, the proof concludes elegantly. We have $q^{-1}(\mathbb{E}) \supset \overline{\mathbb{E}}\backslash\{1\}$, since $|1 + w| < 2$ and hence $|q(w)| < 1$ for all $w \in \overline{\mathbb{E}}\backslash\{1\}$. The function $g = f \circ q \in \mathcal{O}(q^{-1}(\mathbb{E}))$ is thus holomorphic at every point of $\overline{\mathbb{E}}\backslash\{1\}$. Now, if $c \in A$, then g is also holomorphic at 1 since

$q(1) = c$. The Taylor series $\sum b_\nu w^\nu$ of g and a fortiori the sequence of sections $t_{(p+1)m_k}(w)$ then converge in an open disc $B \supset \overline{\mathbb{E}}$. By $(*)$, the sequence $s_{m_k}(z)$ now converges compactly in $q(B)$. Since $q(B)$ is a domain containing $c = q(1)$, the sequence $s_{m_k}(z)$ thus converges compactly in a neighborhood of any point $c \in A$. □

3. Hadamard's gap theorem. A power series $\sum a_\nu z^\nu$ is called an *Hadamard lacunary series* if there exist a $\delta > 0$ and a sequence $m_0, m_1, \ldots$ in $\mathbb{N}$ such that

$$\text{(L)} \quad m_{\nu+1} - m_\nu > \delta m_\nu,\ \nu \in \mathbb{N}; \quad a_j = 0 \text{ if } m_\nu < j < m_{\nu+1}, \quad a_{m_\nu} \neq 0.$$

Every Hadamard lacunary series is a lacunary series in the sense of 1.1 and also an Ostrowski series (with $n_\nu := m_{\nu+1}$). The converse is not true. For lacunary series as in 1.1, only $\lim(m_{\nu+1} - m_\nu) = \infty$ is required; for Ostrowski series, gaps need appear only "here and there"; but (L) requires that a gap lie between any two successive terms that actually appear.

Hadamard's gap theorem. *Every Hadamard lacunary series $f = \sum a_\nu z^\nu$ with radius of convergence $R > 0$ has the disc $B_R(0)$ as domain of holomorphy.*

Proof. The sequence of partial sums $s_n(z)$ is the sequence $s_{m_k}(z)$ (whose terms, however, appear repeatedly). The sequence s_{m_k} thus diverges at every point $\zeta \notin \overline{B_R(0)}$. Hence, by the overconvergence theorem, all the points of $\partial B_R(0)$ are singular points of f. □

The gap theorem is in some sense a paradox: power series that, because of their gaps, converge especially fast in the interior of their disc of convergence have singularities almost everywhere on the boundary precisely on account of these gaps.

Hadamard's gap theorem is both broader and narrower than Ostrowski's gap theorem 1.1: broader, because the sequence a_ν need not be bounded; narrower, because Ostrowski gets by with a weaker gap condition than (L). Hadamard's theorem shows once again that $\mathbb{E}$ is the domain of holomorphy of $\sum z^{2^\nu}$ and $\sum z^{\nu!}$, but is not strong enough to show that the same is true for the theta series $1 + 2\sum z^{\nu^2}$.

An instructive example. *The series $f(z) = 1 + 2z + \sum b_\nu z^{2^\nu}$, with $b_\nu := 2^{-\nu^2}$, defines a function that is one-to-one and continuous on $\overline{\mathbb{E}}$ and holomorphic in $\mathbb{E}$. This function is real differentiable to arbitrarily high order at every point of the circle $\partial\mathbb{E}$, but cannot be continued holomorphically to any point of $\partial\mathbb{E}$.*

Proof. Since $\lim(\nu^2/2^\nu) = 0$, we have $\sqrt[2^\nu]{|b_\nu|} = 1$. Since $2^{\nu k} b_\nu \le 2^{-\nu}$ for $\nu > k$, the sequence and all its derivatives converge uniformly in $\overline{\mathbb{E}}$; hence

f is differentiable to arbitrarily high order on $\overline{\mathbb{E}}$. For all $w, z \in \overline{\mathbb{E}}$, $w \neq z$,

$$\begin{aligned} \left|\frac{f(w)-f(z)}{w-z}\right| &= \left|2+\sum_1^\infty b_\nu(w^{2^\nu-1}+w^{2^\nu-2}z+\cdots+z^{2^\nu-1})\right| \\ &\geq 2-\sum_1^\infty b_\nu 2^\nu = 2-\sum_1^\infty \frac{1}{2^{\nu(\nu-1)}} > 0; \end{aligned}$$

hence $f(w) \neq f(z)$. Thus $\mathbb{E}$ is the domain of holomorphy of f by the gap theorem. □

The surprising thing about this example is that singularities on $\partial\mathbb{E}$ are quite compatible with "smooth and bijective" mapping behavior of the function there. For experts in Riemann mapping theory, however, nothing sensational is going on; all we have done is explicitly give a biholomorphic map $\mathbb{E} \xrightarrow{\sim} G$ that can be extended to a C^∞ diffeomorphism $\overline{\mathbb{E}} \to \overline{G}$. But ∂G is an infinitely differentiable closed path that is not real analytic anywhere, since $f|\partial\mathbb{E}$ cannot be real analytic anywhere (this follows immediately from the Schwarz reflection principle, which is not discussed in this book).

4. Porter's construction of overconvergent series. Let the following be chosen in some way:

- a polynomial $q \neq 0$ of degree d with $q(0) = 0$ that has at least one zero $\neq 0$;
- a lacunary series $f = a_{m_\nu} z^{m_\nu}$ with $m_{\nu+1} > dm_\nu$ and radius of convergence $R \in (0, \infty)$.

Set

$$\begin{aligned} g(z) &:= f(q(z)) = \sum a_{m_\nu} q(z)^{m_\nu}, \\ V &:= \{z \in \mathbb{C} : |q(z)| < R\}, \\ r &:= d(0, \partial V) \in (0, \infty). \end{aligned}$$

Proposition. *The Taylor series $\sum b_\nu z^\nu$ of $g \in \mathcal{O}(V)$ about $0 \in V$ is overconvergent: It has radius of convergence r and its sequence of sections $t_{dm_k}(z) = \sum_0^{dm_k} b_\nu z^\nu$ converges compactly in V. The component $\widehat{V}$ of V containing 0 properly contains $B_r(0)$ and is the domain of holomorphy of $g|\widehat{V}$.*

Proof. The key (as in the overconvergence theorem) is the equation

$$(*) \qquad t_{dm_k}(z) = \sum_{\nu=0}^{k} a_{m_\nu} q(z)^{m_\nu}, \quad k \in \mathbb{N},$$

which follows since $t_{dm_k}(z)$ is a polynomial of degree $\leq dm_k$ and $q(z)^{m_{k+1}}$ has only terms az^j with $j \geq m_{k+1} > dm_k$. The compact convergence of the sequence t_{dm_k} in V is clear from $(*)$.

The Taylor series $\sum b_\nu z^\nu$ of $g \in \mathcal{O}(V)$ converges in $B_r(0) \subset V$. If its radius of convergence were greater than $r = d(0, \partial V)$, there would exist points $v \notin \overline{V}$ such that $\sum b_\nu v^\nu$ converged. Then, by $(*)$, $\sum_0^\infty a_{m_\nu} q(v)^{m_\nu}$ would be convergent. But this is impossible because $|q(v)| > R$. Hence r is the radius of convergence of $\sum b_\nu z^\nu$.

We have $\widehat{V} \supset B_r(0)$ (trivially) and $\widehat{V} \neq B_r(0)$ (by Corollary 9.3.2, since q has distinct zeros). By the gap theorem, $B_R(0)$ is the domain of holomorphy of f; hence, by Theorem 5.3.2, $\widehat{V}$ is the domain of holomorphy of $g = f \circ q$. □

Ostrowski's example in 1 is a special case of the proposition just proved.

5. On the history of the gap theorem. The phenomenon of the existence of power series with natural boundaries, discovered by Weierstrass and Kronecker in the 1860s, was given a natural explanation in 1892 by Hadamard. He proves the gap theorem in [Ha] (p. 72 ff.); a simpler proof was given in 1921 by Szegö ([Sz], pp. 566–568). In 1927 J. L. Mordell, substituting polynomials $w^p(1+w)$, gave a particularly elegant argument [M]. M. B. Porter, however, had already had this beautiful idea in 1906, when he gave the construction in 4 and, in passing, proved the gap theorem for the case $m_{\nu+1} > 2m_\nu$ by means of the substitution $w(1+w)$ ([Por], pp. 191–192). Porter's work remained unnoticed until 1928; see the next subsection. Ostrowski, in 1921, saw Hadamard's theorem as a corollary of his overconvergence theorem ([O], p. 150).

In 1890 the Swedish mathematician I. Fredholm, a student of Mittag-Leffler, pointed out an example like that in 3. For fixed a, $0 < |a| < 1$, he considers the power series

$$g(z) = \sum a^\nu z^{\nu^2} = 1 + az + a^2 z^4 + a^3 z^9 + \cdots;$$

cf. [Fr]. Because $\lim \sqrt[\nu^2]{|a|^\nu} = 1$, Fabry's gap theorem (see Subsection 7) shows that $\mathbb{E}$ is the domain of holomorphy of g. Since $\sum \nu^{2k}|a|^\nu < \infty$ for every k and since $\sum_{\nu=2}^\infty \nu^2 |a|^\nu < |a|$ for small a, g also has the properties that were shown in 3 to hold for f.

In a letter to Poincaré in 1891, Mittag-Leffler called Fredholm's construction "un résultat assez remarquable"; cf. *Acta Math.* 15, 279–280 (1891). Fredholm's example was also discussed by Hurwitz in 1897 ([Hu], p. 478).

Lacunary series, in the form of Fourier series, appear early in real analysis. Weierstrass reports in 1872 ([W], p. 71) that Riemann, "in 1861 or perhaps even earlier," had presented the lacunary series

$$\sum_1^\infty \frac{\sin n^2 x}{n^2}$$

to his students as an example of a nowhere-differentiable function on $\mathbb{R}$. "Leider ist der Beweis hierfür von Riemann nicht veröffentlicht worden und scheint sich auch nicht in seinen Papieren oder durch mündliche Überlieferung erhalten zu haben." (Unfortunately, Riemann's proof of this was not published, nor does it seem to have survived in his papers or to have been passed down orally.) It is now known that Riemann's function is differentiable only at the points $\pi(2p+1)/(2q+1)$, $p, q \in \mathbb{Z}$, and that its derivative there is always $-\frac{1}{2}$ (cf. [Ge] and [Sm]).[3]

Since Weierstrass was unable to prove Riemann's claim, he gave his famous series

$$\sum b^n \cos(a^n x\pi), \quad a \geq 3,\ a \text{ odd},\ 0 < b < a, \quad ab > 1 + \frac{3}{2}\pi,$$

in 1872 as a simple example of a continuous nowhere-differentiable function ([W], pp. 72–74).

6. On the history of overconvergence. In 1921, Erhard Schmidt presented the paper [O], pp. 13–21, to the Prussian Academy; in this paper A. Ostrowski proves his overconvergence theorem with the aid of Hadamard's three-circle theorem. Ostrowski wrote at that time (footnote 2) on p. 14) that R. Jentzsch had discovered overconvergence in 1917 ([J], p. 255 and pp. 265–270). Ostrowski's theorem attracted attention at once. Ostrowski extended his result in several papers (cf. [O], pp. 159–172, and the bibliography given there on p. 159); the "very elegantly constructed examples" of Jentzsch were highly regarded until 1928.

But it had escaped the notice of interested mathematicians that M. B. Porter had already clearly described the phenomenon of overconvergence in 1906. Porter's examples $\sum a_{m_\nu}[z(1+z)]^{m_\nu}$ ([Por], pp. 191–192), which we discussed in 4, are more natural than the "rather artificially constructed" examples of Jentzsch. Porter's series — surprisingly, this too remained hidden from the experts — were also studied in the same year, 1906, by G. Faber in Munich, but Faber did not particularly emphasize the property of overconvergence. Porter's examples were rediscovered by E. Goursat, who discussed them in the fourth edition of his *Cours d'analyse* (vol. 2, p. 284). All this first became known in 1928, when Ostrowski published an addendum; cf. [O], p. 172.

Ostrowski's proof of the overconvergence theorem is complicated. In 1932 T. Estermann, in [E], saw that Porter's trick of exploiting the polynomials $w^p(1+w)$ leads to a direct proof.

7. Glimpses. The sharpest nonextendibility theorem that contains both Hadamard's gap theorem and criterion 1.4 was already discovered in 1899 by E. Fabry (1856–1944). A series $\sum a_\nu z^{m_\nu}$ is called a *Fabry series* if $\lim m_\nu/\nu = \infty$. We have the following deep result.

---

[3] Riemann's example is also discussed in the article "Riemann's example of a continuous 'nondifferentiable' function," by E. Neuenschwander, in *Math. Int.* 1, 40–44 (1978), which is considerably supplemented by S. L. Segal in the same volume, pp. 81–82.

Fabry's gap theorem. *If $f = \sum a_\nu z^{m_\nu}$ is a Fabry series with radius of convergence R, then the disc $B_R(0)$ is the domain of holomorphy of f.*

Proofs can be found in [L] (pp. 76–84) and in [D] (pp. 127–133). Fabry, incidentally, stated his theorem only for lacunary series as defined in 1.1 ([Fabry], p. 382); the formulation given here first appeared in 1906 in [Faber] (p. 581). The following exercise shows that this version is sharper.

Exercise. Prove that every general lacunary series is a Fabry series. Give examples of Fabry series that are not lacunary.

Pólya noticed in 1939 that the converse of Fabry's theorem is true; he shows ([Pól], p. 698):

Let m_ν be a sequence of natural numbers with $m_0 < m_1 < \cdots$. Suppose that every series $\sum a_\nu z^{m_\nu}$ has its disc of convergence as domain of holomorphy. Then $\lim m_\nu/\nu = \infty$.

Bibliography for Section 2

[D] DINGHAS, A.: *Vorlesungen über Funktionentheorie*, Grundlehren, Springer, 1961.

[E] ESTERMANN, T: On Ostrowski's gap theorem, *Journ. London Math. Soc.* 6, 19–20 (1932).

[Faber] FABER, G.: Über Potenzreihen mit unendlich vielen verschwindenden Koeffizienten, *Sitz. Ber. Königl. Bayer. Akad. Wiss., Math.-Phys. Kl.* 36, 581–583 (1906).

[Fabry] FABRY, E.: Sur les points singuliers d'une fonction donnée par son développement en série et l'impossibilité du prolongement analytique dans les cas très généraux, *Ann. Sci. Éc. Norm. Sup.* 13, 3. Sér., 367–399 (1896).

[Fr] FREDHOLM, I.: Om en speciell klass of singulära linier, *Öfversigt K. Vetenskaps-Akad. Förhandl.* 3, 131–134 (1890).

[Ge] GERVER, J.: The differentiability of the Riemann function at certain multiples of π, *Am. Journ. Math.* 92, 33–35 (1970) and : More on the differentiability of the Riemann function at certain multiples of π, *Am. Journn. Math.* 93, 33–41 (1971).

[Ha] HADAMARD, J.: Essai sur l'étude des fonctions données par leur développement de Taylor, *Journ. Math. Pur. Appl.* 8, 4. Sér., 101–186 (1892), *Œuvres* 1, 7–92.

[Hu] HURWITZ, A.: Über die Entwicklung der allgemeinen Theorie der analytischen Funktionen in neuerer Zeit, *Proc. of the 1st International Congress of Mathematicians*, Zurich 1897, Leipzig 1898, 91–112; *Math. Werke* 1, 461–480.

[J] JENTZSCH, R.: Fortgesetzte Untersuchungen über Abschnitte von Potenzreihen, *Acta Math.* 41, 253–270 (1918).

[L] LANDAU, E.: *Darstellung und Begründung einiger neuerer Ergebnisse der Funktionentheorie*, 2nd ed., Julius Springer, 1929; 3rd ed., with supplements by D. GAIER, 1986.

[M] MORDELL, L. J.: On power series with the circle of convergence as a line of essential singularities, *Journ. London Math. Soc.* 2, 146–148 (1927).

[O] OSTROWSKI, A.: *Coll. Math. Pap.* 5.

[Pól] PÓLYA, G.: Sur les séries entières lacunaires non prolongeables, *C.R. Acad. Sci. Paris* 208, 709–711 (1939); *Coll. Pap.* 1, 698–700.

[Por] PORTER, M. B.: On the polynomial convergents of a power series, *Ann. Math.* 8, 189–192 (1906).

[Sm] SMITH, A.: The differentiability of Riemann's function, *Proc. Amer. Math. Soc.* 34, 463–468 (1972).

[Sz] SZEGÖ, G.: Tschebyscheffsche Polynome und nichtfortsetzbare Potenz-reihen, *Math. Ann.* 87, 90–111 (1922); *Coll. Pap.* 1, 563–586.

[W] WEIERSTRASS, K.: Über continuirliche Functionen eines reellen Arguments. die für keinen Werth des letzteren einen bestimmten Differentialquotienten besitzen, *Math. Werke* 2, 71–74.

§3. A Theorem of Fatou-Hurwitz-Pólya

> Für eine beliebige Potenzreihe läßt sich der Konvergenzkreis zur natürlichen Grenze machen, bloß durch geeignete Änderung der Vorzeichen der Koeffizienten. (For an arbitrary power series, the circle of convergence can be turned into the natural boundary just by a suitable change in the signs of the coefficients.)
>
> —G. Pólya, 1916

The natural boundary of Hadamard lacunary series is their circle of convergence. This knowledge now leads to the surprising insight that such series are by no means necessary to specify uncountably many functions with discs as their domain of holomorphy. We will prove:

Theorem (Fatou-Hurwitz-Pólya). *Let B be the disc of convergence of the power series $f = \sum a_\nu z^\nu$. Then the set of all functions of the form $\sum \varepsilon_\nu a_\nu z^\nu$, $\varepsilon_\nu \in \{-1, 1\}$, whose domain of holomorphy is B has the cardinality of the continuum.*[4]

---

[4]It is known that the set of all sequences $\varepsilon : \mathbb{N} \to \{+1, -1\}$ has the cardinality of the continuum (binary number system); hence, in any case, there exist "continuously many" functions of the form $\sum \varepsilon_\nu a_\nu z^\nu$, $\varepsilon = \pm 1$.

There is something paradoxical about this lovely theorem: Although there do exist conditions on the absolute values of the coefficients that guarantee nonextendibility (Hadamard gaps, for example), there is *no* condition that refers only to the absolute values of the coefficients and implies extendibility.

The theorem does *not* assert that there exist at most countably many functions $\sum \varepsilon_\nu a_\nu z^\nu$, $\varepsilon = \pm 1$, whose domain of holomorphy is not the disc B. But F. Hausdorff showed that this always does occur if $\overline{\lim} \sqrt[\nu]{|a_\nu|} = \underline{\lim} \sqrt[\nu]{|a_\nu|}$ (cf. [H], p. 103).

1. Hurwitz's proof. We may assume that $B = \mathbb{E}$. Then $\overline{\lim} \sqrt[\nu]{|a_\nu|} = 1$ and there exists a subseries $h = \sum a_{m_\nu} z^{m_\nu}$ of f such that $m_{\nu+1} > 2m_\nu$ and $\lim \sqrt[m_\nu]{|a_{m_\nu}|} = 1$. From this Hadamard lacunary series $h \in \mathcal{O}(\mathbb{E})$, we construct infinitely many series $h_n \in \mathcal{O}(\mathbb{E})$, $n \in \mathbb{N}$, such that none of the series is finite and every term $a_{m_\nu} z^{m_\nu}$ appears in exactly one of them. Then

$$h = h_0 + h_1 + h_2 + \cdots \text{ in } \mathbb{E} \text{ (normal convergence of power series).}$$

We set $g := f - h$ and assign to every sequence $\eta : \mathbb{N} \to \{+1, -1\}$, $\nu \mapsto \eta_\nu$, the series

$$f_\eta := g + \eta_0 h_0 + \eta_1 h_1 + \cdots + \eta_n h_n + \cdots \in \mathcal{O}(\mathbb{E}).$$

By normal convergence, the Taylor series for every function f_η about 0 has the form $\sum \varepsilon_\nu a_\nu z^\nu$, $\varepsilon_\nu = \pm 1$. Thus it suffices to show that at most countably infinitely many functions f_η do not have the unit disc $\mathbb{E}$ as domain of holomorphy. If this were not true, there would exist an uncountable set of sequences δ such that every function f_δ could be extended holomorphically to a root of unity. Since the set of all roots of unity is *countable*, there would thus exist two *distinct* sequences δ and δ' such that f_δ and $f_{\delta'}$ could be extended holomorphically to the *same* root of unity. Then

$$f_\delta - f_{\delta'} = \alpha_0 h_0 + \alpha_1 h_1 + \cdots, \quad \text{where} \quad \alpha_\nu = \delta_\nu - \delta'_\nu \in \{-2, 0, 2\},$$

would not have the unit disc as domain of holomorphy. But since not all the α_ν vanish (because $\delta \neq \delta'$), and since by construction all the h_n are *infinite series*, the Taylor series $\sum b_\nu z^\nu$ of $f_\delta - f_{\delta'} \in \mathcal{O}(\mathbb{E})$ about 0 is an Hadamard lacunary series (as a subseries of such a series). Moreover, $\overline{\lim} \sqrt[\nu]{|b_\nu|} = 1$ since $\lim \sqrt[m_\nu]{|a_{m_\nu}|}$. By Theorem 1, $\mathbb{E}$ is the domain of holomorphy of $f_\delta - f_{\delta'}$. Contradiction! □

Historical note. P. Fatou conjectured the theorem in 1906 ([F], p. 400) and proved it for the case $\lim a_\nu = 0$ and $\sum |a_\nu| = \infty$; he wrote at that time: "Il est infiniment probable, que cela a lieu dans tous les cas." (It is extremely probable that this occurs in all cases.) A. Hurwitz and G. Pólya gave different proofs of the full theorem in 1916 [HP].

2. Glimpses. In 1896, E. Fabry and E. Borel already thought that *almost all* power series are singular at *all* points of their circle of convergence, hence that holomorphic extendibility at certain boundary points is the exception. Borel saw in this a problem of probabilities. In 1929, H. Steinhaus made these notions precise and proved [St]:

Suppose that the power series $\sum a_n z^n$ has radius of convergence 1. *Furthermore, let $(\varphi_n)_{n\geq 0}$ be a sequence of independent random numbers that are uniformly distributed in the interval* $[0,1]$. *Then, with probability* 1, *the power series $\sum a_n e^{2\pi i \varphi_n} z^n$ has the unit disc as domain of holomorphy (in other words, the set of sequences $(\varphi_n)_{n\geq 0} \in [0,1]^{\mathbb{N}}$ for which the series can somewhere be extended holomorphically beyond $\partial\mathbb{E}$ has measure zero).*

In 1929, it was not at all clear what "probability" and "independent random numbers" meant mathematically, and Steinhaus first had to make these concepts precise. He did so by constructing a *product measure* on the infinite-dimensional unit cube $[0,1]^{\mathbb{N}}$ with the (bounded) Lebesgue measure on each factor $[0,1]$.

A very transparent proof of Steinhaus's theorem is due to H. Boerner [Bo]; in 1938 he gave the theorem the following suggestive form:

Almost all power series have their circle of convergence as natural boundary (here "almost all" means *"all up to a set of measure zero"* in $[0,1]^{\mathbb{N}}$).

But one can also interpret the concept "almost all" topologically and ask whether a corresponding precise formulation of the notions of Fabry and Borel is possible. This idea was successfully worked out by Pólya in 1918, in [P]. He showed that there is a *natural topology* on the *space of all power series* with radius of convergence 1 such that the set of nowhere-extendible power series is open and dense in this topological space and hence, in this sense, contains "almost all" the power series of the space in question.

Thus the domain of holomorphy of a "general power series" is always its disc of convergence. Further results from this circle of ideas can be found in [Bi] (pp. 91–104).

Bibliography for Section 3

[Bi] BIEBERBACH, L.: *Analytische Fortsetzung*, Erg. Math. Grenzgeb. 3, Springer, 1965.

[Bo] BOERNER, H.: Über die Häufigkeit der nicht analytisch fortsetzbaren Potenzreihen, *Sitz. Ber. Bayer. Akad. Wiss., Math.-Nat. Abt.* 1938, 165–174.

[F] FATOU, P.: Séries trigonométriques et séries de Taylor, *Acta Math.* 30, 335–400 (1906).

[H] HAUSDORFF, F.: Zur Verteilung der fortsetzbaren Potenzreihen, *Math. Zeitschr.* 4, 98–103 (1919).

[HP] HURWITZ, A. and G. PÓLYA: Zwei Beweise eines von Herrn Fatou vermuteten Satzes, *Acta Math.* 40, 179-183 (1917); also in HURWITZ's *Math. Werke* 1, 731–734, and in PÓLYA's *Coll. Pap.* 1, 17–21.

[P] PÓLYA, G.: Über die Potenzreihen, deren Konvergenzkreis natürliche Grenze ist, *Acta Math.* 41, 99–118 (1918); Coll. Pap. 1, 64–83.

[St] STEINHAUS, H.: Über die Wahrscheinlichkeit dafür, daß der Konvergenzkreis einer Potenzreihe ihre natürliche Grenze ist, *Math. Zeitschr.* 31, 408–416 (1930).

§4. An Extension Theorem of Szegö

Geometric series $\sum z^{m\nu}$, $m \geq 1$ fixed, and Hadamard lacunary series $\sum z^{m_\nu}$ have radius of convergence 1, but the corresponding holomorphic functions behave completely differently as the boundary $\partial\mathbb{E}$ is approached: although the former can be extended to rational functions with poles at the roots of unity, the latter have $\mathbb{E}$ as domain of holomorphy. This situation is significant for power series with only finitely many distinct coefficients.

Szegö's theorem. *Let $f = \sum a_\nu z^\nu$ be a power series with only finitely many distinct coefficients. Then either $\mathbb{E}$ is the domain of holomorphy of f or f can be extended to a rational function $\widehat{f}(z) = p(z)/(1 - z^k)$, where $p(z) \in \mathbb{C}[z]$ and $k \in \mathbb{N}$.*

This beautiful theorem is proved and discussed in this section. We may assume that f is not a polynomial. Then $\overline{\lim} \sqrt[\nu]{|a_\nu|} = 1$ and the series therefore has radius of convergence 1. It suffices to prove the following:

(Sz) *If $\mathbb{E}$ is not the domain of holomorphy of f, then from some coefficient on the coefficients are periodic; that is, there exist indices λ and μ, $\lambda < \mu$, satisfying*

$$a_{\lambda+j} = a_{\mu+j} \quad \textit{for all} \quad j \in \mathbb{N}.$$

Setting $P := \sum_0^{\lambda-1} a_\nu z^\nu$ and $Q := \sum_\lambda^{\mu-1} a_\nu z^\nu$, we then have

$$f = P + Q + Qz^{\mu-\lambda} + Qz^{2(\mu-\lambda)} + \cdots = P + Q/(1 - z^{\mu-\lambda}), \quad z \in \mathbb{E}.$$

Preliminaries for the proof of (Sz) are given in Subsection 1; Runge's little theorem (12.2.1) is used there. In Subsection 2 we prove a lemma that will yield, in Subsection 3, a surprising proof of (Sz). In Subsection 4, as an application of Szegö's theorem, we characterize the roots of unity (Kronecker's theorem).

1. Preliminaries for the proof of (Sz). Let φ, ψ, $s \in \mathbb{R}$ be prescribed numbers with $0 < \psi - \varphi < 2\pi$, $s > 1$; let $\delta \in [0, 1)$ be a variable. We denote by G_δ a star-shaped domain with center 0 whose boundary $\Gamma(\delta)$ consists of two concentric circular arcs $\gamma_1(t) = se^{it}$, $\varphi \leq t \leq \psi$, and $\gamma_3(t) = (1 - \delta)e^{it}$, $\psi \leq t \leq 2\pi + \varphi$, and the two line segments $\gamma_2(t) = te^{i\psi}$, $1 - \delta \leq t \leq s$, and $\gamma_4(t) = te^{i\varphi}$, $1 - \delta \leq t \leq s$, connecting their endpoints (see Figure 11.2). We need the following:

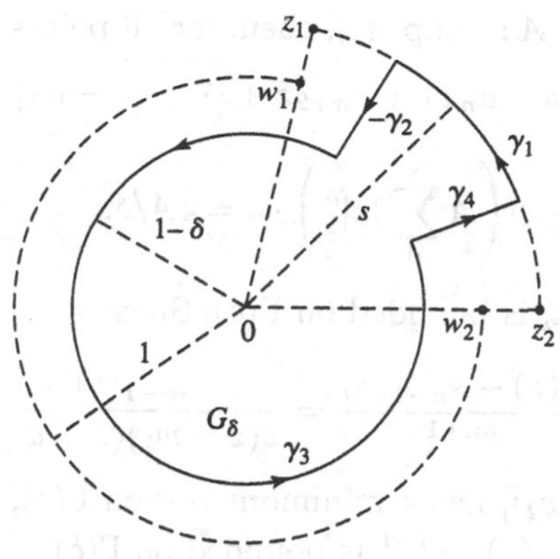

FIGURE 11.2.

a) (*Approximation theorem*) *There exists a* $\delta_0 > 0$ *such that, for every* $\eta > 0$, *there is a function (depending on* η*)* $R(z) = c_0 + c_1/z + c_2/z^2 + \cdots + c_{q-1}/z^{q-1} + 1/z^q$, $q \in \mathbb{N}$, *satisfying*

$$|R|_{\Gamma(\delta)} \le \eta \quad \textit{for all } \delta \textit{ with } 0 \le \delta \le \delta_0.$$

b) (*Variant of Riesz's lemma*) *Suppose that the power series* $f = \sum a_\nu z^\nu$ *has bounded coefficients and that there exists a* $\delta > 0$ *such that* f *has a holomorphic extension* $\widehat{f}$ *to a neighborhood of* G_δ. *Then there exists an* $M > 0$ *such that*

$$\left|\frac{\widehat{f}(z) - s_{n-1}(z)}{z^{n+1}}\right|_{\Gamma(\delta)} \le M \quad \textit{for all } n \ge 1.$$

Proof. ad a). Since $\Gamma(0) \cap \partial\mathbb{E}$ is a compact set $\neq \partial\mathbb{E}$, there exist by 12.2.2(1′) a neighborhood U of $\Gamma(0) \cap \partial\mathbb{E}$ and a function $Q(z) = b_0 + b_1/z + \cdots + b_{k-1}/z^{k-1} + 1/z^k$ such that $|Q|_U < 1$ (Runge's little theorem!). Choose $r > 1$ such that $\Gamma(0) \cap B_r(0) \subset U$ and fix $l \in \mathbb{N}$ such that, for $\widetilde{Q}(z) := z^{-l}Q(z)$,

$$|\widetilde{Q}(z)| < 1 \quad \text{for all } z \in \Gamma(0) \text{ with } |z| \ge r.$$

Then $|\widetilde{Q}|_{\Gamma(0)} < 1$. Now let V be a neighborhood of $\Gamma(0)$ with $|\widetilde{Q}|_V < 1$. For every $\eta > 0$ there exists an $m \in \mathbb{N}$ such that $|R|_V < \eta$ for $R := \widetilde{Q}^m$. Now choose $\delta_0 > 0$ so small that $\Gamma(\delta) \subset V$ for all $\delta \le \delta_0$.

ad b). There exists a compact circular sector S with vertex at 0 and corners z_1, z_2 such that γ_1, γ_2, and γ_4 lie in S (see Figure 11.2) and $\widehat{f}$ is still holomorphic in S. By Riesz's lemma 1.2, the sequence

$$g_n(z) := \frac{\widehat{f}(z) - s_n(z)}{z^{n+1}}(z - w_1)(z - w_2), \quad n \in \mathbb{N},$$

is bounded in S. Set $A := \sup |a_\nu|$; then, for all points $z \in \gamma_3$,

$$\begin{aligned} |g_n(z)| &= |a_{n+1} + a_{n+2}z + \cdots| \cdot |z - w_1| \cdot |z - w_2| \\ &\leq \left(A \sum_0^\infty |z|^\nu \right) \cdot 4 = 4A/\delta. \end{aligned}$$

Thus the sequence g_n is bounded on $\Gamma(\delta)$. Since

$$\frac{\widehat{f}(z) - s_{n-1}(z)}{z^{n+1}} = \frac{g_{n-1}(z)}{z(z - w_1)(z - w_2)}$$

and $|z(z - w_1)(z - w_2)|$ has a minimum > 0 on $\Gamma(\delta)$, this shows that the sequence $[\widehat{f}(z) - s_{n-1}(z)]/z^{n+1}$ is bounded on $\Gamma(\delta)$. □

2. A lemma. *Let $f = \sum a_\nu z^\nu$ be a power series with bounded coefficients and radius of convergence* 1 *that does not have* $\mathbb{E}$ *as domain of holomorphy. Then for every $\varepsilon > 0$ there exist a $q \in \mathbb{N}$ and numbers $c_0, c_1, \ldots, c_{q-1} \in \mathbb{C}$ such that*

$$|c_0 a_n + c_1 a_{n+1} + \cdots + c_{q-1} a_{n+q-1} + a_{n+q}| \leq \varepsilon \quad \textit{for all } n \geq 1.$$

Proof. We choose $\delta_0 > 0$ as in 1,a). Since $\partial\mathbb{E}$ is not the natural boundary of f, there exists a domain G_δ of the kind described in Subsection 1 such that f has a holomorphic extension $\widehat{f}$ to a neighborhood of $G_\delta \cup \Gamma(\delta)$. We may assume that $\delta < \delta_0$. We choose M as in 1,b) and determine the function $R(z) = c_0 + c_1/z + \cdots + c_{q-1}/z^{q-1} + 1/z^q$ as in 1,a) such that $|R|_{\Gamma(\delta)} \leq 2\pi\varepsilon/ML$, where L denotes the Euclidean length of $\Gamma(\delta)$. Then

$$(*) \qquad \left| R(z) \frac{\widehat{f}(z) - s_{n-1}(z)}{z^{n+1}} \right|_{\Gamma(\delta)} \leq 2\pi\varepsilon/L \quad \text{for all } n \geq 1.$$

It is now immediately clear from the equation

$$\begin{aligned} &R(z)\frac{\widehat{f}(z) - s_{n-1}(z)}{z^{n+1}} \\ &\qquad = \left(c_0 + \frac{c_1}{z} + \cdots + \frac{c_{q-1}}{z^{q-1}} + \frac{1}{z^q} \right) \left(\frac{a_n}{z} + a_{n+1} + a_{n+2}z + \cdots \right) \end{aligned}$$

that the number $c_0 a_n + c_1 a_{n+1} + \cdots + c_{q-1} a_{n+q-1} + a_{n+q}$ is the residue at zero of the function on the left-hand side. This function is holomorphic in $\overline{G}_\delta \backslash \{0\}$. It follows (from the residue theorem, since $\Gamma(\delta)$ is a simple closed path) that

$$c_0 a_n + \cdots + c_{q-1} a_{n+q-1} + a_{n+q} = \frac{1}{2\pi i} \int_{\Gamma(\delta)} R(\zeta) \frac{\widehat{f}(\zeta) - s_{n-1}(\zeta)}{\zeta^{n+1}} d\zeta, \quad n \geq 1.$$

Because of $(*)$, the standard estimate for integrals implies the assertion. □

The lemma says that, from the qth coefficient on, no coefficient of the Taylor series for $(1 + c_{q-1}z + \cdots + c_0 z^q)f(z)$ has absolute value greater than ε.

3. Proof of (Sz). Let $d_1, \ldots, d_k$ be the pairwise distinct numbers that occur as values of the coefficients $a_0, a_1, \ldots$. Since (Sz) is trivial for $k = 1$ (geometric series), we may assume that $k \geq 2$. Then

$$d := \min_{\kappa \neq \lambda} |d_\kappa - d_\lambda| > 0.$$

Since the sequence $a_0, a_1, \ldots$ is bounded, Lemma 2 can be applied with $\varepsilon := d/3$. Hence there exist $q \in \mathbb{N}$ and numbers $c_0, c_1, \ldots, c_{q-1} \in \mathbb{C}$ such that

$$(\#) \qquad |c_0 a_n + c_1 a_{n+1} + \cdots + c_{q-1} a_{n+q-1} + a_{n+q}| \leq \frac{1}{3}d \quad \text{for all } n \in \mathbb{N}.$$

We now consider all q-tuples $(a_n, a_{n+1}, \ldots, a_{n+q-1})$, $n \in \mathbb{N}$. Since only *finitely many distinct* q-tuples (namely k^q) can be formed from the k numbers $d_1, \ldots, d_k$, there exist numbers $\lambda, \mu \in \mathbb{N}$ with $\lambda < \mu$ such that

$$(a_\lambda, a_{\lambda+1}, \ldots, a_{\lambda+q-1}) = (a_\mu, a_{\mu+1}, \ldots, a_{\mu+q-1}).$$

Since $a_{\lambda+j} = a_{\mu+j}$ for $0 \leq j < q$, it now follows from the inequality (#) that

$$|a_{\lambda+q} - a_{\mu+q}| = \left| \sum_0^{q-1} c_j a_{\lambda+j} + a_{\lambda+a} - \left(\sum_0^{q-1} c_j a_{\mu+j} + a_{\mu+q} \right) \right| \leq \frac{1}{3}d + \frac{1}{3}d < d.$$

By the choice of d, this implies that $a_{\lambda+q} = a_{\mu+q}$ and hence also that

$$(a_{\lambda+1}, a_{\lambda+2}, \ldots, a_{\lambda+q}) = (a_{\mu+1}, \ldots, a_{\mu+q}).$$

From this it follows, as before, that $a_{\lambda+q+1} = a_{\mu+q+1}$. Thus we see (by induction) that $a_{\lambda+j} = a_{\mu+j}$ for all $j \in \mathbb{N}$. This proves (Sz) and hence Szegö's theorem.

Historical note. P. Fatou, in 1906, was the first to investigate power series with finitely many distinct coefficients [F]. Around 1918, papers of F. Carlson, R. Jentzsch, and G. Pólya appeared; cf. [B], p. 114 ff. In 1922 G. Szegö settled the questions, in a certain sense, with his extension theorem ([Sz], pp. 555–560).

4. An application. We first prove a theorem of Fatou.

Theorem (Fatou, 1905). *Let R be a rational function with the following properties:*

1) *R is holomorphic in $\mathbb{E}$ and has exactly k poles on $\partial\mathbb{E}$, all of first order, $k \geq 1$.*

2) *The set* $\{a_0, a_1, \ldots\}$ *of coefficients of the Taylor series* $\sum a_\nu z^\nu$ *for* R *about* 0 *has no accumulation point in* $\mathbb{C}$,

Then $R(z) = P(z)/(1 - z^k)$, *where* $P(z) \in \mathbb{C}[z]$.

Proof. If $\lambda_1^{-1}, \ldots, \lambda_k^{-1} \in \partial\mathbb{E}$ are the poles of R on $\partial\mathbb{E}$, then the equation

$$R(z) = \frac{B_1}{1 - z\lambda_1} + \cdots + \frac{B_k}{1 - z\lambda_k} + \sum b_\nu z^\nu$$

holds, where the series on the right-hand side has radius of convergence > 1. Thus $\lim b_\nu = 0$. Since $1/(1 - z\lambda) = \sum \lambda^\nu z^\nu$, it follows (by comparing coefficients) that

$$a_\nu = B_1\lambda_1^\nu + \cdots + B_k\lambda_k^\nu + b_\nu; \quad \text{thus} \quad |a_\nu| \leq |B_1| + \cdots + |B_k| + |b_\nu|, \ \nu \in \mathbb{N}.$$

Hence the set $\{a_0, a_1, \ldots\}$ is bounded and therefore, since it has no accumulation points, finite. The assertion follows from Szegö's theorem. □

We note a surprising result.

Corollary (Kronecker's theorem). *If the zeros of the polynomial* $Q(z) = z^n + q_1 z^{n-1} + \cdots + q_{n-1}z + q_n \in \mathbb{Z}[z]$, $n \geq 1$, *all have absolute value* 1, *then all the zeros of* Q *are roots of unity.*

Proof. We may assume that Q is irreducible over $\mathbb{Z}$ (Gauss's lemma). Then Q has only first-order zeros (division with remainder of Q by Q' in $\mathbb{Q}[z]$), and the rational function $1/Q$ has only first-order poles, which all lie on $\partial\mathbb{E}$. Since $\pm q_n$ is the product of all the zeros of Q, we have $|q_n| = 1$; hence $q_n = \pm 1$. Thus all the Taylor coefficients of $1/Q$ are integers (geometric series for $1/(\pm 1 + v)$ with $v := q_{n-1}z + \cdots + z^n$). Our theorem therefore yields

$$\frac{1}{Q(z)} = \frac{P(z)}{1 - z^k}, \quad \text{i.e.} \quad 1 - z^k = P(z)Q(z).$$

Thus $Q(\alpha) = 0$ only if $\alpha^k = 1$. □

Kronecker published his theorem in 1857 ([K], p. 105). In algebra, the theorem is usually formulated as follows:

An algebraic integer $\neq 0$ *which, together with all its conjugates, has absolute value* ≤ 1 *is a root of unity.*

Of course, there are simple algebraic proofs; the reader may refer to [K] and also, for instance, to [PSz] (Part VIII, Problem 200, p. 145 and p. 346). An especially elegant variation of Kronecker's proof was given by L. Bieberbach in 1953, in "Über einen Satz Pólyascher Art," *Arch. Math.* 4, 23–27 (1953).

5. Glimpses. Besides power series with finitely many distinct coefficients, power series with *integer* coefficients have fascinated many mathematicians — beginning with Eisenstein, in 1852. Can having integer coefficients exert a tangible effect on the behavior of the function? The next theorem gives an unexpected answer.

Pólya-Carlson theorem. *Let $f = \sum a_\nu z^\nu$ be a power series with integer coefficients and radius of convergence $R = 1$. Then either $\mathbb{E}$ is the domain of holomorphy of f or f can be extended to a rational function of the form $p(z)/(1-z^m)^n$, where $p(z) \in \mathbb{Z}[z]$ and $m, n \in \mathbb{N}$.*

This theorem was formulated in 1915 by G. Pólya ([P], p. 44), and proved in 1921 by F. Carlson [C]. The hypothesis $R = 1$ is the proper one: when $R > 1$, f is obviously a polynomial since $\overline{\lim} \sqrt[\nu]{|a_\nu|} = R^{-1} < 1$ and $a_\nu \in \mathbb{Z}$; when $R < 1$, the series

$$\frac{1}{\sqrt{1-4z^m}} = \sum_0^\infty \binom{2\nu}{\nu} z^{m\nu}, \quad \text{where } R = 1/\sqrt[m]{4},\ m = 1, 2, \ldots,$$

show that f can have extensions that are not rational. The Pólya-Carlson theorem was generalized in 1931 by H. Petersson to *power series with algebraic integer coefficients* (cf. *Abh. Math. Sem. Univ. Hamburg* 8, 315–322). Further contributions are due to W. Schwarz: Irrationale Potenzreihen, *Arch. Math.* 17, 435–437 (1966).

F. Hausdorff had already pointed out in 1919, "as support for Pólya's conjecture," that there are only countably many power series with integer coefficients that converge in $\mathbb{E}$ and do not have $\mathbb{E}$ as domain of holomorphy ([H], p.103). In 1921 G. Szegö gave a new proof of the Pólya-Carlson theorem ([Sz], pp. 577–581); in the same year, Pólya finally gave the theorem the following definitive form ([P], p. 176):

Pólya's theorem. *Let G be a simply connected domain with $0 \in G$, and let f be a function that is holomorphic in G up to isolated singularities and whose Taylor series about 0 has only integer coefficients. Let $\rho(G)$ denote the mapping radius of G with respect to 0* (cf. 8.4.3).

1) *If $\rho(G) > 1$, then f can be extended to a rational function.*
2) *If $\rho(G) = 1$ and ∂G is a simple closed path, then either f cannot be extended holomorphically beyond ∂G anywhere or else f can be extended to a rational function.*

From this deep theorem, which combines such heterogeneous properties as the rationality of a function, the integrality of its Taylor coefficients, and the conformal equivalence of domains, one easily obtains the Pólya-Carlson theorem by observing that $\rho(G) < \rho(G')$ when $G \subsetneq G'$ (cf. 8.4.3). A beautiful presentation of this circle of problems can be found in [P] (pp. 192–198).

Bibliography for Section 4

[B] Bieberbach, L.: *Analytische Fortsetzung*, Erg. Math. Grenzgeb. 3, Springer, 1955.

[C] Carlson, F.: Über Potenzreihen mit ganzzahligen Koeffizienten, *Math. Zeitschr.* 9, 1–13 (1921).

[F] Fatou, P.: Séries trigonométriques et séries de Taylor, *Acta Math.* 30, 335–400 (1906).

[H] Hausdorff, F.: Zur Verteilung der fortsetzbaren Potenzreihen, *Math. Zeitschr.* 4, 98–103 (1919).

[K] Kronecker, L.: Zwei Sätze über Gleichungen mit ganzzahligen Coeffizienten, *Journ. reine angew. Math.* 53, 173–175 (1857); *Werke* 1, 103–108.

[P] Pólya, G: *Coll. Pap.* 1.

[PSz] Pólya, G. and G. Szegö: *Problems and Theorems in Analysis*, vol. 2, trans. C. E. Billigheimer, Springer, New York, 1976.

[Sz] Szegö, G.: *Coll. Pap.* 1.

12

Runge Theory for Compact Sets

Runge approximation theory charms by its wonderful balance between freedom and necessity.

In discs B, all holomorphic functions are approximated compactly by their Taylor polynomials. In particular, *for every $f \in \mathcal{O}(B)$ and every compact set K in B, there exists a sequence of polynomials p_n such that* $\lim |f - p_n|_K = 0$. In arbitrary domains, polynomial approximation is not always possible; in $\mathbb{C}^\times$, for example, there is no sequence of polynomials p_n that approximates the holomorphic function $1/z$ uniformly on a circle γ, for it would then follow that

$$2\pi i = \int_\gamma \frac{d\zeta}{\zeta} = \lim \int_\gamma p_n(\zeta) d\zeta = 0.$$[1]

The problem of polynomial approximation is contained in a more general approximation problem. Let $K \subset \mathbb{C}$ be a compact set. A function $f : K \to \mathbb{C}$ is called *holomorphic on* K if there exist an open neighborhood U of K and a function g that is holomorphic in U and satisfies $g|K = f$. For regions $D \supset K$, we pose the following question:

When can all functions holomorphic on K be approximated uniformly by functions holomorphic in D?

---

[1]More generally: *A function f that is holomorphic in a neighborhood of a circle γ about c can be approximated uniformly by polynomials on γ if and only if there exist a disc B about c with $\gamma \subset B$ and a function $\widehat{f} \in \mathcal{O}(B)$ such that $\widehat{f}|\gamma = f$.* The proof is left to the reader.

The example $K = \partial\mathbb{E}$, $D = \mathbb{C}$, shows that such approximation is not always possible: Runge theory, named after the Göttingen mathematician Carl Runge, answers the question definitively. Our starting point is the following classical theorem.

Runge's approximation theorem. *Every function holomorphic on K can be approximated uniformly on K by rational functions with poles outside K.*

Since the location of the poles can be well controlled, this leads to a surprising answer to the question above (cf. Theorem 2.3).

Every function holomorphic on K can be approximated uniformly by functions holomorphic in D if and only if the topological space $D\backslash K$ has no connected component that is relatively compact in D.

This contains, in particular:

Runge's little theorem. *If $\mathbb{C}\backslash K$ is connected, then every function holomorphic on K can be approximated uniformly on K by polynomials.*

This approximation theorem, odd at first glance, will be obtained in Section 2 from a Cauchy integral formula for compact sets by means of a "pole-shifting method." These techniques will be set up in Section 1.

Runge's little theorem already permits surprising applications; we use it in Section 3 to prove, among other things,

- the existence of sequences of polynomials that converge pointwise to functions that are not continuous everywhere;
- the existence of a *holomorphic imbedding* of the unit disc in $\mathbb{C}^3$.

§1. Techniques

The Cauchy integral formula for *discs* is sufficient to prove almost all the fundamental theorems of local function theory. For approximation theory, however, we need a Cauchy integral formula for compact sets in arbitrary regions. Our starting point is the following formula, which was mentioned in I.7.2.2.

Cauchy integral formula for rectangles. *Let R be a compact rectangle in a region D. Then, for every function $f \in \mathcal{O}(D)$,*

$$\frac{1}{2\pi}\int_{\partial R}\frac{f(\zeta)}{\zeta - z}d\zeta = \begin{cases} f(z) & \text{if } z \in \mathring{R}, \\ 0 & \text{if } z \notin R. \end{cases}$$

In what follows, we apply this formula to rectangles *parallel to the axes.* In Subsection 1 we obtain the Cauchy integral formula for compact sets, which is fundamental for Runge theory. The structure of the Cauchy kernel in the integral formula suggests that we try to approximate holomorphic functions by linear combinations of functions of the form $(z - w_\mu)^{-1}$. The approximation lemma 2 describes how to proceed. In Subsection 3, we finally show how the poles of the approximating functions can be "shifted out of the way."

1. Cauchy integral formula for compact sets. Let $K \neq \emptyset$ be a compact subset of D. Then in $D \backslash K$ there exist finitely many distinct oriented horizontal or vertical line segments $\sigma^1, \ldots, \sigma^n$ of equal length such that, for every function $f \in \mathcal{O}(D)$,

$$f(z) = \frac{1}{2\pi i} \sum_{\nu=1}^{n} \int_{\sigma^\nu} \frac{f(\zeta)}{\zeta - z} d\zeta, \quad z \in K. \tag{1}$$

Proof. We may assume that $D \neq \mathbb{C}$. Then $\delta := d(K, \partial D) > 0$.[2] We lay a lattice on the plane that is parallel to the axes, consists of *compact* squares, and has "mesh width" d satisfying $\sqrt{2}d < \delta$. Since K is compact, it intersects only finitely many squares of the lattice (see Figure 12.1); we denote them by $Q^1, \ldots, Q^k$. We claim that

$$K \subset \bigcup_{\kappa=1}^{k} Q^\kappa \subset D.$$

The first inclusion is clear. To show that $Q^\kappa \subset D$, let us fix a point $c_\kappa \in Q^\kappa \cap K$. Then $B_\delta(c_\kappa) \subset D$ by the definition of δ. Since the square Q^κ has

[2]The *distance* between two sets $A, B \neq \emptyset$ is denoted by $d(A, B) := \inf\{|a - b| : a \in A, b \in B\}$. If A is compact and B is closed in $\mathbb{C}$, then $d(A, B) > 0$ whenever $A \cap B = \emptyset$.

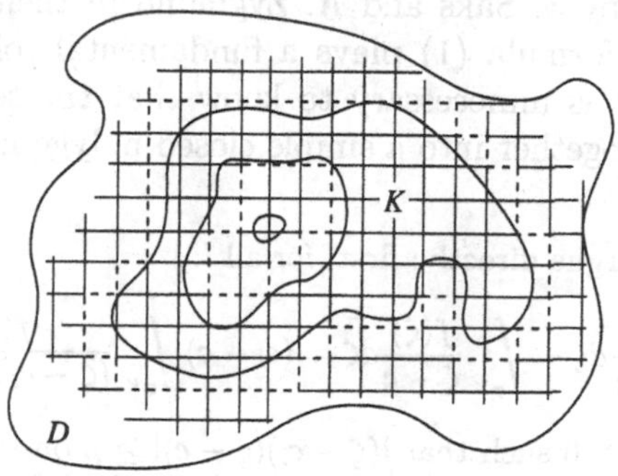

FIGURE 12.1.

diameter $\sqrt{2}d$, the distance from c_κ of every point in Q^κ is $\leq \sqrt{2}d$. Since $\sqrt{2}d < \delta$, it follows that $Q^\kappa \subset B_\delta(c_\kappa) \subset D$, $1 \leq \kappa \leq k$.

We now consider those line segments, say $\sigma^1, \ldots, \sigma^n$, that are part of the boundaries ∂Q^κ but are not common sides of two squares Q^p, Q^q, $p \neq q$. We claim that

$$(*) \qquad \bigcup_{\nu=1}^{n} |\sigma^\nu| \subset D \backslash K.$$

For if K intersected a segment σ^j, then the two squares of the lattice with σ^j as a side would have points in common with K, contradicting the choice of the segments $\sigma^1, \ldots, \sigma^n$.

Since the common sides of different squares of the lattice occur in their boundaries with opposite orientations, it follows that

$$\sum_{\kappa=1}^{k} \int_{\partial Q^\kappa} \frac{f(\zeta)}{\zeta - z} d\zeta = \sum_{\nu=1}^{n} \int_{\sigma^\nu} \frac{f(\zeta)}{\zeta - z} d\zeta \quad \text{for all } z \in D \backslash \bigcup_{\kappa=1}^{k} \partial Q^\kappa.$$

If c is an interior point of some square, say $c \in \mathring{Q}^\iota$, then

$$\int_{\partial Q^\iota} \frac{f(\zeta)}{\zeta - c} d\zeta = 2\pi i f(c) \quad \text{and} \quad \int_{\partial Q^\kappa} \frac{f(\zeta)}{\zeta - c} d\zeta = 0 \text{ for all } \kappa \neq \iota$$

by the integral formula for rectangles. Thus (1) has already been proved for all points of the set $\cup \mathring{Q}^\kappa$. Now let $c \in K$ be a boundary point of a square Q^j. Because of $(*)$, c does not lie on any line segment σ^ν. Hence the integrals on the right-hand side in (1) are also well defined in this case. We choose a sequence $c_l \in \mathring{Q}^j$ with $\lim c_l = c$. By what has already been proved, equation (1) holds for all points $z := c_l$. That it holds for $z := c$ follows by continuity, once we observe that the νth summand on the right-hand side of (1) is a continuous function at $z \in D \backslash |\sigma^\nu|$.[3] □

Remark. The theorem was probably first presented and used as a basis for Runge theory by S. Saks and A. Zygmund in their textbook [SZ], p. 155. The integral formula (1) plays a fundamental role in what follows. For the present, it is unnecessary to know that the segments $\sigma^1, \ldots, \sigma^n$ automatically fit together into a simple closed polygon; see §4.

[3]We could also argue directly: first, for all l,

$$\int_{\sigma^\nu} \frac{f(\zeta)}{\zeta - c_l} d\zeta - \int_{\sigma^\nu} \frac{f(\zeta)}{\zeta - c} d\zeta = (c_l - c) \int_{\sigma^\nu} \frac{f(\zeta)}{(\zeta - c_l)(\zeta - c)} d\zeta.$$

If we now choose $\rho > 0$ such that $|(\zeta - c_l)(\zeta - c)| \geq \rho$ on $|\sigma^\nu|$ for all l, then the absolute value of the difference of the integrals on the left-hand side is bounded above by $|c_l - c| \cdot |f|_{\sigma^\nu} \cdot \rho^{-1} \cdot L(\sigma^\nu)$; thus it tends to zero as l increases.

2. Approximation by rational functions. We begin with a lemma.

Lemma. *Let σ be a line segment in $\mathbb{C}$ disjoint from K and let h be continuous on $|\sigma|$. Then the function $\int_\sigma h(\zeta)(\zeta - z)^{-1}d\zeta$, $z \in \mathbb{C}\backslash|\sigma|$, can be approximated uniformly on K by rational functions of the form*

$$\sum_{\mu=1}^{m} \frac{c_\mu}{z - w_\mu}, \quad c_1, \ldots, c_m \in \mathbb{C}, \quad w_1, \ldots, w_m \in |\sigma|.$$

Proof. The function $v(\zeta, z) := h(\zeta)/(\zeta - z)$ is continuous on $|\sigma| \times K$. Since $|\sigma| \times K$ is compact, v is uniformly continuous on $|\sigma| \times K$; hence for every $\varepsilon > 0$ there exists a $\delta > 0$ such that

$$|v(\zeta, z) - v(\zeta', z)| \leq \varepsilon \quad \text{for all } (\zeta, \zeta', z) \in |\sigma| \times |\sigma| \times K \text{ with } |\zeta - \zeta'| \leq \delta.$$

We subdivide σ into segments $\pi^1, \ldots, \pi^m$ of length $\leq \delta$ and choose $w_\mu \in |\pi^\mu|$. With $c_\mu := -h(w_\mu)\int_{\pi^\mu} d\zeta$, we then have, for $z \in K$ (standard estimate),

$$\left|\int_{\pi^\mu} v(\zeta, z)d\zeta - \frac{c_\mu}{z - w_\mu}\right| = \left|\int_{\pi^\mu} v((\zeta, z) - v(w_\mu, z))d\zeta\right| \leq \varepsilon \cdot L(\pi^\mu).$$

For $q(z) := \sum_{\mu=1}^{m} c_\mu(z - w_\mu)^{-1} \in C(K)$, it now follows, since $L(\sigma) = \sum_{\mu=1}^{m} L(\pi^\mu)$, that

$$\left|\int_\sigma v(\zeta, z)d\zeta - q(z)\right| \leq L(\sigma) \cdot \varepsilon \quad \text{for all } z \in K. \qquad \square$$

The lemma and the integral formula (1) now easily imply the basic

Approximation lemma. *For every compact set K in a region D, there exist finitely many line segments $\sigma^1, \ldots, \sigma^n$ in $D\backslash K$ such that every function $f \in \mathcal{O}(D)$ can be approximated uniformly on K by rational functions of the form*

$$\sum_{\kappa=1}^{k} \frac{c_\kappa}{z - w_\kappa}, \quad c_\kappa \in \mathbb{C}, \quad w_\kappa \in \bigcup_{\nu=1}^{n} |\sigma^\nu|.$$

Proof. We choose line segments $\sigma^1, \ldots, \sigma^n$ in $D\backslash K$, as in Theorem 1, such that 1(1) holds. By the lemma, for a given $\varepsilon > 0$ there exist functions

$$q_\nu(z) = \sum_{\mu=1}^{m_\nu} \frac{c_{\mu\nu}}{z - w_{\mu\nu}}, \quad w_{\mu\nu} \in |\sigma^\nu|, \quad 1 \leq \nu \leq n,$$

such that

$$\frac{1}{2\pi i}\left|\int_{\sigma^\nu} \frac{f(\zeta)}{\zeta - z}d\zeta - q_\nu(z)\right|_K \leq \frac{\varepsilon}{n}, \quad 1 \leq \nu \leq n.$$

For $q := q_1 + \cdots + q_n$, we then have $|f - q|_K \leq \varepsilon$. By construction, q is a finite sum of terms of the form $c_\kappa/(z - w_\kappa)$, where $w_\kappa \in \bigcup|\sigma^\nu|$. □

The set $\bigcup|\sigma^\nu|$ in which the poles of q lie is — independently of the quality of the approximation — determined only by D and K (it does, of course, depend on the choice of the lattice in the proof of Theorem 1). It is true that if ε is decreased, the approximating function q will have more poles w_κ on $\bigcup|\sigma^\nu|$, but they will not get any closer to K. We show in the next subsection that, in addition, these poles can be shifted by admitting polynomials in $(z - w)^{-1}$ instead of the functions $c/(z - w)$.

3. Pole-shifting theorem. Every topological space X can be uniquely represented as the union of its *components* (= maximal connected subspaces); for more on this, see I.0.6.4 and Subsection 1 of the appendix to Chapter 13. In this subsection, X is a space $\mathbb{C}\backslash K$, where K denotes a compact set in $\mathbb{C}$. Every component of $\mathbb{C}\backslash K$ is then a *domain* in $\mathbb{C}$. There is exactly one *unbounded* component.

Pole-shifting theorem. *Let a and b be arbitrary points in a component Z of $\mathbb{C}\backslash K$. Then $(z-a)^{-1}$ can be approximated uniformly on K by polynomials in $(z - b)^{-1}$. In particular, if Z is the unbounded component of $\mathbb{C}\backslash K$, then $(z - a)^{-1}$ can be approximated uniformly on K by polynomials.*

Proof. For $w \notin K$, let L_w denote the set of all $f \in \mathcal{O}(K)$ that can be approximated uniformly on K by polynomials in $(z-w)^{-1}$. Then the following is clear.

(∗) *If $(z - s)^{-1} \in L_c$ and $(z - c)^{-1} \in L_b$, then $(z - s)^{-1} \in L_b$ (transitivity).*

The first assertion of the theorem is that $S := \{s \in Z : (z-s)^{-1} \in L_b\} = Z$. Since $b \in S$, it suffices to prove the following:

If $c \in S$ and $B \subset Z$ is a disc about c, then $B \subset S$.[4]

Let $s \in B$. The (geometric) series $\sum(s - c)^\nu/(z - c)^{\nu+1}$ converges normally in $\mathbb{C}\backslash B$ to $(z - s)^{-1}$. Since $K \cap B = \emptyset$, the sequence of partial sums converges uniformly on K to $(z - s)^{-1}$. Hence $(z - s)^{-1} \in L_c$. Since $c \in S$, it now follows from (∗) that $(z - s)^{-1} \in L_b$, i.e. $s \in S$.

If Z is unbounded, then there exists a $d \in Z$ such that $K \subset B_{|d|}(0)$. Then all the functions $(z-d)^{-n}$ are approximated uniformly on K by their Taylor polynomials. Hence, by what has already been proved (transitivity!), $(z - a)^{-1}$ can also be approximated uniformly on K by polynomials.

□

[4]We use the following assertion, whose justification is left to the reader: *Let $S \neq \emptyset$ be a subset of a domain G such that every disc $B \subset G$ about a point $c \in S$ lies in S. Then $S = G$.*

The statement about the unbounded components of $\mathbb{C}\backslash K$ is often interpreted as saying that the pole can be shifted to ∞. (Polynomials are viewed as rational functions with poles at ∞.)

Remark. The statement that a and b lie in the same component of $\mathbb{C}\backslash K$ is necessary for the pole-shifting theorem to hold. For instance, if we choose $K := \partial\mathbb{E}$, $a \in \mathbb{E}$, and $b \in \mathbb{C}\backslash\overline{\mathbb{E}}$, then $(z-a)^{-1}$ cannot be approximated uniformly on $\mathbb{E}$ by polynomials in $(z-b)^{-1}$, since such functions g are holomorphic in a neighborhood of $\overline{\mathbb{E}}$ and therefore

$$2\pi i = \int_{\partial\mathbb{E}} \left[\frac{1}{\zeta - a} - g(\zeta)\right] d\zeta, \quad \text{whence} \quad \left|\frac{1}{z-a} - g(z)\right|_{\partial\mathbb{E}} \geq 1.$$

In general:

Let Z be a component of $\mathbb{C}\backslash K$, $a \in Z$, $b \notin K \cup Z$, and $\delta = |z-a|_K > 0$. Then $|(z-a)^{-1} - g(z)|_K \geq \delta^{-1}$ for every holomorphic function g that is a nonconstant polynomial in $(z-b)^{-1}$ and for which $\lim_{z\to\infty} g(z) = 0$.

Proof. First observe that $\partial Z \subset K$ (cf. 2.3(1)). Now, if there were a g with $|(z-a)^{-1} - g(z)|_K < \delta^{-1}$, we would have $|1-(z-a)g(z)|_K < 1$. Since $\partial Z \subset K$ and $\lim g(z) = 0$, it would follow from the maximum principle that $|1-(z-a)g(z)|_Z < 1$, which is absurd because $a \in Z$. □

§2. Runge Theory for Compact Sets

For every compact set $K \subset \mathbb{C}$, the set $\mathcal{O}(K)$ of all functions holomorphic in K is a $\mathbb{C}$-algebra. (Prove this.) We first prove that every function in $\mathcal{O}(K)$ can be *approximated uniformly on K by rational functions with poles outside K:* the location of these poles is given more precisely in Subsection 1. We obtain as a special case that, if the space $\mathbb{C}\backslash K$ is connected, every function in $\mathcal{O}(K)$ can be approximated uniformly by polynomials. With the techniques of §1, the proofs are easy.

In Subsection 3 we prove the main theorem of Runge theory for compact sets.

1. Runge's approximation theorem. For every set $P \subset \mathbb{C}$, the collection $\mathbb{C}_P[z]$ of rational functions whose poles all lie in P is a $\mathbb{C}$-algebra, and $\mathbb{C}[z] \subset \mathbb{C}_P[z] \subset \mathcal{O}(\mathbb{C}\backslash\overline{P})$. The significance of the algebras $\mathbb{C}_P[z]$ for Runge theory lies in the following theorem.

Approximation theorem (*Version* 1). *If $P \subset \mathbb{C}\backslash K$ intersects every bounded component of $\mathbb{C}\backslash K$, then every function in $\mathcal{O}(K)$ can be approximated uniformly on K by functions in $\mathbb{C}_P[z]$.*

Proof. Let $f \in \mathcal{O}(K)$ and let $\varepsilon > 0$. Since f is holomorphic in an open neighborhood of K, by the approximation lemma 1.2 there exist a compact set L in $\mathbb{C}$, disjoint from K, and a function

$$q(z) = \sum_{\kappa=1}^{k} \frac{c_\kappa}{z - w_\kappa}, \quad \text{with } w_1, \ldots, w_k \in L,$$

such that $|f - q|_K \leq \varepsilon/2$. Let Z_κ be the component of $\mathbb{C}\backslash K$ containing w_κ. If Z_κ is bounded, there exists by hypothesis a $t_\kappa \in P \cap Z_\kappa$. By the pole-shifting theorem 1.3, there is then a polynomial g_κ in $(z - t_\kappa)^{-1}$ such that

$$(*) \qquad \left| \frac{c_\kappa}{z - w_\kappa} - g_\kappa(z) \right| \leq \frac{1}{2k}\varepsilon \quad \text{for all } z \in K.$$

But if Z_κ is unbounded, then $(*)$ holds, by the same theorem, even with a polynomial g_κ in z. The function $g := \sum_{\kappa=1}^{k} g_\kappa$ is now rational, all its poles lie in P, and

$$|f - g|_K \leq |f - q|_K + |q - g|_K \leq \frac{1}{2}\varepsilon + \sum_{\kappa=1}^{k} \left| \frac{c_\kappa}{z - w_\kappa} - g_\kappa(z) \right|_K \leq \varepsilon. \quad \square$$

The set P can be infinite, e.g. for $K := \{0\} \cup \bigcup_{n=1}^{\infty} \partial B_{1/n}(0)$. — The next result follows immediately from the theorem just proved.

Approximation theorem (*Version* 2). *If K is a compact set in the region D and if every bounded component of $\mathbb{C}\backslash K$ intersects the set $\mathbb{C}\backslash D$, then every function in $\mathcal{O}(K)$ can be approximated uniformly on K by rational functions that are all holomorphic in D.*

Proof. We can choose the set $P \subset \mathbb{C}\backslash K$ to lie outside D. $\square$

The next theorem follows for the special case $D = \mathbb{C}$.

Runge's little theorem on polynomial approximation. *If $\mathbb{C}\backslash K$ is connected, then every function in $\mathcal{O}(K)$ can be approximated uniformly on K by polynomials.*

The sufficient conditions for approximability given in this subsection are also necessary, as will become apparent in Subsection 3.

Glimpses. Runge's little theorem stands at the beginning of a chain of theorems on approximation by polynomials in the complex plane. The hypothesis that f is holomorphic on K can be weakened. If f is to be approximated uniformly on K by polynomials, it must certainly be continuous on K and holomorphic at all interior points of K. In 1951, after Walsh, Keldych, and Lavrentieff had settled special cases, Mergelyan proved that this necessary condition for polynomial approximation also suffices if $\mathbb{C}\backslash K$ is again assumed to be connected. The book [Ga$_1$] of D. Gaier is recommended to readers who would like to go more deeply into this circle of problems; see also [Ga₂], where the construction of approximating polynomials is considered.

2. Consequences of Runge's little theorem. The next result follows at once.

(1) *If $K \neq \partial\mathbb{E}$ is a nonempty compact set in $\partial\mathbb{E}$, there exists a polynomial P such that $P(0) = 1$ and $|P|_K < 1$.*

Proof. Since $\mathbb{C}\backslash K$ is connected, there exists by Runge a polynomial $\widetilde{P}$ such that $|\widetilde{P} + 1/z|_K < 1$. Then $P := 1 + z\widetilde{P}$ is a polynomial with the desired properties. □

Remark. Every polynomial $P(z) = 1 + b_1 z + \cdots + b_n z^n$, $b_n \neq 0$, $n \geq 1$, assumes values of magnitude > 1 on $\partial\mathbb{E}$ by the maximum principle. This does not contradict (1).

The following variant of (1) was used in 11.4.1:

(1′) *If $K \neq \partial\mathbb{E}$ is a compact set in $\partial\mathbb{E}$, then there exist a neighborhood U of K and a function $Q(z) = b_0 + b_1/z + \cdots + b_{k-1}/z^{k-1} + 1/z^k$, with $k \geq 1$, such that $|Q|_U < 1$.*

Proof. By (1), there exists a polynomial $P(z) = 1 + a_1 z + \cdots + a_k z^k$ with $|P|_K < 1$. Set $Q(z) := P(z)/z^k$; then the inequality $|Q|_K < 1$ also holds. By continuity, there exists a neighborhood U of K such that $|Q|_U < 1$. □

We will need the following in 3.2:

(2) *Let $A_1, \ldots, A_k$, $B_1, \ldots, B_l$ be pairwise disjoint compact sets in $\mathbb{C}$ such that $\mathbb{C}\backslash(A_1 \cup \cdots \cup B_l)$ is connected, and let $u_1, \ldots, u_k$, $v_1, \ldots, v_l$ be entire functions. Then for any two real numbers $\varepsilon > 0$, $M > 0$, there exists a polynomial p such that*

$$|u_\kappa + p|_{A_\kappa} \leq \varepsilon, \quad 1 \leq \kappa \leq k,$$

and

$$\min\{|v_\lambda(z) + p(z)| : z \in B_\lambda\} \geq M, \quad 1 \leq \lambda \leq l.$$

Proof. Let $K := A_1 \cup \cdots \cup B_l$. Since $A_1, \ldots, B_l$ are pairwise disjoint, the function h defined to be u_κ on A_κ and $v_\lambda - M - \varepsilon$ on B_λ is holomorphic on K. Hence, by Runge, there exists a polynomial p such that $|h + p|_K \leq \varepsilon$. This means that $|u_\kappa + p| \leq \varepsilon$, $1 \leq \kappa \leq k$. Moreover, since $v_\lambda + p = M + \varepsilon + h + p$ on B_λ, it follows for all $z \in B_\lambda$, $1 \leq \lambda \leq l$, that

$$|v_\lambda(z) + p(z)| \geq M + \varepsilon - |h(z) + p(z)| \geq M. \qquad \square$$

One can also approximate and interpolate *simultaneously*. For example, let $\mathfrak{d}$ be any positive divisor on $\mathbb{C}$ with *finite* support A, and let K denote a compact set with $A \subset K$ such that $\mathbb{C}\backslash K$ is connected. Then the following holds.

(3) *If* $\varepsilon > 0$, *then for every* $f \in \mathcal{O}(K)$ *there exists a polynomial* p *such that*

$$|f - p|_K \le \varepsilon \quad \textit{and} \quad o_a(f - p) \ge \mathfrak{d}(a) \quad \textit{for all } a \in A.$$

Proof. Choose $\widetilde{p} \in \mathbb{C}[z]$ such that $o_a(f - \widetilde{p}) \ge \mathfrak{d}(a)$, $a \in A$. Then, setting $q(z) := \prod_{a \in A}(z - a)^{\mathfrak{d}(a)}$, we have $F := (f - \widetilde{p})/q \in \mathcal{O}(K)$. We may assume that $|q|_K \neq 0$, since the case $A = K$ is trivial. Hence, by Runge, there exists a $\widehat{p} \in \mathbb{C}[z]$ with $|F - \widehat{p}|_K \le \varepsilon |q|_K^{-1}$. Now $p := \widetilde{p} + q\widehat{p}$ is a polynomial with the desired properties. □

3. Main theorem of Runge theory for compact sets. The approximation theorem 1 (version 2) will be strengthened. We first note two simple statements:

(1) *For every component* Z *of* $D \backslash K$, *we have* $D \cap \partial Z \subset K$. *If, in addition,* Z *is relatively compact*[5] *in* D, *then* $|f|_Z \le |f|_K$ *for all* $f \in \mathcal{O}(D)$.

Proof. If there were a $c \in D \cap \partial Z$ with $c \notin K$, then there would exist a disc $B \subset D \backslash K$ about c. Since $Z \cap B \neq \emptyset$, B would then be contained in Z (since Z is a component of $D \backslash K$), contradicting $c \in \partial Z$. Thus $D \cap \partial Z \subset K$.

If Z is relatively compact in D, then $\partial Z \subset D$; hence $\partial Z \subset K$. The estimate now follows from the maximum principle for bounded domains. □

(2) *Every component* Z_0 *of* $\mathbb{C} \backslash K$ *that lies in* D *is a component of* $D \backslash K$. *If, in addition,* Z_0 *is bounded, then* Z_0 *is relatively compact in* D.

Proof. Since Z_0 is a domain in $D \backslash K$, there exists a component Z_1 of $D \backslash K$ such that $Z_0 \subset Z_1$. Since Z_1 is a domain in $\mathbb{C} \backslash K$, it follows from the maximality of Z_0 that $Z_0 = Z_1$.

If Z_0 is bounded, then $\overline{Z_0} := Z_0 \cup \partial Z_0$ is compact. Since $\partial Z_0 \subset K$ (by (1), with $D := \mathbb{C}$), it follows that $\overline{Z_0} \subset D \cup K \subset D$. □

We now prove:

Theorem. *The following statements about a compact set* K *in* D *are equivalent.*

i) *No component of the space* $D \backslash K$ *is relatively compact in* D.
ii) *Every bounded component of* $\mathbb{C} \backslash K$ *intersects* $\mathbb{C} \backslash D$.
iii) *Every function in* $\mathcal{O}(K)$ *can be approximated uniformly on* K *by rational functions without poles in* D.
iv) *Every function in* $\mathcal{O}(K)$ *can be approximated uniformly on* K *by functions holomorphic in* D.
v) *For every* $c \in D \backslash K$ *there exists a function* $h \in \mathcal{O}(D)$ *such that* $|h(c)| > |h|_K$.

---

[5] A subset M of D is *relatively compact* in D if there exists a compact set $L \subset D$ with $M \subset L$.

Proof. i) $\Rightarrow$ ii). Clear by (2). — ii) $\Rightarrow$ iii). This is approximation theorem 1 (version 2). — iii) $\Rightarrow$ iv). Trivial. — iv) $\Rightarrow$ i). If $D\backslash K$ has a component Z that is relatively compact in D, let $a \in Z$ and $\delta := |z-a|_K \in (0, \infty)$. For $(z-a)^{-1} \in \mathcal{O}(K)$, there exists a $g \in \mathcal{O}(D)$ such that

$$|(z-a)^{-1} - g(z)|_K < \delta^{-1}; \quad \text{hence} \quad |1-(z-a)g(z)|_K < 1.$$

By (1), it follows that $|1-(z-a)g(z)| < 1$ for all $z \in Z$, which is absurd for $z = a$.

i) $\Rightarrow$ v). The region $D\backslash(K \cup \{c\})$ has the same components as $D\backslash K$, except one from which c has been removed. Hence i) holds and thus, by what has already been proved, iv) also holds for $K \cup \{c\}$ instead of K. Thus, corresponding to the function $g \in \mathcal{O}(K \cup \{c\})$ defined by

$$g(z) := 0 \quad \text{for } z \in K, \quad g(c) := 1,$$

there is an $h \in \mathcal{O}(D)$ such that $|h|_K < \frac{1}{2}$ and $|1-h(c)| < \frac{1}{2}$. This implies that $|h(c)| > \frac{1}{2}$ and proves v).

v) $\Rightarrow$ i). If $D\backslash K$ had a component Z that was relatively compact in D, then, by (1), v) would fail for every point $c \in Z$. □

For $D = \mathbb{C}$, we obtain a more precise form of Runge's little theorem.

Corollary 1. *Every function in $\mathcal{O}(K)$ can be approximated uniformly on K by polynomials if and only if $\mathbb{C}\backslash K$ is connected. This occurs if and only if for every $c \in \mathbb{C}\backslash K$ there exists a polynomial p such that $|p(c)| > |p|_K$.*

Proof. $\mathbb{C}\backslash K$ is connected if and only if $\mathbb{C}\backslash K$ has no component that is relatively compact in $\mathbb{C}$. Thus the corollary follows immediately from the theorem, since every entire function in $\mathbb{C}$ can be approximated compactly by Taylor polynomials. □

We now also have the converse of the first version of the approximation theorem 1.

Corollary 2. *If $P \subset \mathbb{C}\backslash K$ is such that every function in $\mathcal{O}(K)$ can be approximated uniformly on K by functions in $\mathbb{C}_P[z]$, then P intersects every bounded component of $\mathbb{C}\backslash K$.*

Proof. By the pole-shifting theorem 1.3, we may assume that *P intersects every component of $\mathbb{C}\backslash K$ in at most one point.* Then $D := \mathbb{C}\backslash P$ is a region with $K \subset D$. Since $\mathbb{C}_P[z] \subset \mathcal{O}(D)$, every bounded component of $\mathbb{C}\backslash K$ intersects $\mathbb{C}\backslash D = P$ because of the implication iv) $\Rightarrow$ ii). □

With the aid of (2), the equivalence i) $\Leftrightarrow$ ii) can immediately be improved. For a set $B \subset \mathbb{C}$, the following statements are equivalent:

– B is a component of $D\backslash K$ that is relatively compact in D;

– B is a bounded component of $\mathbb{C}\backslash K$ that lies in D.

The implication iv) ⇒ i) can be strengthened. A necessary condition for approximability is the following (the proof, using (1), is left to the reader):

If $f \in \mathcal{O}(K)$ can be approximated uniformly on K by functions in $\mathcal{O}(D)$, then for every component Z of $D\backslash K$ that is relatively compact in D there exists exactly one function $\widehat{f}$ that is continuous in $Z \cup K$ and satisfies $\widehat{f}|K = f$ and $\widehat{f}|Z \in \mathcal{O}(Z)$ (holomorphic extendibility of f to Z).

§3. Applications of Runge's Little Theorem

Runge's theorem belongs in every analyst's bag of tricks.
— L. A. Rubel

It is easy to construct sequences of *continuous* functions that converge *pointwise* to functions with *points of discontinuity.* But it is hard to find a *pointwise convergent sequence of holomorphic functions* whose limit function is *not holomorphic.* Osgood's theorem (see 7.1.5*) could even serve as an indication that such pathological (?) functions do not exist and that pointwise convergence is the appropriate notion of convergence for function theory. Using Runge's little theorem, one is easily convinced of the opposite; with this, we fulfill a promise made in the first volume (p. 92). We prove in Subsection 1 that there exist sequences of polynomials that converge pointwise in $\mathbb{C}$ to discontinuous limit functions and for which the sequences of derivatives converge everywhere in $\mathbb{C}$ — though not compactly — to the zero function.

In Subsection 2, we use Runge's little theorem to map the unit disc biholomorphically onto a *complex curve in* $\mathbb{C}^3$.

1. Pointwise convergent sequences of polynomials that do not converge compactly everywhere. We set $\mathbb{R}^+ := \{x \in \mathbb{R} : x \geq 0\}$ and prove:

Proposition. *There exists a sequence of polynomials p_n with the following properties:*

1) $\lim\limits_{n\to\infty} p_n(0) = 1,\ \lim\limits_{n\to\infty} p_n(z) = 0$ *for every point* $z \in \mathbb{C}^\times$.
2) $\lim\limits_{n\to\infty} p_n^{(k)}(z) = 0$ *for every point* $z \in \mathbb{C}$ *and every* $k \geq 1$.
3) *Every sequence* $p_1^{(k)}, p_2^{(k)}, \ldots, p_n^{(k)}, \ldots,\ k \in \mathbb{N}$, *converges compactly in* $\mathbb{C}\backslash\mathbb{R}^+$, *but none of these sequences converges compactly in any neighborhood of any point in* $\mathbb{R}^+$.

Proof. We set

$$I_n := \left\{z \in \overline{B}_n(0) \text{ and } d(z, \mathbb{R}^+) \geq \frac{1}{n}\right\}, \quad K_n := \{0\} \cup [\tfrac{1}{n}, n] \cup I_n, \quad n \geq 1.$$

K_n is compact and every set $\mathbb{C}\backslash K_n$ is connected (see Figure 12.2). We place compact rectangles R_n and S_n around 0 and $[\frac{1}{n}, n]$, respectively, in such a way that the following holds.

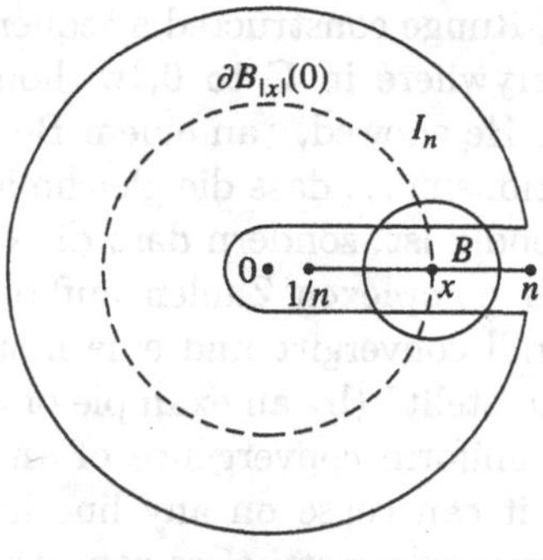

FIGURE 12.2.

The sets R_n, S_n, and I_{n+1} are pairwise disjoint. The compact set $L_n := R_n \cup S_n \cup I_{n+1}$ is a neighborhood of K_n; that is, $K_n \subset \mathring{L}_n$. The set $\mathbb{C}\backslash L_n$ is connected.

The function g_n defined by

$$g_n(z) := 0 \text{ for } z \in L_n\backslash R_n, \quad g_n(z) := 1 \text{ for } z \in R_n$$

is holomorphic in L_n. Hence, by Runge, there exists a polynomial p_n such that

$$(*) \qquad |p_n - g_n|_{L_n} \leq \frac{1}{n} \quad n = 1, 2, \ldots.$$

Since $g_n' \equiv 0$ on $\mathring{L}_n$, we can even approximate so closely in $(*)$ that p_n also satisfies

$$(**) \qquad |p_n^{(k)}|_{K_n} \leq \frac{1}{n} \quad \text{for } k = 1, \ldots, n; \quad n = 1, 2 \ldots.^6$$

Since $\bigcup K_n = \mathbb{C}$, assertions 1) and 2) follow from $(*)$ and $(**)$.

By construction, all the sequences $p_n^{(k)}$ converge compactly in $\mathbb{C}\backslash\mathbb{R}^+$. Certainly the sequence p_n does not converge compactly in any disc about 0. If it converged compactly in a disc B about $x > 0$, then the sequence would be uniformly convergent on $\partial B_{|x|}(0)$. By the maximum principle,

[6]This follows immediately from the Cauchy estimates for derivatives in compact sets; cf. I.8.3.1.

it would also be compactly convergent in $B_{|x|}(0)$, which is impossible. It now follows that none of the sequences of derivatives converges compactly about points of $\mathbb{R}^+$; cf. I.8.4.4. □

Historical note. In 1885, Runge constructed a sequence of polynomials that converges pointwise everywhere in $\mathbb{C}$ to 0, without the convergence being compact in all of $\mathbb{C}$. He showed, "an einem Beispiel einer Summe von ganzen rationalen Functionen ..., dass die gleichmässige Convergenz eines Ausdrucks nicht nothwendig ist, sondern dass dieselbe auf irgend welchen Linien in der Ebene der complexen Zahlen aufhören kann, während der Ausdruck dennoch überall convergirt und eine monogene analytische [= holomorphe] Function darstellt" (by an example of a sum of rational entire functions ..., that the uniform convergence of an expression is not necessary, but rather that it can cease on any line in the plane of complex numbers, while the expression nevertheless converges everywhere and represents a monogenic analytic [= holomorphic] function) ([Run], p. 245). Runge approximates the functions $1/[n(nz-1)]$ by polynomials: his sequence "convergirt ungleichmässig auf dem positiven Theil der imaginären Achse" (converges nonuniformly on the positive part of the imaginary axis). Runge was the first to ask whether, in the Weierstrass convergence theorem, "die gleichmässige Convergenz nothwendig ist, damit eine monogene analytische Function dargestellt werde" (uniform convergence is necessary for a monogenic analytic function to be represented). His counterexample was hardly noticed at the time.

I have yet to learn who gave the first example of a sequence of holomorphic functions that converges pointwise and has a discontinuous limit function. This was already mathematical folklore by around 1901; cf., for example, [O], p. 32. Since 1904 at the latest, it has been easy to give such sequences explicitly. It was then that Mittag-Leffler constructed an entire function F with the following paradoxical property (cf. [ML], théorème E, p. 263):

$$(1) \qquad F(0) \neq 0, \quad \lim_{r\to\infty} F(re^{i\varphi}) = 0 \quad \text{for every (fixed) } \varphi \in [0, 2\pi).$$

The sequence $g_n(z) := F(nz)$ converges pointwise in $\mathbb{C}$ to a discontinuous limit function. — The Mittag-Leffler function F can be written down explicitly as a "closed analytic expression"; see, for example, [PS] (vol. 1, III, Problem 158, and vol. 2, IV, Problem 184). Montel, in 1907, systematically investigated the properties of such limit functions, which he called "fonctions de première classe" ([Mo], p. 315 and p. 326). F. Hartogs and A. Rosenthal, in 1928, obtained very detailed results [HR].

A particularly simple construction of an entire function satisfying (1) was given in 1976 by D. J. Newman [N]. He considers the *nonconstant* entire function

$$G(z) := g(z+4i), \quad \text{where} \quad g(z) := \int_0^\infty e^{zt} t^{-t} dt, \quad z \in \mathbb{C},$$

and proves:

The function G is bounded on every real line through $0 \in \mathbb{C}$.

Then, if $G - G(0)$ vanishes to order k at 0, it is obvious that $F(z) := [G(z) - G(0)]/z^k$ is an entire function for which (1) holds.

Remark. "Explicit" sequences in $\mathcal{O}(\mathbb{C})$ can also be given which vanish pointwise everywhere in $\mathbb{C}$ but do not converge compactly to 0. Let F denote the "Mittag-Leffler function" given by (1).

The sequence $f_n(z) := F(nz)/n \in \mathcal{O}(\mathbb{C})$ *converges pointwise to* 0 *in* $\mathbb{C}$, *but this convergence is not uniform in any neighborhood of the origin.*

Proof. By (1), the sequence f_n converges pointwise in $\mathbb{C}$ to 0. If it were uniformly convergent in a neighborhood of 0, then, in particular, it would be bounded in a neighborhood of 0. There would then exist $r > 0$ and $M > 0$ such that $|F(z)| \leq nM$ for all $n \geq 1$ and all $z \in \mathbb{C}$ with $|z| \leq nr$. For the Taylor coefficients $a_0, a_1, \ldots$ of F about 0, it would follow that

$$|a_\nu| \leq (nM)/(nr)^\nu \quad \text{for all } \nu \geq 0 \text{ and all } n \geq 1.$$

Since $\lim_{n\to\infty}(nM)/(nr)^\nu = 0$ for $\nu \geq 2$, F would be at most linear and hence, since $\lim_{r\to\infty} F(re^{i\varphi}) = 0$, identically zero, contradicting $F(0) \neq 0$. □

Exercise. Let $f \in \mathcal{O}(\mathbb{C})$. Construct a sequence of polynomials p_n that converges pointwise to the function

$$g(z) := \begin{cases} 0 & \text{for } z \in \mathbb{C}\backslash\mathbb{R} \\ f(z) & \text{for } z \in \mathbb{R}. \end{cases}$$

Do this in such a way that the convergence in $\mathbb{C}\backslash\mathbb{R}$ is compact and all the sequences of derivatives $p_n^{(k)}$ converge pointwise everywhere in $\mathbb{C}$.

2. Holomorphic imbedding of the unit disc in $\mathbb{C}^3$. If $f_1, \ldots, f_n \in \mathcal{O}(D)$, $1 \leq n < \infty$, the map $D \to \mathbb{C}^n$, $z \mapsto (f_1(z), \ldots, f_n(z))$, is called *holomorphic.* Such a map is called *smooth* if $(f_1'(z), \ldots, f_n'(z)) \neq (0, \ldots, 0)$ for all $z \in D$. Holomorphic maps that are injective, closed,[7] and smooth are called *holomorphic imbeddings* of D into $\mathbb{C}^n$. It can be shown that the image set of D in $\mathbb{C}^n$ is then a smooth (with no singularities) closed complex curve in $\mathbb{C}^n$ (= complex submanifold of $\mathbb{C}^n$ of real dimension 2). — Regions $D \neq \mathbb{C}$ admit no holomorphic imbeddings in $\mathbb{C}$, since every holomorphic imbedding $D \to \mathbb{C}$ is biholomorphic. Our goal is to prove the following theorem.

Imbedding theorem. *There exists a holomorphic imbedding* $\mathbb{E} \to \mathbb{C}^3$.

To prove this, it suffices to construct two functions $f, g \in \mathcal{O}(\mathbb{E})$ such that $\lim(|f(z_n)| + |g(z_n)|) = \infty$ for every sequence $z_n \in \mathbb{E}$ converging to a point in $\partial\mathbb{E}\backslash\{1\}$. Then the three functions f, g, and h, where $h(z) := 1/(z-1)$,

[7] A map $X \to Y$ between topological spaces is called *closed* if every closed set in X has a closed image in Y; see also 9.4*.1.

give a *closed holomorphic* map $\mathbb{E} \to \mathbb{C}^3$. Since $h : \mathbb{E} \to \mathbb{C}$ is injective and $h'(z) \neq 0$ for all z, this map is also injective and smooth.

The functions f and g are constructed with the aid of Runge's little theorem. We choose a "horseshoe sequence" K_n of compact pairwise disjoint sets in $\mathbb{E}$ such that $\mathbb{C}\backslash(K_0 \cup \cdots \cup K_n)$ is always connected and the sets K_n tend to the boundary $\partial\mathbb{E}$ and taper toward 1 (as in Figure 12.3).

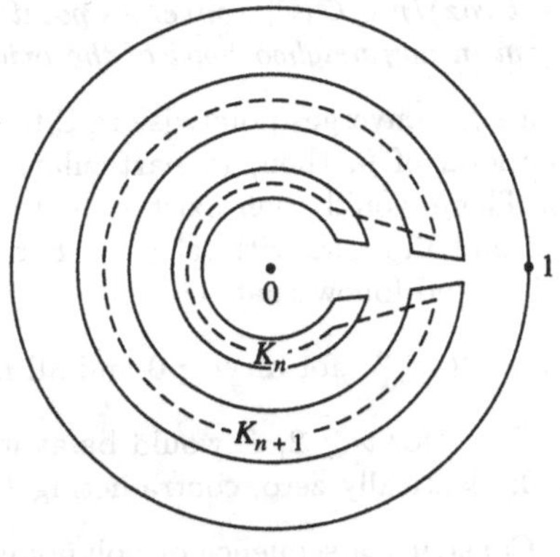

FIGURE 12.3.

Lemma. *There exists a holomorphic function* $f \in \mathcal{O}(\mathbb{E})$ *such that* $\min\{|f(z)| : z \in K_n\} \geq 2^n$ *for all* $n \in \mathbb{N}$.

Proof. We choose an increasing sequence $V_0 \subset V_1 \subset \cdots$ of compact discs about 0 such that $\bigcup V_\nu = \mathbb{E}$ and

$$K_0 \cup K_1 \cup \cdots \cup K_{n-1} \subset V_n, \quad V_n \cap K_n = \emptyset, \quad n \in \mathbb{N}.$$

Then $\mathbb{C}\backslash(V_n \cup K_n)$ is connected. We now recursively construct a sequence (p_n) of polynomials. We set $p_0 := 3$. Let $n \geq 1$ and assume that $p_0, p_1, \ldots, p_{n-1}$ have already been constructed. By 2.2(2) (with $A_1 := V_n$, $B_1 := K_n$, $u_1 := 0$, $v_1 := p_0 + \cdots + p_{n-1}$), there exists a polynomial p_n such that

$$(*) \quad |p_n|_{V_n} \leq 2^{-n}, \quad \min\{|p_0(z) + \cdots + p_{n-1}(z) + p_n(z)| : z \in K_n\} \geq 2^n + 1.$$

The series $\sum p_\nu$ converges *normally* in $\mathbb{E}$ to a function $f \in \mathcal{O}(\mathbb{E})$, since

$$|p_\nu|_{V_n} \leq |p_\nu|_{V_\nu} \leq 2^{-\nu} \quad \text{for all } \nu \geq n$$

and therefore

$$\sum_{\nu=0}^{\infty} |p_\nu|_{V_n} < \infty \quad \text{for all } n \geq 1.$$

Since $K_n \subset V_\nu$ for $\nu > n$, we have $|p_\nu|_{K_n} \le 2^{-\nu}$ for $\nu > n$ by (*). It follows from the second inequality in (*) that, for all $z \in K_n$,

$$\begin{aligned} |f(z)| &\ge |p_0(z) + \cdots + p_{n-1}(z) + p_n(z)| - \sum_{\nu>n}^{\infty} |p_\nu|_{K_n} \\ &\ge 2^n + 1 - \sum_{\nu>n}^{\infty} 2^{-\nu} \ge 2^n. \end{aligned}$$

□

Similarly, we choose a second horseshoe sequence L_n in $\mathbb{E}$ such that L_n covers the ring between K_n and K_{n+1} except for a trapezoid with the real axis as midline (L_n is drawn with a dashed line in the figure). Then, by the lemma, there exists a $g \in \mathcal{O}(\mathbb{E})$ such that $\min\{|g(z)| : z \in L_n\} \ge 2^n$ for all n. By the construction of f and g, it now follows immediately that, for every sequence $z_n \in \mathbb{E}$ with $\lim z_n \in \partial\mathbb{E}\setminus\{1\}$,

$$(**) \qquad \lim(|f(z_n)| + |g(z_n)|) = \infty.$$

This completes the proof of the imbedding theorem. □

Remark. The construction of the functions f and g can be refined so that (**) also holds for all sequences $z_n \in \mathbb{E}$ with $\lim z_n = 1$. H. Cartan noticed this as early as 1931 ([C], p. 301); cf. also [Rub], pp. 187–190. Thus there exist *finite* holomorphic maps $\mathbb{E} \to \mathbb{C}^2$ (for the concept of a finite map, see 9.3). It can be shown that there even exist holomorphic imbeddings of $\mathbb{E}$ into $\mathbb{C}^2$; see, for example, [Gau]. It can also be proved that *every* domain in $\mathbb{C}$ can be imbedded in $\mathbb{C}^3$.

§4. Discussion of the Cauchy Integral Formula for Compact Sets

It is aesthetically unsatisfying that in the Cauchy integral formula, which underlies Runge theory, integrals are taken over line segments rather than closed paths. In this section we give Theorem 1.1 a more attractive form. It turns out that, although we cannot get by with a *single* closed path in general, there always exist finitely many such paths with nice properties. — To formulate the theorem conveniently, we introduce some terminology that is also used in the next chapter. Every (formal) linear combination

$$\gamma = a_1\gamma^1 + \cdots + a_n\gamma^n, \quad a_\nu \in \mathbb{Z}, \quad \gamma^\nu \text{ a closed path in } D, \quad 1 \le \nu \le n,$$

is called a *cycle in D*. The *support* $|\gamma| = \bigcup |\gamma^\nu|$ of γ is compact. *Integrals over cycles* are defined by

$$\int_\gamma f d\zeta := \sum_{\nu=1}^{n} a_\nu \int_{\gamma^\nu} f d\zeta, \quad f \in C(|\gamma|).$$

The *index function*

$$\operatorname{ind}_\gamma(z) := \frac{1}{2\pi i}\int_\gamma \frac{d\zeta}{\zeta - z} \in \mathbb{Z}, \quad z \in \mathbb{C}\backslash|\gamma|,$$

is locally constant. The *interior* and *exterior* of γ are defined by

$$\operatorname{Int}\gamma := \{z \in \mathbb{C}\backslash|\gamma| : \operatorname{ind}_\gamma(z) \neq 0\}, \quad \operatorname{Ext}\gamma := \{z \in \mathbb{C}\backslash|\gamma| : \operatorname{ind}_\gamma(z) = 0\};$$

these sets are *open* in $\mathbb{C}$, and the exterior is *never empty*.

A closed polygon $\tau = [p_1 p_2 \dots p_k p_1]$ composed of k segments $[p_1, p_2]$, $[p_2, p_3], \dots, [p_k, p_1]$ is called a *step polygon* if every segment is horizontal or vertical in $\mathbb{C}$ and all the "vertices" $p_1, p_2, \dots, p_k$ are pairwise distinct. Then every point in τ except p_1 is traversed *exactly once*.

1. Final form of Theorem 1.1. We use the notation of 1.1. Let $\sigma^1, \dots, \sigma^n \in D\backslash K$ be oriented segments of a square lattice, parallel to the axes in $\mathbb{C}$, for which Theorem 1.1 holds. *We may assume that* $\sigma^\mu \neq \pm\sigma^\nu$ *for* $\mu \neq \nu$. The integral formula 1.1(1), applied to $f(z)(z-c)$ with $c \in K$, gives the integral theorem

$$\text{(1)} \qquad \sum_{\nu=1}^{n} \int_{\sigma^\nu} f(\zeta)d\zeta = 0 \quad \textit{for all } f \in \mathcal{O}(D).$$

One consequence of this formula is that the segments σ^ν *automatically* fit together into a step polygon. We claim:

Cauchy integral formula for compact sets (final form). *For every compact set $K \neq \emptyset$ in a region D, there exist finitely many step polygons $\tau^1, \dots, \tau^m$ in $D\backslash K$ such that, for the cycle $\gamma := \tau^1 + \dots + \tau^m$,*

$$\text{(2)} \qquad f(z) = \frac{1}{2\pi i}\int_\gamma \frac{f(\zeta)}{\zeta - z} d\zeta \quad \textit{for all } f \in \mathcal{O}(D) \textit{ and all } z \in K.$$

In particular,

$$\text{(3)} \qquad K \subset \operatorname{Int}\gamma \subset D \quad \textit{and} \quad \operatorname{ind}_\gamma(z) = 1 \quad \textit{for all } z \in K.$$

Proof. We write $\sigma^\nu = [a_\nu, b_\nu]$ and first prove:

(*) *Every point $c \in \mathbb{C}$ is the initial point a_ν exactly as many times as it is the terminal point b_μ of one of the segments $\sigma^1, \dots, \sigma^n$.*

Let k and ℓ denote the *multiplicity* with which c occurs in the n-tuples $(a_1, \dots, a_n)$ and $(b_1, \dots, b_n)$, respectively. We choose a polynomial p such that

$$p(c) = 1, \quad p(a_\nu) = 0 \text{ for } a_\nu \neq c, \quad p(b_\mu) = 0 \text{ for } b_\mu \neq c.$$

Then $\sum_{\nu=1}^{n} p(a_\nu) = \ell$ and $\sum_{\mu=1}^{n} p(b_\mu) = \ell$. By (1),

$$0 = \sum_{\nu=1}^{n} \int_{\sigma^\nu} p'(\zeta)d\zeta = \sum_{\nu=1}^{n} [p(b_\nu) - p(a_\nu)]; \quad \text{hence } \ell = \ell, \text{ i.e. } (*).$$

The next statement now follows easily.

($\ddagger$) *There exist an enumeration of the σ^ν and natural numbers $0 = k_0 < k_1 < \cdots < k_{m+1} = n$ such that $\tau^\mu := \sigma^{k_\mu} + \sigma^{k_\mu+1} + \cdots + \sigma^{k_{\mu+1}}$ is a step polygon, $1 \leq \mu \leq m$.*

Because of $(*)$, finitely many closed polygons can be formed from the σ^ν. Let $\delta = [d_1 d_2 \ldots d_k d_1]$ be such a polygon. We show by induction on k that δ decomposes into step polygons. The case $k = 4$ is clear. If $k > 4$ and δ is not a step polygon, then there exist indices $s < t$ with $d_s = d_t$. Then $[d_1 d_2 \ldots d_s d_{t+1} \ldots d_k d_1]$ and $[d_s d_{s+1} \ldots d_t]$ are closed polygons with fewer than k segments to which the induction hypothesis can be applied. Hence ($\ddagger$) holds. By Theorem 1.1, this proves the existence of a cycle for which (2) holds.

The integral theorem for γ now follows from (2), and $1/(z-c) \in \mathcal{O}(D)$ for all $c \notin D$; hence $\text{Int}\,\gamma \subset D$. Setting $f \equiv 1$ in (2) gives $\text{ind}_\gamma(K) = 1$. □

Warning. In general, there is no closed path γ in $D \backslash K$ with $K \subset \text{Int}\,\gamma \subset D$ and a fortiori no closed path for which formula (2) holds. If K is a circle about 0 in $D := \mathbb{E}^\times$, for instance, then every closed path γ with $K \subset \text{Int}\,\gamma$ lies in the exterior of K. The disc bounded by K then lies in $\text{Int}\,\gamma$, so that $\text{Int}\,\gamma \not\subset \mathbb{E}^\times$. In this example, the theorem holds for all cycles $\gamma^1 + \gamma^2$, where γ^1 and γ^2 are oppositely oriented circles about 0 in $\mathbb{E}^\times$, one of which lies outside and one inside K.

Remark. That the segments $\sigma^1, \ldots, \sigma^n$ in Theorem 1.1 automatically fit together into closed polygons was observed in 1979 by R. B. Burckel; in [B], pp. 259–260, he derives $(*)$ somewhat differently.

2. Circuit theorem. It is intuitively clear that one can go around any connected compact set K in D by following a closed path in $D \backslash K$. This statement is made precise in what follows; in addition to Theorem 1, we need the Jordan curve theorem for step polygons:

Lemma. *Every step polygon divides $\mathbb{C}$ into exactly two domains:*

$$\mathbb{C} \backslash |\tau| = \text{Int}\,\tau \cup \text{Ext}\,\tau \quad \textit{and} \quad \text{ind}_\tau(\text{Int}\,\tau) = \pm 1.$$

We carry out the *proof* in three steps. Let $\tau = [p_1 p_2 \ldots p_k p_1]$, with successive segments $\pi_\kappa = [p_\kappa, p_{\kappa+1}]$, $1 \leq \kappa \leq k$, where $p_{k+1} := p_1$. Around each segment π_κ we place the open rectangle R_κ of length d (= mesh width of the lattice) and width $d/4$, with π_κ as midline. In one subrectangle of

$R_1\backslash|\pi_1|$ we choose a point p, and in the other a point q. We denote by U and V the *components* of $\mathbb{C}\backslash|\tau|$ with $p \in U$, $q \in V$, and first prove that

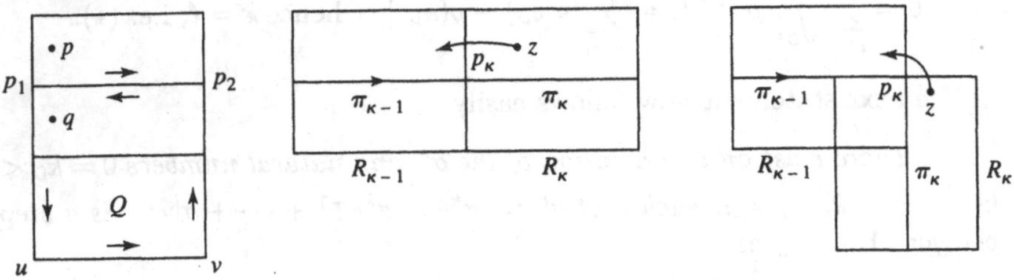

FIGURE 12.4.

a) $\text{ind}_\tau(U) = \text{ind}_\tau(V) \pm 1$; *hence, in particular,* $U \cap V = \emptyset$.

If Q is the square that has $|\pi_1|$ as one side and contains q, let $\gamma \in \{\pm\partial Q\}$ be chosen in such a way that $-\pi_1$ is a segment of γ (first part of Figure 12.4). With the auxiliary polygon $\delta := [p_1uvp_2 \dots p_kp_1]$, we then have

$$\text{ind}_\delta(z) = \text{ind}_\tau(z) + \text{ind}_\gamma(z) \quad \text{for all } z \in \mathbb{C}\backslash(|\delta| \cup [p_1, p_2]).$$

Since the index function is locally constant and $[p, q] \subset \mathbb{C}\backslash|\delta|$, it follows that $\text{ind}_\delta(p) = \text{ind}_\delta(q)$. Since $\text{ind}_\gamma(p) = 0$ and $\text{ind}_\gamma(q) = \pm 1$, this yields $\text{ind}_\tau(p) = \text{ind}_\tau(q) \pm 1$ and therefore a). — We next observe that

b) $R_\kappa\backslash|\pi_\kappa| \subset U \cup V$ *for all* $\kappa = 1, \dots, k$.

Every point in $R_\kappa\backslash|\pi_\kappa|$ can be joined to a point in $R_{\kappa-1}\backslash|\pi_{\kappa-1}|$ by a path in $\mathbb{C}\backslash|\tau|$, $2 \le \kappa \le k$ (the two possible situations are illustrated in the second and third parts of Figure 12.4; since τ is a step polygon, neither of the segments of the grid distinct from $\pi_{\kappa-1}$ and π_κ that intersect at p_κ belongs to τ). Thus b) follows by induction since $R_1\backslash|\pi_1| \subset U \cup V$ trivially, by the choice of U and V.

After these preliminaries, it is easy to complete the proof of the lemma. Let $w \in \mathbb{C}\backslash|\tau|$ be arbitrary. We pass a line g through w such that $g \cap |\tau| \neq \emptyset$ and g intersects no vertex p_k of τ. On g, there is a point $w' \in |\tau|$ that is closest to w. Since w' is not a vertex, some point of the rectangle R_κ lies on the segment $[w, w']$. Since $[w, w') \subset \mathbb{C}\backslash|\tau|$, it follows from b) that $\mathbb{C}\backslash|\tau| \subset U \cup V$. The domains U and V are thus the only components of $\mathbb{C}\backslash|\tau|$. Since $U \cap V = \emptyset$ by a), it follows that either $U = \text{Int}\,\tau$ and $V = \text{Ext}\,\tau$ or vice versa. In both cases, a) implies that $\text{ind}_\tau(\text{Int}\,\tau) = \pm 1$. □

Remark. The proof can be modified so as to hold for all simply closed polygons; it is also easy to see that the polygon is the *common* boundary of its interior and exterior. — The proof above is probably due to A. Pringsheim ([Pr], pp. 41–43).

The theorem now follows from Theorem 1 and the lemma.

Circuit theorem. *If K is a compact set in D, then for every connected subset L of K there exists a closed path (step polygon) τ in $D\backslash K$ such that* $\mathrm{ind}_\tau(L) = 1$.

Proof. Choose $\gamma = \tau^1 + \cdots + \tau^m$ as in Theorem 1. Let $c \in L$. Since $1 = \mathrm{ind}_\gamma(c) = \sum_{\mu=1}^m \mathrm{ind}_{\tau^\mu}(c)$, there exists a k such that $\mathrm{ind}_{\tau^k}(c) \neq 0$. By the lemma, $\mathrm{ind}_\tau(c) = 1$ for $\tau := \tau^k$ or $\tau := -\tau^k$. That $\mathrm{ind}_\tau(L) = 1$ follows because L is connected. □

Remark. One cannot always find τ such that $\mathrm{Int}\,\tau \subset D$; for example, no such τ exists in the case $L = K := \partial\mathbb{E}$ and $D := \mathbb{C}^\times$.

Bibliography

[B] BURCKEL, R. B.: *An Introduction to Classical Complex Analysis*, vol. 1, Birkhäuser, 1979.

[C] CARTAN, H.: *Œuvres 1*, Springer, 1979.

[Gai$_1$] GAIER, D.: *Vorlesungen über Approximation im Komplexen*, Birkhäuser, 1980.

[Gai$_2$] GAIER, D.: Approximation im Komplexen, *Jber. DMV* 86, 151–159 (1984).

[Gau] GAUTHIER, P.: Un plongement du disque unité, *Sém. F. Norguet*, Lect. Notes 482, 333–336 (1975).

[HR] HARTOGS, F. and A. ROSENTHAL: Über Folgen analytischer Funktionen, *Math. Ann.* 100, 212–263 (1928).

[ML] MITTAG-LEFFLER, G.: Sur une classe de fonctions entières, *Proc. 3rd Int. Congr. Math.*, Heidelberg 1904, 258–264, Teubner, 1905.

[Mo] MONTEL, P.: Sur les suites infinies de fonctions, *Ann. Sci. Ec. Norm. Sup.* (3)24, 233–334 (1907).

[N] NEWMAN, D. J.: An entire function bounded in every direction, *Amer. Math. Monthly* 83, 192–193 (1976).

[O] OSGOOD, W. F.: Note on the functions defined by infinite series whose terms are analytic functions of a complex variable; with corresponding theorems for definite integrals, *Ann. Math.* (2)3, 25–34 (1901–1902).

[Pr] PRINGSHEIM, A.: Über eine charakteristische Eigenschaft sogenannter Treppenpolygone und deren Anwendung auf einen Fundamentalsatz der Funktionentheorie, *Sitz. Ber. Math.-Phys. Kl. Königl. Bayrische Akad. Wiss.* 1915, 27–57.

[PS] PÓLYA, G. and G. SZEGÖ: *Problems and Theorems in Analysis*, 2 volumes, trans. C. E. BILLIGHEIMER, Springer, New York, 1976.

[Rub] RUBEL, L. A.: How to use Runge's theorem, *L'Enseign. Math.* (2)22, 185–190 (1976) and Errata ibid. (2)23, 149 (1977).

[Run] RUNGE, C.: Zur Theorie der analytischen Functionen, *Acta Math.* 6, 245–248 (1885).

[Sz] SAKS, S. and A. ZYGMUND: *Analytic Functions*, trans. E. J. SCOTT, 3rd ed., Elsevier, 1971.

13

Runge Theory for Regions

> Jede eindeutige analytische Function kann durch eine einzige unendliche Summe von rationalen Functionen in ihrem ganzen Gültigkeitsbereich dargestellt werden. (Every single-valued analytic function can be represented by a single infinite sum of rational functions in its whole region of validity.)
>
> — C. Runge, 1884

In Chapter 12 we proved approximation theorems for compact sets; we now prove their analogues for regions. We pose the following question:

When are regions D, D' with $D \subset D'$ a Runge pair? That is, when can every function holomorphic in D be approximated compactly by functions holomorphic in D'?

The pair $\mathbb{E}^{\times}$, $\mathbb{C}$ is *not* Runge since $1/z \in \mathcal{O}(\mathbb{E}^{\times})$ cannot be approximated on circles about 0. Topological constraints must be imposed on how D lies in D'. Our starting point is the following theorem.

Runge's approximation theorem. *Every function holomorphic in D can be approximated in D by rational functions that have no poles in D* (Theorem 1.2).

Since, as in the case of compact sets, the location of the poles in $\mathbb{C}\backslash D$ can be well controlled, a bit more thought shows that

D and D′ form a Runge pair if and only if the space $D'\backslash D$ has no compact components (Theorem 2.1).

This contains:

If D has no holes (= compact components of $\mathbb{C}\backslash D$), then every function holomorphic in D can be approximated compactly by polynomials.

The proofs will be carried out in Sections 1 and 2. They involve no function-theoretic difficulties, but some topological obstacles must be overcome. An important tool is Šura-Bura's theorem, proved in the appendix to this chapter, on compact components of locally compact spaces. (References to this appendix will be indicated by A.)

In 1.3, as an application of the approximation theorem, we give a short proof of the main theorem of Cauchy function theory. In Section 2 we prove, among other things, that D, D' form a Runge pair if and only if every cycle in D that is null homologous in D' is already null homologous in D (the Behnke-Stein theorem). In Section 3 we introduce the concept of the holomorphically convex hull; this leads to another characterization of Runge pairs.

§1. Runge's Theorem for Regions

Let $\mathcal{A}$ and $\mathcal{B}$, where $\mathcal{A} \subset \mathcal{B}$, be subalgebras of the $\mathbb{C}$-algebra of all $\mathbb{C}$-valued functions that are continuous in D. The function algebra $\mathcal{A}$ is said to be *dense in* $\mathcal{B}$ if *every* function in $\mathcal{B}$ can be *approximated compactly in D by functions in* $\mathcal{A}$; that is, if for every $g \in \mathcal{B}$ there exists a sequence $f_n \in \mathcal{A}$ that converges compactly to g. We observe immediately:

A function $g \in \mathcal{B}$ can be approximated compactly in D by functions in $\mathcal{A}$ if it can be approximated uniformly on every compact set in D by functions in $\mathcal{A}$.

Proof. Every set $K_n := \{z \in D : |z| \le n \text{ and } d(z, \partial D) \ge 1/n\} \subset D$, $n \ge 1$, is compact. Choose $f_n \in \mathcal{A}$ such that $|g - f_n|_{K_n} \le 1/n$, $n \ge 1$. Since $K_m \subset K_n$ for $m \le n$, we have

$$|g - f_n|_{K_m} \le |g - f_n|_{K_n} \le \frac{1}{n}, \quad 1 \le m \le n.$$

The sequence (f_n) thus converges uniformly to g on every K_m. Since every compact set in D is contained in some K_m (prove this), the sequence f_n converges compactly to g in D. □

In what follows, the components of the (locally compact) space $\mathbb{C}\backslash D$ play a crucial role. They are closed in $\mathbb{C}\backslash D$ (see the appendix) and hence also in $\mathbb{C}$. There may be *uncountably many* bounded and unbounded components;

the reader should sketch examples (Cantor sets, see the appendix). The bounded components are *compact*; we call them — in accord with intuition — the *holes of D (in* $\mathbb{C}$).

Warning. The "theory of holes" is more complicated than one might think at first. Thus isolated boundary points are obviously one-point holes; but it is not at all clear that isolated one-point holes are also isolated boundary points. (Such holes could still be accumulation points of unbounded components of $\mathbb{C}\backslash D$; Šura-Bura's theorem shows that this is impossible.)

1. Filling in compact sets. Runge's proof of Mittag-Leffler's theorem. For the approximation theorem 12.2.1 (version 2), every bounded component of $\mathbb{C}\backslash K$ must intersect the set $\mathbb{C}\backslash D$. By enlarging K, we can ensure that this happens.

(1) *For every compact set K in D, there exists a compact set $K_1 \supset K$ in D such that every bounded component of $\mathbb{C}\backslash K_1$ contains a hole of D.*

Proof. If $D = \mathbb{C}$, let $K_1 \supset K$ be a compact disc. Then $\mathbb{C}\backslash K_1$ has no bounded component. Thus we may assume that $D \neq \mathbb{C}$. We choose ρ with $0 < \rho < d(K, \partial D)$ and set $M := \{z \in D : d(z, \partial D) \geq \rho\}$; then $K \subset M$. It follows immediately from the definition of M that

$$(*) \qquad \mathbb{C}\backslash M = \bigcup_{w \in \mathbb{C}\backslash D} B_\rho(w).$$

Thus M is closed in $\mathbb{C}$. We choose a compact disc $\overline{B}$ with $K \subset \overline{B}$ and set $K_1 := M \cap \overline{B}$. Then K_1 is compact and $K \subset K_1 \subset D$.

Now let Z be a *bounded* component of $\mathbb{C}\backslash K_1$. Since $\mathbb{C}\backslash \overline{B}$ is connected and *unbounded*, and since $\mathbb{C}\backslash K_1 = (\mathbb{C}\backslash \overline{B}) \cup (\mathbb{C}\backslash M)$, it follows that $Z \subset \mathbb{C}\backslash M$. For every disc $B_\rho(w) \subset \mathbb{C}\backslash M$, either $B_\rho(w) \subset Z$ or $B_\rho(w) \cap Z = \emptyset$. Hence, by $(*)$,

$$Z = \bigcup_{w \in Z\backslash D} B_\rho(w).$$

Therefore Z intersects the set $\mathbb{C}\backslash D$. Thus there exists a component S of $\mathbb{C}\backslash D$ such that $Z \cap S \neq \emptyset$. Because $\mathbb{C}\backslash D \subset \mathbb{C}\backslash K_1$, S is a connected subset of $\mathbb{C}\backslash K_1$; hence S lies in the maximal connected subset Z of $\mathbb{C}\backslash K_1$. Since Z is bounded, S is also bounded and therefore a hole of D. □

The construction shows that K_1 is *not uniquely* determined by K. We will see in 3.1 that K_1 can always be chosen to be the *holomorphically convex hull* $\widetilde{K}_D$, and that $\widetilde{K}_D$ is the smallest compact set $K_1 \supset K$ in D with property (1).

Mittag-Leffler's theorem 6.2.2 can be elegantly derived from (1) and the approximation theorem 12.2.1. We use the notation of Chapter 6. Let a principal part distribution (d_ν, q_ν) be given, with support T in $D := \mathbb{C}\backslash T'$. We first choose a sequence $K_1 \subset K_2 \subset \cdots$ of compact sets K_n in D such that $T \cap K_1 = \emptyset$ and every compact subset of D lies in some K_m (exhaustion sequence for D). By (1), we can arrange that every bounded component of $\mathbb{C}\backslash K_n$ intersects the set $\mathbb{C}\backslash D$, $n \geq 1$. Since T is locally finite in D, every set $T_n := T \cap (K_{n+1}\backslash K_n)$, $n \geq 1$, is finite. We may assume that $T_n \neq \emptyset$. Let k_n be the number of points in T_n. Since, for every point $d_\nu \in T_n$, the principal part q_ν is holomorphic in K_n, there exists by 12.2.1 a $g_\nu \in \mathcal{O}(D)$ such that $|q_\nu - g_\nu|_{K_n} \leq 2^{-n}/k_n$. Now $h := \sum_1^\infty (q_\nu - g_\nu)$ is a Mittag-Leffler series for the distribution (d_ν, q_ν): Every compact set in D lies in almost all the K_n and, after omission of finitely many terms, the series is bounded above on K by $\sum 2^{-\nu}$; thus it converges normally in $D\backslash T$. □

Historical note. This proof was given in 1885 by C. Runge ([R], pp. 243–244).

2. Runge's approximation theorem. As an analogue of the approximation theorem 12.2.1, we now have

Runge's theorem on rational approximation. *Let $P \subset \mathbb{C}\backslash D$ be a set whose closure $\overline{P}$ intersects every hole of D. Then the algebra $\mathbb{C}_P[z] \subset \mathcal{O}(D)$ is dense in $\mathcal{O}(D)$.*

Proof. Let K be a compact subset of D. We choose K_1 as in 1(1). Then every bounded component of $\mathbb{C}\backslash K_1$ intersects $\overline{P}$ and hence also P. By the approximation theorem 12.2.1, every function in $\mathcal{O}(D) \subset \mathcal{O}(K_1)$ can therefore be approximated uniformly on K_1 and a fortiori on K by functions in $\mathbb{C}_P[z]$. The claim follows from the observation made in the introduction. □

Setting $P := \emptyset$ gives an immediate corollary.

Runge's polynomial approximation theorem. *If D has no holes, then the polynomial algebra $\mathbb{C}[z]$ is dense in the algebra $\mathcal{O}(D)$.*

We also obtain (even if D has *uncountably many* holes):

For every region D in $\mathbb{C}$, there exists a finite or countable set P of boundary points of D such that the algebra $\mathbb{C}_P[z]$ is dense in the algebra $\mathcal{O}(D)$.

Proof. Every hole of D intersects ∂D (see A.1(3)). Since ∂D is a subspace of $\mathbb{C}$, there exists a countable set $P \subset \partial D$ with $\overline{P} = \partial D$ (second countability of the topology on ∂D). □

3. Main theorem of Cauchy function theory. In I.9.5 it was proved, by an argument due to Dixon, that the Cauchy integral theorem holds for closed paths in D whose interiors lie in D. We alluded then to a short proof using Runge theory. We now give this proof for arbitrary cycles. The following convergence theorem holds for cycles γ just as it does for paths:

(1) *If a sequence $f_n \in C(|\gamma|)$ converges uniformly on $|\gamma|$, then*

$$\lim \int_\gamma f_n d\zeta = \int_\gamma (\lim f_n) d\zeta.$$

A cycle γ in D is called *null homologous in* D if its interior lies in D: $\operatorname{Int}\gamma \subset D$. We can now state and prove the

Main theorem of Cauchy function theory. *The following statements about a cycle γ in D are equivalent.*

i) *The integral theorem $\int_\gamma f\, d\zeta = 0$ holds for all $f \in \mathcal{O}(D)$.*

ii) *The integral formula*

$$\operatorname{ind}_\gamma(z) f(z) = \frac{1}{2\pi i} \int_\gamma \frac{f(\zeta)}{\zeta - z} d\zeta, \quad z \in D\backslash|\gamma|,$$

holds for all $f \in \mathcal{O}(D)$.

iii) *γ is null homologous in D.*

Proof. The equivalence i) $\Leftrightarrow$ ii) follows verbatim as for paths. The implication i) $\Rightarrow$ iii) is trivial, since $1/(\zeta - w) \in \mathcal{O}(D)$ for all $w \in \mathbb{C}\backslash D$. The implication iii) $\Rightarrow$ i) is the core of the theorem and is proved as follows: Let $f \in \mathcal{O}(D)$ be given. By Runge's theorem 2, there exists a sequence q_n, all of whose terms are rational functions with poles only in $\mathbb{C}\backslash D$, which converges uniformly to f on the support $|\gamma|$ of γ. By (1),

$$(*) \qquad \int_\gamma f d\zeta = \lim \int_\gamma q_n d\zeta.$$

Now, if P_n denotes the finite pole set of q_n, we have

$$\frac{1}{2\pi i} \int_\gamma q_n d\zeta = \sum_{c \in P_n \cap \operatorname{Int}\gamma} \operatorname{ind}_\gamma(c) \operatorname{res}_c q_n \quad \text{(residue theorem)}.$$

But $P_n \subset \mathbb{C}\backslash D$ and $\operatorname{Int}\gamma \subset D$; hence $P_n \cap \operatorname{Int}\gamma = \emptyset$ and all the integrals on the right-hand side of $(*)$ vanish.

4. On the theory of holes. Runge's theorems in Subsection 2 lead us to consider the holes of regions. The proof of our first result is quite easy.

(1) *If γ is a cycle in D that is not null homologous, then the interior of γ contains at least one hole of D.*

Proof. Since $\operatorname{Int}\gamma \not\subset D$, there exists a point $a \in \mathbb{C}\backslash D$ with $\operatorname{ind}_\gamma(a) \neq 0$. Since the index function is locally constant and the component L of $\mathbb{C}\backslash D$ containing a is connected, $\operatorname{ind}_\gamma(z) = \operatorname{ind}_\gamma(a)$ for all $z \in L$. Hence $L \subset \operatorname{Int}\gamma$. Since the function $\operatorname{ind}_\gamma$ vanishes when z is large, L is bounded and therefore a hole of D. □

Our next result follows from the contrapositive of (1).

(2) *If D has no holes, then D is (homologically) simply connected.*

The converse of (2) is true but not as easy to prove. We obtain it in 2.4 by a function-theoretic detour. In 14.1.3, we will reverse (1) and show (as is immediately evident) that for every hole L of D there exist closed paths γ with $L \subset \operatorname{Int}\gamma$. The construction of such paths is difficult, since the family of holes of D can be quite unpleasant (Cantor sets). □

The theory of holes is complicated. The definition of holes is based on "how D lies in $\mathbb{C}$"; it is *not* a definition by "intrinsic properties of D," and is not a priori an invariant of D. Perhaps domains with *two* holes could be topologically or even biholomorphically mapped onto domains with *three* holes. We will see in 14.2.2 that this cannot happen.

Although simply connected domains in $\mathbb{R}^2$ ($\simeq \mathbb{C}$), by the discussion above, are precisely those domains that have no holes, domains in $\mathbb{R}^n$, $n \geq 3$, may have holes and still be simply connected: For instance, if (in)finitely many pairwise disjoint compact balls are removed from $\mathbb{R}^n$, $n \geq 3$, what remains is a simply connected domain in $\mathbb{R}^n$ with (in)finitely many holes.

5. On the history of Runge theory. The year 1885 marks the beginning of *complex and real* approximation theory: Runge published his groundbreaking theorem — our Theorem 2 — in *Acta Mathematica* [R]; Weierstrass, in the *Sitzungsberichte der Königlichen Akademie der Wissenschaften zu Berlin*, published his theorem on the approximation of continuous functions by polynomials on real intervals [W]. Runge's approximation theorem did not receive much attention at first; only since the 1920s has it decisively influenced the development of function theory (cf., for example, [G]).

Runge approximates the Cauchy integral by Riemann sums. In doing so, he already obtains rational functions, which, however, still "become infinite at points where the behavior of the function is regular. To remedy this situation," he develops the *pole-shifting technique*, whereby he moves the poles to the boundary. Runge says nothing about approximation by polynomials, although his method also contains the theorem on polynomial approximation. No less a mathematician than D. Hilbert proved this important special case by different means in 1897 [Hi]; Hilbert does not mention Runge's work.

The motive for Runge's investigations was the question whether every domain in $\mathbb{C}$ is a domain of holomorphy. With his approximation theorem, he could give an affirmative answer (see also 5.2.5); he showed moreover that his approximation theorem subsumed that of Mittag-Leffler, proved two years earlier.

Even today, a hundred years later, Runge's method provides the easiest approach to complex approximation theory. Other approaches to Runge theory can be found in [FL], [Hö], and [N].

Runge's theorem was generalized in 1943 by H. Behnke and K. Stein. In their paper [BS], which because of the war did not appear until 1948, they consider arbitrary *noncompact* Riemann surfaces X instead of $\mathbb{C}$; they prove (cf. p. 456 and p. 460):

For every region D in X, there exists a set T of boundary points of D (in X) that is at most countable and has the following property: Every function in $\mathcal{O}(D)$ can be approximated compactly in D by functions meromorphic in X that each have only finitely many poles, all of which lie in T.

This theorem can be used to show, for example, that noncompact Riemann surfaces are Stein manifolds.

§2. Runge Pairs

In this section, D, D' always denote regions in $\mathbb{C}$ with $D \subset D'$. We call D, D' a *Runge pair* if every function in $\mathcal{O}(D)$ can be approximated compactly in D by functions $h|D$, $h \in \mathcal{O}(D')$ (the subalgebra of $\mathcal{O}(D)$ resulting from the restriction $\mathcal{O}(D') \to \mathcal{O}(D)$, $h \mapsto h|D$, is then dense in $\mathcal{O}(D)$). In Subsection 1, we characterize such pairs topologically by properties of the (unpleasant) *locally compact* difference $D'\backslash D$; its components play a crucial role. We need Šura-Bura's theorem from set-theoretic topology (see the appendix to this chapter).

In Subsection 2 we consider *Runge hulls;* in Subsection 3, following Behnke and Stein, we characterize Runge pairs by a *homological* property. In Subsection 4, those functions that can be approximated will be described by an *extension property.*

Runge pairs D, D' with $D' = \mathbb{C}$ are of particular interest; D is then called a *Runge region.* In Subsection 5 we prove, among other things, that every Runge domain $\neq \mathbb{C}$ can be mapped biholomorphically onto the unit disc.

1. Topological characterization of Runge pairs. We begin by proving two lemmas.

(1) *Every component L of $\mathbb{C}\backslash D$ with $L \subset D'$ is a component of $D'\backslash D$.*

Proof. We have $L \subset D'\backslash D$. Since L is connected, there exists a component L' of $D'\backslash D$ with $L' \supset L$. Since $L' \subset \mathbb{C}\backslash D$ and L' is connected, it follows from the definition of components that $L' = L$. □

(2) *For every open compact subset A of $D'\backslash D$, there exists a set V that is open and relatively compact in D' and satisfies $A \subset V$ and $\partial V \subset D$.*

Proof. We have $D'\backslash D = A \cup B$, $A \cap B = \emptyset$, where B is closed in $D'\backslash D$. Since $D'\backslash D$ is closed in D', B is also closed in D'. Hence there exists a covering of A by discs that are relatively compact in $D'\backslash B$. By Heine-Borel there exists a set V that is relatively compact and open in D' and satisfies $A \subset V$, $\overline{V} \cap B = \emptyset$. It follows from $\partial V \cap B = \emptyset = \partial V \cap A$ that $\partial V \cap (D' \backslash D) = \emptyset$. But $\partial V \subset D'$; hence $\partial V \subset D$. □

The next theorem is an easy consequence of (1), (2), and Corollary A.2.1.

Theorem. *The following statements about regions D, D' with $D \subset D'$ are equivalent:*

i) *The space $D' \backslash D$ has no compact component.*
ii) *The algebra of all rational functions without poles in D' is dense in $\mathcal{O}(D)$.*
iii) *D, D' is a Runge pair.*
iv) *In the space $D' \backslash D$, there is no open compact set $\neq \emptyset$.*

Proof. i) ⇒ ii). By (1), the set $P := \mathbb{C} \backslash D'$ intersects every hole of D. Since $P \subset \mathbb{C} \backslash D$, Theorem 1.2 implies that the algebra $\mathbb{C}_P[z]$ is dense in $\mathcal{O}(D)$.

ii) ⇒ iii). This is clear, since rational functions without poles in D' are in $\mathcal{O}(D')$.

iii) ⇒ iv). Let A be an open compact subset of $D' \backslash D$. We choose V as in (2); then ∂V is compact and $\partial V \subset D$. Suppose there were a point $a \in A$. Then $(z-a)^{-1} \in \mathcal{O}(D)$; choose a sequence $g_n \in \mathcal{O}(D')$ with $\lim |(z-a)^{-1} - g_n|_{\partial V} = 0$. But then $\lim |1 - (z-a)g_n(z)|_{\partial V} = 0$. Since $\overline{V} \subset D'$, the sequence $(z-a)g_n(z)$ would converge uniformly to 1 by the maximum principle, which is impossible because $a \in A \subset V$. Hence A is empty.

iv) ⇒ i). This is clear by Corollary A.2.1. □

Behind the purely topological implication iv) ⇒ i) lies Šura-Bura's theorem; see the appendix. In general, compact components of $D' \backslash D$ are *not open in $D' \backslash D$*, as is shown by the examples $D := D' \backslash$ *Cantor sets*. □

We now also have the converse of Theorem 1.2.

Corollary. *If $P \subset \mathbb{C} \backslash D$ is such that the algebra $\mathbb{C}_P[z]$ is dense in $\mathcal{O}(D)$, then $\overline{P}$ intersects every hole of D.*

Proof. $D \subset \mathbb{C} \backslash \overline{P}$ and $\mathbb{C}_P[z] \subset \mathcal{O}(\mathbb{C} \backslash \overline{P})$; hence D, $\mathbb{C} \backslash \overline{P}$ form a Runge pair. It follows from the implication iii) ⇒ i) that $(\mathbb{C} \backslash \overline{P}) \backslash D = \mathbb{C} \backslash (\overline{P} \cup D)$ has no compact component. But this means that $\overline{P}$ intersects every hole of D. □

2. Runge hulls. Let D' be a *given fixed* region in $\mathbb{C}$ (total space). For every subregion D of D', we set

$$\widetilde{D} := D \cup R_D, \text{ where } R_D := \textit{union of all open compact subsets of } D' \backslash D.$$

The set R_D can be quite pathological, e.g. a Cantor set; see A.1.4). By

Corollary 1 of A.2, R_D is the union of *all the compact components of* $D'\backslash D$. The next statement now follows from A.3(1).

(1) *The set* $\widetilde{D}$ *is a subregion of* D'. *The difference* $D'\backslash\widetilde{D}$ *contains no open compact set.*

$\widetilde{D}$ is called the *Runge hull of* D *(in* D'*)*. When $D' = \mathbb{C}$, we obtain $\widetilde{D}$ from D by "plugging up all the holes of D." We now justify the choice of the term "Runge hull."

Proposition. *The pair* $\widetilde{D}$, D' *is Runge.* $\widetilde{D} = D$ *if and only if* D, D' *is a Runge pair. For every Runge pair* E, D' *with* $D \subset E$, *we have* $\widetilde{D} \subset E$.

Proof. The first two statements are clear by (1) and Theorem 1. — If $D \subset E \subset D'$, then $D'\backslash E$ is closed in $D'\backslash D$. Thus $(D'\backslash E) \cap K$ is compact and open in $D'\backslash E$ for every open compact subset K of $D'\backslash D$. Since E, D' is a Runge pair, Theorem 1 implies that $(D'\backslash E) \cap K$ is empty. It follows that $K \subset E$; therefore $R_D = \bigcup K \subset E$ and $\widetilde{D} = D \cup R_D \subset E$. □

We see that $\widetilde{D}$ is the "smallest region between D and D' that is relatively Runge in D'." It follows immediately from the proposition that

$$\widetilde{\widetilde{D}} = \widetilde{D}, \quad D \subset E(\subset D') \Rightarrow \widetilde{D} \subset \widetilde{E}.$$

3. Homological characterization of Runge hulls. The Behnke-Stein theorem. Let the regions D, D' be given, with $D \subset D'$. We denote by $\mathscr{S}$ the family of all cycles γ in D that are null homologous in D', hence for which $\operatorname{Int}\gamma$ is a subregion of D'.

Theorem. $\widetilde{D} = \bigcup_{\gamma\in\mathscr{S}} \operatorname{Int}\gamma$.

Proof. Let $\widetilde{D} = D \cup R_D$. If $w \in D$, then $w \in \operatorname{Int}\gamma$ for every small circle $\gamma \in \mathscr{S}$ about w. If $w \in R_D$, then w lies in an *open* compact subset K of $D'\backslash D$. Then $D_0 := D \cup K$ is a subregion of D' (see A.3(1)). By 12.4.1(3), there exists a cycle γ in $D = D_0\backslash K$ such that $K \subset \operatorname{Int}\gamma \subset D_0$. In particular, $\gamma \in \mathscr{S}$ and $w \in \operatorname{Int}\gamma$.

For the opposite inclusion, let $w \in \operatorname{Int}\gamma$ with $\gamma \in \mathscr{S}$. In order to show that $w \in \widetilde{D}$, we may assume that $w \notin D$. Then, since $|\gamma| \subset D$, we have $Z \cap |\gamma| = \emptyset$ for the component Z of $D'\backslash D$ containing w. The index function $\operatorname{ind}_\gamma$ is now well defined and constant on Z. Since $w \in \operatorname{Int}\gamma$, it follows that $Z \subset \operatorname{Int}\gamma$. Hence $Z \subset (\operatorname{Int}\gamma \cup |\gamma|)\backslash D$. Thus Z is compact. It follows that $Z \subset R_D$ and therefore that $w \in R_D \subset D$. □

Corollary (Behnke-Stein theorem). *A pair* D, D' *of regions with* $D \subset D'$ *is Runge if and only if every cycle* γ *in* D *that is null homologous in* D' *is already null homologous in* D.

Proof. D, D' is a Runge pair if and only if $D = \widetilde{D}$, i.e. if and only if $\operatorname{Int}\gamma \subset D$ for all $\gamma \in \mathscr{S}$. □

In the theorem and the corollary, we *must* admit cycles, not just closed paths. If $D := \mathbb{C}^\times \setminus \partial\mathbb{E}$, $D' := \mathbb{C}^\times$, then $\widetilde{D} = \mathbb{C}^\times$ but $\operatorname{Int}\gamma \subset D$ for *every path* $\gamma \in \mathscr{S}$. All cycles $\gamma^1 + \gamma^2$ consisting of oppositely oriented circles γ^1 in $\mathbb{E}^\times$ and $\gamma^2 \in \mathbb{C}\setminus\overline{\mathbb{E}}$ about 0 are null homologous in D' but not in D.

If D is a domain G, we can get by with paths $\gamma \in \mathscr{S}$ in the theorem; every cycle in G can then be turned into a closed path by joining the components of the cycle by paths in G.

Historical note. H. Behnke and K. Stein proved their theorem in 1943 in [BS] for arbitrary *noncompact* Riemann surfaces; they state the homology condition in such a way that D *is simply connected relative to* D' (loc. cit., pp. 444–445).

4. Runge regions. A region D in $\mathbb{C}$ is called a Runge region if every function holomorphic in D can be approximated compactly in D by polynomials. Since $\mathbb{C}[z]$ is dense in $\mathcal{O}(\mathbb{C})$, D is a Runge region if and only if D, $\mathbb{C}$ is a Runge pair. The statements of Theorem 1 and the Behnke-Stein theorem make possible a simple characterization of Runge regions.

Theorem. *The following statements about a region D in $\mathbb{C}$ are equivalent:*

i) *D has no holes.*
ii) *D is a Runge region.*
iii) *There is no open compact set $\neq \emptyset$ in the space $\mathbb{C}\setminus D$.*
iv) *D is homologically simply connected.*
v) *Every component of D is a simply connected domain.*

Proof. i) ⇒ ii) ⇒ iii) ⇒ i). This is Theorem 1, i) ⇒ iii) ⇒ iv) ⇒ i), with $D' = \mathbb{C}$.

ii) ⇔ iv). This is the Behnke-Stein theorem with $D' = \mathbb{C}$, since *all* cycles are null homologous in $\mathbb{C}$.

iv) ⇔ v). Since D is homologically simply connected if and only if every component of D is, the assertion follows from Theorem 8.2.6, i) ⇔ ix). □

The equivalence i) ⇔ v) says in particular that the simply connected domains in $\mathbb{R}^2$ are precisely the domains without holes; see 1.4. The next statement follows immediately.

A bounded domain G in $\mathbb{C}$ is simply connected if and only if $\mathbb{C}\setminus G$ is connected.

There are, however, *unbounded* simply connected domains G in $\mathbb{C}$ for which $\mathbb{C}\setminus G$ has (in)finitely many components (the reader should sketch examples).

With the aid of i) ⇒ v), the injection theorem 8.2.2 can be generalized:

For every domain G in $\mathbb{C}$ whose complement $\mathbb{C}\backslash G$ has a component with more than one point, there exists a holomorphic injection $G \to \mathbb{E}$.

Proof. If $\mathbb{C}\backslash G$ has an *unbounded* component Z, then $\mathbb{C}\backslash Z \neq \mathbb{C}$ is a domain (!) without holes and can therefore be mapped biholomorphically onto $\mathbb{E}$. Hence $G \subset \mathbb{C}\backslash Z$ can be mapped biholomorphically and injectively into $\mathbb{E}$.

Now let M be a component of $\mathbb{C}\backslash G$ that has more than one point. Then $z \mapsto 1/(z-a)$, $a \in M$, maps $\mathbb{C}\backslash M$ biholomorphically onto a domain G_1 whose complement has at least one unbounded component. Thus G_1 admits a holomorphic injection into $\mathbb{E}$; hence so does $G \subset \mathbb{C}\backslash M \simeq G_1$. □

5*. Approximation and holomorphic extendibility. If D, D' is not a Runge pair, then not all functions in $\mathcal{O}(D)$ can be approximated in D by functions in $\mathcal{O}(D')$. We prove:

Theorem. *The following statements about a function $f \in \mathcal{O}(D)$ are equivalent:*

i) *f can be approximated compactly in D by functions in $\mathcal{O}(D')$.*
ii) *There exists a single-valued holomorphic extension $\widetilde{f}$ of f to the Runge hull $\widetilde{D}$.*

Proof. ii) ⇒ i). Clear, since $\widetilde{f}|D = f$ and $\widetilde{D}$, D' is a Runge pair by Theorem 2.

i) ⇒ ii). We define $\mathscr{S}$ as in Subsection 3 and consider, in D', all the subregions $D_\gamma := D \cup \operatorname{Int}\gamma$, $\gamma \in \mathscr{S}$. Let a pair D_γ, f_γ with $f_\gamma \in \mathcal{O}(D_\gamma)$, $\gamma \in \mathscr{S}$, be called an *f-extension* if $f_\gamma|D = f$. We claim:

a) *For every $\gamma \in \mathscr{S}$, there exists an f-extension.*
b) *If D_γ, f_γ and D_δ, f_δ are f-extensions, then $f_\gamma = f_\delta$ on $D_\gamma \cap D_\delta$.*

It follows immediately from a) and b) that there exists exactly one holomorphic function $\widetilde{f}$ on $\bigcup_{\gamma \in \mathcal{S}} D_\gamma$ such that $\widetilde{f}|D_\gamma = f_\gamma$, $\gamma \in \mathcal{S}$. Theorem 3 then implies ii).

For the proof of a), we choose a sequence $g_n \in \mathcal{O}(D')$ that converges compactly to f in D. Then, since $\lim |f - g_n|_{|\gamma|} = 0$, g_n is a Cauchy sequence on $|\gamma|$ for every $\gamma \in \mathscr{S}$. Since $\operatorname{Int}\gamma$ is a bounded region with $\partial(\operatorname{Int}\gamma) \subset |\gamma|$, the maximum principle implies that g_n is also a Cauchy sequence in $\operatorname{Int}\gamma$ and hence in D_γ. Let $f_\gamma := \lim_{n\to\infty}(g_n|D_\gamma)$; then D_γ, f_γ is an f-extension.

For the proof of b), it suffices (by the identity theorem) to prove that every component Z of $(\operatorname{Int}\gamma) \cap (\operatorname{Int}\delta)$ intersects the region D. But this is clear since

$$\partial Z \subset \partial(\operatorname{Int}\gamma) \cup \partial(\operatorname{Int}\delta) \subset |\gamma| \cup |\delta| \subset D. \qquad \square$$

§3. Holomorphically Convex Hulls and Runge Pairs

If M is a subset of a region D in $\mathbb{C}$, the set

$$\widehat{M}_D := \{z \in D : |f(z)| \leq |f|_M \text{ for all } f \in \mathcal{O}(D)\}$$

is called the *holomorphically convex hull of M in D*. This is often written $\widehat{M}$ instead of $\widehat{M}_D$. For the circle $S := \partial\mathbb{E}$, we have $\widehat{S}_{\mathbb{C}} = \overline{E}$ and $\widehat{S}_{\mathbb{C}^\times} = S$. The next statement follows immediately from the maximum principle for bounded domains.

If V is relatively compact and open in D, then $(\widehat{\partial V})_D \supset \overline{V}$.

Let α, β, $r \in \mathbb{R}$, $z = x + iy$. Then, since $|e^{(\alpha+i\beta)z}| \leq e^r$ if and only if $\alpha x - \beta y \leq r$, $\widehat{M}$ always lies in the intersection of all the closed half-planes containing M. This implies:

The holomorphically convex hull $\widehat{M}$ is contained in the linearly convex hull of M.

Runge pairs D, D' can be characterized analytically by the property that every compact set $K \subset D$ has the same holomorphically convex hull in D and in D'; this equivalence is the main result of this final section on Runge theory.

1. Properties of the hull operator. The following properties are immediate from the definition:

(1) *$\widehat{M}$ is closed in D. Moreover,*

$$M \subset \widehat{M} = \widehat{\widehat{M}} \subset D, \ M \subset M' \Rightarrow \widehat{M} \subset \widehat{M'}, \text{ and } D \subset D' \Rightarrow \widehat{M}_D \subset \widehat{M}_{D'}.$$

(2) *$c \in D\backslash\widehat{M}$ if and only if there exists an $h \in \mathcal{O}(D)$ such that $|h|_M < 1 < |h(c)|$.*

The next property is important for what follows.

(3) *It is always true that $d(M, \mathbb{C}\backslash D) = d(\widehat{M}, \mathbb{C}\backslash D)$. If M is compact, so is $\widehat{M}$.*

Proof. Since $M \subset \widehat{M}$, we need only prove that $d(M, \mathbb{C}\backslash D) \leq d(\widehat{M}, \mathbb{C}\backslash D)$. Let $\zeta \notin D$. Since $(z-\zeta)^{-1} \in \mathcal{O}(D)$, we have $|w-\zeta|^{-1} \leq \sup\{|z-\zeta|^{-1} : z \in M\}$ for all $w \in \widehat{M}$. It follows that $d(\widehat{M}, \zeta) = \inf\{|z-\zeta| : z \in M\} \leq |w-\zeta|$ for all $w \in \widehat{M}$; hence $d(M,\zeta) \leq d(\widehat{M},\zeta)$ for all $\zeta \in \mathbb{C}\backslash D$. — If M is compact, then $\widehat{M}$ is *closed in* $\mathbb{C}$ because $0 < d(M, \mathbb{C}\backslash D) = d(\widehat{M}, \mathbb{C}\backslash D)$. Since $\widehat{M}$ lies in the linearly convex hull of M, which is bounded if M is, $\widehat{M}$ is also bounded. □

By (2), $M = \widehat{M}_D$ if and only if, for every $c \in D\backslash M$, there exists an $h \in \mathcal{O}(D)$ such that $|h(c)| > |h|_M$. Our next result thus follows directly from Theorem 12.2.3.

Theorem. *The following statements about a compact set K in D are equivalent:*

i) $K = \widehat{K}_D$.
ii) *The space $D\backslash K$ has no component that is relatively compact in D.*
iii) *Every bounded component of $\mathbb{C}\backslash K$ contains a hole of D.*
iv) *The algebra $\mathcal{O}(D)$ is dense in $\mathcal{O}(K)$.*

For the *proof*, it suffices to observe that a bounded component of $\mathbb{C}\backslash K$ contains a hole of D if and only if it intersects $\mathbb{C}\backslash D$. □

The equivalence i) ⇔ iii) immediately yields a more precise form of 1.1(1):

For every compact set K in D, there exists a smallest compact set $K_1 \supset K$ in D, namely $\widehat{K}_D$, such that every bounded component of $\mathbb{C}\backslash K_1$ contains a hole of D.

The implication i) ⇒ ii) can be generalized as follows.

(4) *For every compact set K in D, $\widehat{K}_D$ is the union of K and all the components of $D\backslash K$ that are relatively compact in D.*

Proof. Let A denote the union of all the components of $D\backslash K$ that are relatively compact in D; let B denote the union of all the remaining components of $D\backslash K$. Since each component is a domain,

$$A \text{ and } B \text{ are open in } D, \quad D\backslash K = A \cup B, \quad \text{and} \quad A \cap B = \emptyset.$$

We set $M := K \cup A$. Then, by 12.2.3(1) and (1), $M \subset \widehat{K}_D \subset \widehat{M}_D$. Hence $M = \widehat{K}_D$ if $M = \widehat{M}_D$. Since $D\backslash M = B$ is open, M is closed in D and therefore compact, since $\widehat{K}_D$ is compact by (3). By the definition of B, no component of $D\backslash M$ is relatively compact in D; hence the theorem implies that $M = \widehat{M}_D$. □

Thus passage from K to $\widehat{K}_D$ "fills in the holes of $D\backslash K$ in D." The purely topological description of holomorphically convex hulls given by (4) is occasionally used in the literature as the definition of $\widehat{K}_D$; cf. [N], pp. 112–113, and [FL], p. 204.

By a simple argument, one can also show (cf. [N], pp. 112–113):

If $K = \widehat{K}_D$, then $\mathbb{C}\backslash K$ has only finitely many components.

The concept of the holomorphically convex hull comes from the function theory of several variables; it was developed in the 1930s (cf. [BT], Chapter VI, and the appendix by O. Forster). For regions D in $\mathbb{C}^n$, $n > 1$, the hull $\widehat{K}_D$ of a compact set $K \subset D$, which is defined word for word as above, is in general *no longer compact*. The following holds:

A domain G in $\mathbb{C}^n$, $1 \le n < \infty$, is a domain of holomorphy if and only if, for every compact set $K \subset G$, the holomorphically convex hull $\widehat{K}_G$ has only compact components (weak holomorphic convexity of G); this occurs if and only if $\widehat{K}_G$ is always compact (holomorphic convexity of G).

2. Characterization of Runge pairs by means of holomorphically convex hulls. *The following statements about regions D, D' with $D \subset D'$ are equivalent:*

i) *D, D' is a Runge pair.*
ii) *For every compact set $K \subset D$, $\widehat{K}_{D'} = \widehat{K}_D$.*
iii) *For every compact set $K \subset D$, $D \cap \widehat{K}_{D'} = \widehat{K}_D$.*
iv) *For every compact set $K \subset D$, $D \cap \widehat{K}_{D'}$ is compact.*

Proof. We argue according to the following diagram:

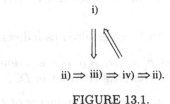

i)

$\Downarrow \quad \nwarrow$

ii) $\Rightarrow$ iii) $\Rightarrow$ iv) $\Rightarrow$ ii).

FIGURE 13.1.

The implications ii) $\Rightarrow$ iii) $\Rightarrow$ iv) are clear, the last one because of 1(3).

i) $\Rightarrow$ iii). Since $\widehat{K}_D \subset \widehat{K}_{D'}$ and $\widehat{K}_D \subset D$, all we need to prove is that $D \cap \widehat{K}_{D'} \subset \widehat{K}_D$, or, equivalently, that $D\backslash\widehat{K}_D \subset D\backslash\widehat{K}_{D'}$. Let $c \in D\backslash\widehat{K}_D$. By 1(2), there exists an $h \in \mathcal{O}(D)$ with $|h|_K < 1 < |h(c)|$. Since h can be approximated uniformly on $K \cup \{c\}$ by functions in $\mathcal{O}(D')$, there exists a $g \in \mathcal{O}(D')$ with $|g|_K < 1 < |g(c)|$. It follows from 1(2) that $c \notin \widehat{K}_D$.

iv) $\Rightarrow$ i) *and* ii). Not only is $K' := D \cap \widehat{K}_{D'}$ compact, but (as the intersection of a closed set with a compact set) so is $K'' := (\mathbb{C}\backslash D) \cap \widehat{K}_{D'}$. Let $f \in \mathcal{O}(D)$ be arbitrary. Since $K' \cap K'' = \emptyset$, setting

$$h(z) := f(z) \text{ for } z \in K' \quad \text{and} \quad h(z) := 2 \text{ for } z \in K''$$

defines a function $h \in \mathcal{O}(K' \cup K'')$. Since $\widehat{K}_{D'} = K' \cup K''$, Theorem 1 — applied to $\widehat{K}_{D'}$ in D' — implies that h can be approximated uniformly by functions in $\mathcal{O}(D')$. This proves i) since $K \subset K'$. In particular, choosing $f = 0$ shows that there exists a function $g \in \mathcal{O}(D')$ with $|g|_K < 1 < |g(w)|$ for all $w \in K''$. It follows that $K'' = \emptyset$, since otherwise 1(2) gives the contradiction $w \notin \widehat{K}_{D'}$. Hence $\widehat{K}_{D'} = D \cap \widehat{K}_{D'}$. Since iii) also holds by i) and what has already been proved, we see that $\widehat{K}_D = D \cap \widehat{K}_{D'} = \widehat{K}_{D'}$. □

The theorem also holds verbatim for regions in $\mathbb{C}^n$, $1 \le n < \infty$, if D and D' are assumed to be *regions of holomorphy*; cf. [Hö], p. 91. A purely topological characterization of Runge pairs — for example, by analogy with the Behnke-Stein theorem — is no longer possible when $n > 1$. For more on this, see the appendix by O. Forster to Chapter VI of [BT], where general Runge pairs of Stein spaces are considered.

Every set M in $\mathbb{C}$ is assigned its *polynomially convex hull*

$$M' := \{z \in \mathbb{C} : |p(z)| \le |p|_M \text{ for all polynomials } p\};$$

we have $M' = \widehat{M}_{\mathbb{C}}$. A region D in $\mathbb{C}$ is called *polynomially convex* if, for every compact set $K \subset D$, the (compact) set K' lies in D. The theorem contains the following statement:

A region in $\mathbb{C}$ is Runge if and only if it is polynomially convex.

This equivalence also remains valid for domains in $\mathbb{C}^n$.

Appendix: On the Components of Locally Compact Spaces. Šura-Bura's Theorem

Theorem 2.1 says in particular that a difference $D' \backslash D$ has compact connected components if and only if it contains nonempty open compact sets. This is a theorem of set-theoretic topology (which does not appear in current textbooks). We prove it here in a more general situation; cf. Theorem 2 and Corollary 2.1. — X always denotes a topological space.

1. Components. We consider connected subspaces of X; every one-point subspace $\{x\}$, $x \in X$, is connected. If $A_\imath$, $\imath \in J$, is a family of connected subspaces of X such that any two have a nonempty intersection, then the union $\bigcup A_\imath$ is connected. Hence the union of all the connected subspaces of X containing a fixed point is a *maximal connected* subspace of X. Every such subspace is called a *component* — more precisely, a *connected component* — *of* X.

(1) *Distinct components of X are disjoint. Every component of X is closed in X.*

The last statement follows because the closure $\overline{A}$ of A is connected whenever A is. — In general, components of X are *not open* in X.

(2) *Every subspace of X that is both open and closed is the union of components of X.*

Examples. 1) Let $X := \mathbb{Q} \subset \mathbb{R}$, equipped with the relative topology. The components are the points of X; no component of X is open; there are no open compact sets $\ne \emptyset$ in X.

2) Let $X := \{0, 1, 1/2, \dots, 1/n, \dots\} \subset \mathbb{R}$, equipped with the relative topology. The components of X are the points of X; every component $\neq \{0\}$ is open in X. The component $\{0\}$ is the intersection of all the open compact subsets of X containing it.

3) Let X be a *Cantor set* in $[0, 1]$. The components are the points of X; no component of X is open; there are open compact sets $\neq \emptyset$ in X.

4) For regions $D \subset D' \subset \mathbb{C}$, the difference $D' \backslash D$ is *locally compact.* In the case $D := D' \backslash$ *Cantor set*, $D' \backslash D$ has *uncountably many* compact components, and no union of D with finitely many of these components is a region.

The following was used in 1.2.

(3) *If D' is a domain, then every component Z of $D' \backslash D$ intersects the boundary of D.*

Proof. Let $a \in Z$, let $b \in D$, and let γ be a path in D' from a to b. On γ, there is a "first" point $c \in \partial D$. Since Z is connected and the subpath $\widehat{\gamma}$ from a to c passes through $D' \backslash D$, we have $|\widehat{\gamma}| \subset Z$; hence $c \in Z$. □

2. Existence of open compact sets. *Every compact component A of a locally compact (Hausdorff) space X has a neighborhood base in X consisting of open compact subsets of X.*

This theorem was proved in 1941 by M. Šura-Bura for (bi)compact spaces; cf. [SB]. The theorem appears implicitly in N. Bourbaki; cf. [Bo], p. 205, *corollary and its proof.* The significance of the theorem for function theory was pointed out by R. B. Burckel; cf. [Bu]. We prove Šura-Bura's theorem in Subsection 4, but we first derive a few of its consequences here and in Subsection 3. Since open compact sets, by 1(2), are always the union of compact components, our first result is immediate.

Corollary 1. *A locally compact space X has compact components if and only if there exist nonempty open compact sets in X. The union of all the compact components of X coincides with the union of all the open compact subsets of X, and in particular is open in X.*

This contains the equivalence statement i) $\Leftrightarrow$ iv) of Theorem 2.1.

Corollary 2. *If the locally compact space X has only finitely many compact components, then each of these components is open in X.*

Proof. Let A be a compact component of X. If $A_1, \dots, A_k$ are the remaining compact components, then $U := X \backslash (A_1 \cup \dots \cup A_k)$ is a neighborhood of A which intersects no other compact component of X. By the theorem, there exists an open compact set B in X such that $A \subset B \subset U$. Since B is the union of compact components, it follows that $A = B$. □

Corollary 3. *Let X be a connected compact space containing more than one point, and let $p \in X$. Then p is an accumulation point of every component of $X\backslash\{p\}$.*

Proof. Let A be a component of $X\backslash\{p\}$. Since A is closed in $X\backslash\{p\}$, if p were not in $\overline{A}$ then A would also be closed in X and therefore compact. Since $X\backslash\{p\}$ is locally compact, there would exist an open compact subset B of $X\backslash\{p\}$ such that $B \supset A$. But then X would not be connected, since $X = B \cup (X\backslash B)$, with B and $X\backslash B$ nonempty closed subsets of X. □

3. Filling in holes. We apply Šura-Bura's theorem to the difference $D'\backslash D$ of two regions $D \subset D' \subset \mathbb{C}$; see Example 1.4). We first observe:

(1) *If M is open in $D'\backslash D$, then $D \cup M$ is a subregion of D'. If M is also a union of components of $D'\backslash D$, then the components of $D'\backslash(D \cup M)$ are exactly those components of $D'\backslash D$ that do not lie in M.*

For the proof, it suffices to show that $D \cup M$ is open in D'. Since there exists a set U that is open in D' and satisfies $M = (D'\backslash D) \cap U$, it follows that

$$D \cup M = D \cup [(D'\backslash D) \cap U] = D \cup (U\backslash D) = D \cup U. \qquad \square$$

An important application of (1) and Corollary 2.1 was already given in 13.2.2(1). Here we note further:

(2) *If $D'\backslash D$ has exactly n compact components $L_1, \dots, L_n$, $1 \le n \le \infty$, then $D \cup L_1$ is a subregion of D'; the space $D'\backslash(D \cup L_1)$ has exactly the $(n-1)$ compact components $L_2, \dots, L_n$.*

(3) *If L is a compact component of $\mathbb{C}\backslash D$ (a hole of D), N is closed in D, and $L \cap N = \emptyset$, then there exists a compact subset K of $\mathbb{C}\backslash D$ such that $L \subset K \subset \mathbb{C}\backslash N$ and $D \cup K$ is a region in $\mathbb{C}$.*

Proof. (2) is clear by (1) and Corollary 2.2. — *ad* (3). Since $(\mathbb{C}\backslash N) \cap (\mathbb{C}\backslash D)$ is a neighborhood of L in $\mathbb{C}\backslash D$, there exists by Šura-Bura a compact set K that is *open* in $\mathbb{C}\backslash D$ and satisfies $L \subset K \subset \mathbb{C}\backslash N$. By (1), $D \cup K$ is then a region. □

We will need statement (3) in 14.1.3.

4. Proof of Šura-Bura's theorem. We first reduce the claim to the compact case. Thus assume that the theorem has already been proved for compact spaces. Let U be any neighborhood of A in X. Since X is locally compact, there exists an open neighborhood V of A in X whose closure $\overline{V}$ is a compact subset of U. Now A is also a component of the space $\overline{V}$ (every connected subspace of $\overline{V}$ is also connected as a subspace of X). By hypothesis, there exists an open compact subset B of $\overline{V}$ such that $A \subset B \subset V$. Then B is also open in V and hence in X. Thus B is an open compact subset of X such that $A \subset B \subset U$. □

The reduction step can be carried out more easily by passing from X to the Alexandroff compactification $X \cup \{\infty\}$.

Now let X be compact. If A is *any* compact set in X, we denote by $\mathcal{F}$ the family of all *open compact* sets F in X with $F \supset A$. Then $X \in \mathcal{F}$. The intersection B of all the sets in $\mathcal{F}$ is compact and contains A.

(o) *Every set U that is open in X and satisfies $U \supset B$ contains an element of $\mathcal{F}$.*

Proof. We have $(X \backslash U) \cap \bigcap_{F \in \mathcal{F}} F = \emptyset$. Since $X \backslash U$ is compact, there exist *finitely many* sets $F_1, \ldots, F_p \in \mathcal{F}$ such that $(X \backslash U) \cap \bigcap_{j=1}^{p} F_j = \emptyset$.[1] — Since $\bigcap_{j=1}^{p} F_j \in \mathcal{F}$, this proves (o). □

We now prove Theorem 2 for compact spaces X. Let A be a *compact component* of X. We retain our earlier notation. If we prove that B is connected, it will follow from $A \subset B$ that $A = B$ since A is a *maximal connected* subspace of X. Hence it suffices to prove the following:

If $B = B_1 \cup B_2$, where B_1 and B_2 are disjoint sets that are closed in X, then either B_1 or B_2 is empty.

Since $A = (B_1 \cap A) \cup (B_2 \cap A)$ and A is connected, either $A = B_1 \cap A$ or $A = B_2 \cap A$. Suppose that $A \subset B_1$. Since B_1, B_2 are disjoint compact sets, there are sets V_1, V_2 that are *open* in X and satisfy $B_1 \subset V_1$, $B_2 \subset V_2$, and $V_1 \cap V_2 = \emptyset$. Since $B \subset V_1 \cup V_2$, there exists by (o) an $F \in \mathcal{F}$ such that $B \subset F \subset V_1 \cup V_2$. But now (!)

$$F \cap (X \backslash V_2) = F \cap V_1 =: W.$$

Since F and V_1 are open and F and $X \backslash V_2$ are compact, this shows that W is a compact set that is open in X. The inclusions $A \subset B \subset F$ and $A \subset B_1 \subset V_1$ imply that $A \subset W$; thus $W \in \mathcal{F}$ and $B \subset W \subset V_1$. Then $B \cap V_2$ is empty, whence $B_2 = \emptyset$. — This proves Theorem 2. □

Bibliography

[Bo] BOURBAKI, N.: *General Topology*, Chapters I and II, Springer, 1989.

[BS] BEHNKE, H. and K. STEIN: Entwicklung analytischer Funktionen auf Riemannschen Flächen, *Math. Ann.* 120, 430–461 (1947–49).

[1] Here we use the following

Lemma: Suppose that X is compact and N is a collection of closed subsets of X such that $\bigcap_{N \in \mathcal{N}} N_j = \emptyset$. Then there exist finitely many sets $N_1, \ldots, N_p \in \mathcal{N}$ such that $\bigcap_{j=1}^{p} N_j = \emptyset$.

To prove this, observe that the open cover $\{X \backslash N\}_{N \in \mathcal{N}}$ of X contains a *finite* subcover. — We apply the lemma to the family $\mathcal{F} \cup \{X \backslash U\}$.

[BT] BEHNKE, H. and P. THULLEN: *Theorie der Funktionen mehrerer komplexer Veränderlichen*, 2nd expanded ed., Springer, 1970.

[Bu] BURCKEL, R. B.: *An Introduction to Classical Complex Analysis*, vol. 1, Birkhäuser, 1979.

[FL] FISCHER, W. and I. LIEB: *Funktionentheorie*, Vieweg, 1980.

[G] GAIER, D.: Approximation im Komplexen, *Jber. DMV* 86, 151–159 (1984).

[Hi] HILBERT. D.: Ueber die Entwickelung einer beliebigen analytischen Function einer Variabeln in eine unendliche nach ganzen rationalen Functionen fortschreitende Reihe, *Nachr. Königl. Ges. Wiss. Göttingen, Math.-phys. Kl.* 63–70 (1897); *Ges. Abh.* 3, 3–9.

[Hö] HÖRMANDER, L.: *An Introduction to Complex Analysis in Several Variables*, 2nd. ed., North-Holland Publ. Co., 1973.

[N] NARASIMHAN, R.: *Complex Analysis in One Variable*, Birkhäuser, 1985.

[R] RUNGE, C.: Zur Theorie der eindeutigen analytischen Functionen, *Acta Math.* 6, 229–244 (1885).

[SB] ŠURA-BURA, M. R.: Zur Theorie der bikompakten Räume, *Rec. Math. Moscou* [*Math. Sbornik*], (2), 9, 385–388 (Russian with German summary), 1941.

[W] WEIERSTRASS, K.: Über die analytische Darstellbarkeit sogenannter willkürlicher Functionen reeller Argumente, *Sitz. Ber. Königl. Akad. Wiss. Berlin*, 1885; *Math. Werke* 3, 1–37.

14

Invariance of the Number of Holes

Is it intuitively clear that biholomorphically (more generally, topologically) equivalent domains have *the same number of* holes? There is no direct proof of this invariance theorem. The property of "having the same number of holes" is defined by how G lies in $\mathbb{C}$ and at first glance is not an invariant of G. In order to prove the invariance of the number of holes, we assign every domain in $\mathbb{C}$ its *(first) homology group.* The *rank* of this group, called the *Betti number of* G, is a biholomorphic (even topological) invariant of the domain.

The invariance of the number of holes now follows from the equation

number of holes of G = *Betti number of* G.

We carry out the proof in 2.2 with the aid of special families of paths, which we call *orthonormal.* We obtain the (intuitively clear) existence of such families of paths in 1.3, using Šura-Bura's theorem and the circuit theorem 12.4.2.

§1. Homology Theory. Separation Lemma

In Subsection 1, we assign every region in $\mathbb{C}$ its (first) *homology group* (with coefficients in the ring $\mathbb{Z}$ of integers). In Subsection 2 we prove, among other things, that biholomorphically equivalent regions have isomorphic homology groups. In Subsection 3, we make precise the idea that holes in regions can always be "separated" by paths that go around them. — U, V, and W denote regions in $\mathbb{C}$.

1. Homology groups. The Betti number. The set $Z(U)$ of *all* cycles

(1) $\gamma = a_1\gamma_1 + \cdots + a_n\gamma_n$, $a_\nu \in \mathbb{C}$, γ_ν a closed path in U, $n \in \mathbb{N}\backslash\{0\}$,

in U forms a *free abelian group* with respect to (natural) addition, with the closed paths as *basis.* Every cycle (1) defines a $\mathbb{C}$-linear form

(2) $\overline{\gamma} : \mathcal{O}(U) \to \mathbb{C}$, $f \mapsto \overline{\gamma}(f) := \int_\gamma f d\zeta$.

The following holds because of Theorem 13.1.3.

(3) *A cycle* $\gamma \in Z(U)$ *is null homologous in* U *if and only if* $\overline{\gamma} = 0$.

Null homology and holes are related in the following way:

(4) *A cycle* γ *in* U *is null homologous in* U *if and only if its index* $\operatorname{ind}\gamma$ *vanishes identically on every hole* L *of* U, *i.e. if* $\operatorname{ind}_\gamma(L) = 0$.

Proof. $\operatorname{Int}\gamma \subset U$ if and only if $\operatorname{ind}_\gamma(\mathbb{C}\backslash U) = 0$. Since $\operatorname{ind}_\gamma$ *always* vanishes on *all unbounded* components of $\mathbb{C}\backslash U$, (4) follows. □

The set of all $\mathbb{C}$-linear forms

$$H(U) := \{\overline{\gamma} : \gamma \in Z(U)\}$$

defined by (2) is a subgroup of the $\mathbb{C}$-vector space of *all* $\mathbb{C}$-linear forms on $\mathcal{O}(U)$. We call $H(U)$ the *(first) homology group of* U *(with coefficients in* $\mathbb{Z}$*)*. $H(U) = 0$ if and only if U is homologically simply connected. The next assertion is clear because of (3).

(5) *The map* $Z(U) \to H(U)$, $\gamma \mapsto \overline{\gamma}$, *is a group epimorphism with the group* $B(U) := \{\gamma \in Z(U) : \operatorname{Int}\gamma \subset U\}$ *as kernel; it induces a group isomorphism*

$$(*) \qquad Z(U)/B(U) \xrightarrow{\sim} H(U).$$

The left-hand side of $(*)$ gives a *topological* description of $H(U)$. In algebraic topology, null-homologous cycles in U are called *boundaries in* U (intuitively, γ "bounds" the surface $\operatorname{Int}\gamma$, which lies in U). Two cycles γ, γ' in U are called *homologous in* U if $\gamma - \gamma'$ is a boundary in U, i.e. if $\overline{\gamma} = \overline{\gamma}'$. "Being homologous" is an equivalence relation. The set of all cycles homologous to γ is the homology class $\overline{\gamma} \in H(U)$.

The abelian group $H(U)$ has a well-defined *rank* $b(U)$ (= maximal number of $\mathbb{Z}$-linearly independent elements in $H(U)$). This rank $b(U)$ is called the *(first) Betti number* of U.

It can be shown that $H(U)$ is *always* a *free* abelian group whose rank $b(U)$ is *at most countably infinite.*

The vector space $\mathcal{O}'(U)$ of all derivatives f', $f \in \mathcal{O}(U)$, is characterized homologically by

$$(6) \qquad \mathcal{O}'(U) = \{f \in \mathcal{O}(U) : \overline{\gamma}(f) = 0 \textit{ for all } \overline{\gamma} \in H(U)\}.$$

2. Induced homomorphisms. Natural properties. Every holomorphic map $h : U \to V$ induces a group homomorphism

$$h : Z(U) \to Z(V), \quad \gamma = \sum a_\nu \gamma_\nu \mapsto h \circ \gamma := \sum a_\nu (h \circ \gamma_\nu).$$

By linearity, the *substitution rule*

$$\overline{h \circ \gamma}(f) = \overline{\gamma}((f \circ h) \cdot h') \quad \text{for all } \gamma \in Z(U),\ f \in \mathcal{O}(V)$$

holds for cycles as well as for paths.

It is clear that *if $\gamma_1, \gamma_2 \in Z(U)$ are homologous, then $h(\gamma_1), h(\gamma_2) \in Z(V)$ are homologous.* Hence h induces a homomorphism $\widetilde{h} : H(U) \to H(V)$ of the homology groups. We now make this precise (vertical arrows denote passage to homology classes).

Proposition. *Given h, there exists exactly one map $\widetilde{h} : H(U) \to H(V)$ that makes the following a commutative diagram:*

$$\begin{array}{ccc} Z(U) & \xrightarrow{\ h\ } & Z(V) \\ \big\downarrow & & \big\downarrow \\ H(U) & \xrightarrow{\ \widetilde{h}\ } & H(V) \end{array}$$

The map $\widetilde{h}$ is a group homomorphism and

(1) $$\widetilde{h}(\overline{\gamma}) = \overline{h \circ \gamma} \quad \textit{for all } \gamma \in Z(U).$$

Proof. By the discussion above, (1) defines a map $H(U) \to H(V)$ that is clearly additive. This is obviously the only map that makes the diagram commute. □

The correspondence $h \rightsquigarrow \widetilde{h}$ has the following "natural" properties:

(2) *If $id : U \to U$ is the identity map on U, then $\widetilde{id} : H(U) \to H(U)$ is the identity map on $H(U)$. If $h : U \to V$ and $g : V \to W$ are holomorphic, then $\widetilde{g \circ h} = \widetilde{g} \circ \widetilde{h}$.*

Proof. The first statement holds since $\widetilde{id}(\overline{\gamma}) = \overline{id \circ \gamma} = \overline{\gamma}$ by (1). The second statement holds because (1) implies that, for all $\gamma \in Z(U)$,

$$(\widetilde{g \circ h})(\overline{\gamma}) = \overline{(g \circ h) \circ \gamma} \quad \text{and} \quad (\widetilde{g} \circ \widetilde{h})(\overline{\gamma}) = \widetilde{g}(\overline{h \circ \gamma}) = \overline{g \circ (h \circ \gamma)}. \quad □$$

The next result is immediate.

Invariance theorem. *If $h : U \xrightarrow{\sim} V$ is biholomorphic, then $\widetilde{h} : H(U) \to H(V)$ is an isomorphism. In particular, the Betti numbers of U and V are equal.*

Proof. This is clear by (2) for $g := h^{-1}$, since $g \circ h = id_U$ and $h \circ g = id_V$. □

The invariance theorem refines the statement that, for a biholomorphic map $h : U \xrightarrow{\sim} V$, a cycle γ in U is null homologous in U if and only if its image $h \circ \gamma$ is null homologous in V.

Remark. In modern terminology, we have proved:

The correspondences $U \rightsquigarrow H(U)$ and $h \rightsquigarrow \widetilde{h}$ give a covariant functor from the category of all regions in $\mathbb{C}$ (with the holomorphic maps as morphisms) to the category of abelian groups (with the group homomorphisms as morphisms).

The homomorphisms $\widetilde{h}$ can be defined for all *continuous* maps $h : U \to V$. If this is done, the *functorial* property (2) is preserved; the invariance theorem thus holds for all homeomorphisms $U \xrightarrow{\sim} V$.

3. Separation of holes by closed paths. We begin by noting:

(1) *Let $L_1, L_2, \ldots, L_n$ be finitely many holes of a domain G. Then there exists a closed, unbounded, and connected set N in $\mathbb{C}$ that does not intersect L_1 and contains all the remaining holes $L_2, \ldots, L_n$.*

Proof. Let $p \in G$ be fixed. For every component $M \neq L_1$ of $\mathbb{C}\backslash G$, there exist paths in $\mathbb{C}\backslash L_1$ from p to points in $M \cap \partial G$ (there exist a $q \in M \cap \partial G$, cf. Example 13.A.1(4), and a corresponding line segment $[q, \widehat{q}] \subset \mathbb{C}\backslash L_1$ with $\widehat{q} \in G$). We assign such paths $\gamma_2, \gamma_3, \ldots, \gamma_n$ to $L_2, L_3, \ldots, L_n$, and set $N' := L_2 \cup |\gamma_2| \cup L_3 \cup |\gamma_3| \cup \cdots \cup L_n \cup |\gamma_n|$. We also choose a ray σ in $\mathbb{C}\backslash L_1$ with initial point $w \in G$ and a path δ in G from p to w. Then $N := |\sigma| \cup |\delta| \cup N'$ is a set with the desired properties. □

The next result, which is intuitively clear, now follows from (1), 13.A.3(3), and the circuit theorem 12.4.2.

Separation lemma. *Let $L_1, L_2, \ldots, L_n$ be finitely many holes of a domain G. Then there exist closed paths $\gamma_1, \ldots, \gamma_n$ in G such that*

$$\operatorname{ind}_{\gamma_\mu}(L_\nu) = \delta_{\mu\nu} = \begin{cases} 0 & \text{for} \quad \mu \neq \nu \\ 1 & \text{for} \quad \mu = \nu \end{cases} \qquad \textit{(orthonormality relations).}$$

Proof. It suffices to construct the path γ_1. We choose N as in (1). Since $N \cap L_1 = \emptyset$, there exists by 13.A.3(3) a compact set $K \subset \mathbb{C}\backslash G$ such that $L_1 \subset K \subset \mathbb{C}\backslash N$ and $G \cup K$ is open in $\mathbb{C}$. The compact set K lies in the *region* $D := (G \cup K)\backslash N$. Since $L_1 \subset K$, the circuit theorem 12.4.2 implies that there exists a closed path γ_1 in $D\backslash K \subset G$ with $\operatorname{ind}_{\gamma_1}(L_1) = 1$. Since $|\gamma_1| \cap N = \emptyset$ and N is *unbounded and connected*, we have $\operatorname{ind}_{\gamma_1}(N) = 0$. Since $L_2 \cup \cdots \cup L_n \subset N$, it follows that $\operatorname{ind}_{\gamma_1}(L_\nu) = 0$ for $\nu > 1$. □

The first appearance of the separation lemma in the textbook literature is probably in the book of S. Saks and A. Zygmund (cf. [SZ], p. 209). The lemma does not rule out that still other holes $\neq L_\nu$ of G may lie in the interior of the path γ_ν. For example, if there is a Cantor set C in the space of holes of G, then every path in G that encloses a hole of C contains *uncountably many* other holes of C in its interior.

§2. Invariance of the Number of Holes. Product Theorem for Units

In Subsections 1 and 2, the homology group $H(G)$ and the $\mathbb{C}$-vector space $\mathcal{O}(G)/\mathcal{O}'(G)$ of an arbitrary domain G are investigated; one result that emerges is the equality of the Betti number $b(G)$ and the number of holes. In Subsection 3, the multiplicative group $\mathcal{O}(G)^\times$ of all the *units* of $\mathcal{O}(G)$ and its subgroup $\exp \mathcal{O}(G)$ are studied; one result here is that, in domains with finitely many holes, for every function $f \in \mathcal{O}(G)^\times$ there exists a rational function q such that $q|G \in \mathcal{O}(G)^\times$ and qf has a holomorphic logarithm in G (product theorem).

1. On the structure of the homology groups. If $L_1, \ldots, L_n$ are distinct holes of G and $\gamma_1, \ldots, \gamma_n$ form a corresponding *orthonormal* family of paths (as in the separation lemma 1.3), we consider the two maps

$$\varepsilon : H(G) \to H(G), \qquad \overline{\gamma} \mapsto \sum_{\nu=1}^{n} \operatorname{ind}_\gamma(L_\nu)\overline{\gamma}_\nu,$$
$$\eta : \mathcal{O}(G) \to \mathbb{C}^n, \qquad f \mapsto (\overline{\gamma}_1(f), \ldots, \overline{\gamma}_n(f));$$

the first is $\mathbb{Z}$-*linear* (additivity of the index!), the second $\mathbb{C}$-*linear.*

Lemma. $H(G) = \ker \varepsilon \oplus \operatorname{image} \varepsilon$. *The elements* $\overline{\gamma}_1, \ldots, \overline{\gamma}_n$ *form a basis of* $\operatorname{image}\varepsilon$; *for every* $\overline{\gamma} \in \operatorname{image}\varepsilon$, *we have* $\overline{\gamma} = \sum_{\nu=1}^{n} \operatorname{ind}_\gamma(L_\nu)\overline{\gamma}_\nu$.

The map η *is surjective and* $\mathcal{O}'(G) \subset \ker \eta$.

Proof. a) $\varepsilon(\overline{\gamma}_\mu) = \overline{\gamma}_\mu$ by the orthonormality of $\overline{\gamma}_1, \ldots, \overline{\gamma}_n$; hence $\varepsilon^2 = \varepsilon$ (projection) and therefore $H(G) = \ker \varepsilon \oplus \operatorname{image}\varepsilon$. Let $\sum_{\nu=1}^{n} a_\nu \overline{\gamma}_\nu = 0$, $a_\nu \in \mathbb{Z}$. Applying this linear form to a function $(z-c)^{-1} \in \mathcal{O}(G)$ with $c \in L_\mu$ gives $a_\mu = 0$ for all μ. Hence $\overline{\gamma}_1, \ldots, \overline{\gamma}_n$ form a basis of $\operatorname{image}\varepsilon$.

b) By 1.1(6), $\mathcal{O}'(G) \subset \ker \eta$. Since $\eta((z-c)^{-1})$, $c \in L_\nu$, is the νth unit vector $(0, \ldots, 1, \ldots, 0)$, η is also surjective. □

The next result now follows quickly.

Theorem. *If G has exactly n holes $L_1, \ldots, L_n$, $n \in \mathbb{N}$, and $\gamma_1, \ldots, \gamma_n$ is an orthonormal family of paths corresponding to these holes, then*

1) $\overline{\gamma}_1, \ldots, \overline{\gamma}_n$ *is a basis of the group $H(G)$ and*

$$\overline{\gamma} = \sum \operatorname{ind}_\gamma(L_\nu)\overline{\gamma}_\nu \quad \textit{for every homology class } \overline{\gamma} \in H(G);$$

2) $\eta : \mathcal{O}(G) \to \mathbb{C}^n$ *induces a $\mathbb{C}$-vector space isomorphism* $\mathcal{O}(G)/\mathcal{O}'(G) \xrightarrow{\sim} \mathbb{C}^n$.

Proof. By the lemma, it suffices to prove that $\ker \varepsilon = 0$ and $\ker \eta = \mathcal{O}'(G)$.

1) Since $\ker \varepsilon = \{\overline{\gamma} : \operatorname{ind}_\gamma(L_\nu) = 0 \text{ for } \nu = 1, \ldots, n\}$ by the lemma and since $L_1, \ldots, L_n$ are *all* the holes of G, 1.1(3) and (4) imply that $\ker \varepsilon = 0$.

2) By (1), $\ker \eta = \{f \in \mathcal{O}(G) : \overline{\gamma}(f) = 0 \text{ for all } \overline{\gamma} \in H(G)\}$. Hence $\ker \eta = \mathcal{O}'(G)$ by 1.1(6). □

The theorem contains, as a special case:

If A is an annulus with hole L and $\Gamma \subset A$ is a circle around L, then

$$H(A) = \mathbb{Z}\overline{\Gamma} \quad \textit{and} \quad \overline{\gamma} = \operatorname{ind}_\gamma(L)\overline{\Gamma} \textit{ for all } \overline{\gamma} \in H(A).$$

2. The number of holes and the Betti number. A domain G is called $(n+1)$*-connected*, $0 \le n \le \infty$, if it has exactly n distinct holes. (We do not distinguish among infinitely large cardinal numbers.) Simply connected domains are, by 13.2.4, precisely those domains *without holes*; all annuli are examples of doubly connected domains. For more on this, see Subsection 3.

We prove here that the number of holes of G is an invariant and hence a measure of the intrinsic connectivity of G. The next theorem follows immediately from the insights into the structure of $H(G)$ and $\mathcal{O}(G)/\mathcal{O}'(G)$ that we have obtained so far.

Theorem. *For every $(n+1)$-connected domain G, the following statements hold.*

1) *If $n \in \mathbb{N}$, then the groups $H(G)$ and $\mathbb{Z}^n$, as well as the $\mathbb{C}$-vector spaces $\mathcal{O}(G)/\mathcal{O}'(G)$ and $\mathbb{C}^n$, are isomorphic; in particular, $b(G) = n$.*
2) *If $n = \infty$, then $b(G) = \infty = \dim_{\mathbb{C}} \mathcal{O}(G)/\mathcal{O}'(G)$.*

Proof. ad 1). This follows immediately from Theorem 1.

ad 2). For *every* $k \in \mathbb{N}$, there are k distinct holes in G. By Lemma 1,

- $H(G)$ then contains a subgroup isomorphic to $\mathbb{Z}^k$ (namely Image ε);
- there exists a $\mathbb{C}$-epimorphism $\mathcal{O}(G)/\mathcal{O}'(G) \to \mathbb{C}^k$ (induced by η).

The rank of $H(G)$ and the dimension of $\mathcal{O}(G)/\mathcal{O}'(G)$ are thus $\geq k$, $k \geq \mathbb{N}$. □

In particular, we have the equations

$$\text{Betti number of } G = \text{number of holes of } G = \dim_{\mathbb{C}} \mathcal{O}(G)/\mathcal{O}'(G).$$

Since Betti numbers are biholomorphic invariants by 1.2, this implies the

Invariance of the number of holes. *Biholomorphically equivalent domains in $\mathbb{C}$ have the same number of holes.*

Invariance also follows from the right-hand equality, since for every biholomorphic map $h : G \xrightarrow{\sim} G_1$ the map $\mathcal{O}(G_1) \to \mathcal{O}(G)$, $f \mapsto (f \circ h)h'$, is a $\mathbb{C}$-vector space isomorphism which maps $\mathcal{O}'(G_1)$ onto $\mathcal{O}'(G)$; see Exercise I.9.4.4.

Glimpse. The *topological invariance of the number of holes* is contained in a general (and quite deep) *duality theorem* of algebraic topology for compact sets in oriented manifolds. Let G denote a domain in the two-dimensional sphere $S^2 := \mathbb{C} \cup \{\infty\}$. Then

$$H_2(S^2, G; \mathbb{Q}) \simeq \overline{H}^0(S^2 \backslash G; \mathbb{Q}),$$

where the 2nd homology group of the pair S^2, G with coefficients in $\mathbb{Q}$ is on the left-hand side, and the 0th homology group, which appears on the right-hand side, is isomorphic to the *group of locally constant functions* $S^2 \backslash G \to \mathbb{Q}$. (See, for example, E. H. Spanier: *Algebraic Topology*, McGraw-Hill and Springer, 1966. In Theorem 17 on p. 296, set $X := S^2 := \mathbb{C} \cup \{\infty\}$, $A := S^2 \backslash G$, $B := \emptyset$, $G := \mathbb{Q}$, $n := q := 2$; then use the theorem at the bottom of p. 309.) If G now has n holes in $\mathbb{C}$, then $S^2 \backslash G$ has exactly $n+1$ components (all the *unbounded* components of G in $\mathbb{C}$ have ∞ as an accumulation point and thus form — together with ∞ — *one* component of $S^2 \backslash G$). In the case $n < \infty$, it follows that $\overline{H}^0(S^2 \backslash G; \mathbb{Q}) \simeq \mathbb{Q}^{n+1}$. Since $H_2(G, \mathbb{Q}) = H_1(S^2, \mathbb{Q}) = 0$, the exact homology sequence for the pair (S^2, G) shows that $H_2(S^2, G; \mathbb{Q}) \simeq \mathbb{Q} \oplus H_1(G; \mathbb{Q})$ depends only on the domain G and not on the imbedding $G \subset S^2$ (exactness axiom, loc. cit., p. 200). This gives the topological invariance of the number of holes.

One can also prove:

Every $(n+1)$-connected domain in $\mathbb{C}$, where $n \in \mathbb{N}$, is homeomorphic to the n-punctured plane $\mathbb{C} \backslash \{1, 2, \ldots, n\}$.

(Cf. M. H. A. Newman: *Elements of the Topology of Plane Sets of Points*, Cambridge Univ. Press 1951, p. 157.)

Thus domains in $\mathbb{C}$ with the same Betti number in $\mathbb{N}$ are always homeomorphic.

3. Normal forms of multiply connected domains (report). Let G be an $(n+1)$-connected domain, $n \in \mathbb{N}$, and let $L_1, L_2, \ldots, L_n$ be the holes of G. Then, by 3(2) of the appendix to Chapter 13, $G \cup L_1 \cup L_2 \cup \cdots \cup L_n$ is a simply connected domain and can therefore, by the Riemann mapping theorem, be mapped biholomorphically onto $\mathbb{C}$ or $\mathbb{E}$. The domain G itself is thus biholomorphically equivalent to a domain that results from "drilling" n holes out of $\mathbb{C}$ or $\mathbb{E}$, respectively. But much more can be said.

Mapping theorem. *Every $(n+1)$-connected domain in $\mathbb{C}$ can be mapped biholomorphically onto a circle domain, i.e. a disc $B_r(0)$, $0 < r \leq \infty$, from which n pairwise disjoint compact discs (possibly single points) have been removed, $n \in \mathbb{N}$.*

Every biholomorphic map between circle domains is realized by a linear fractional transformation.

Koebe was the first to prove this theorem. We refer the reader interested in details to [Gai] or [Gru], where the problem of mapping arbitrary domains conformally onto slit domains is also treated.

For $n = 1$, a more precise result can be proved. (We write $\simeq$ for "biholomorphically equivalent" and A_{r1} for the annulus $\{z \in \mathbb{C} : r < |z| < 1\}$, $0 < r < 1$.)

If L is the only hole of G, then

a) *L consists of a single point and $G \cup L = \mathbb{C} \Leftrightarrow G \simeq \mathbb{C}^\times$; or*
b) *L consists of a single point and $G \cup L \neq \mathbb{C} \Leftrightarrow G \simeq \mathbb{E}^\times$; or*
c) *L consists of more than one point and $G \cup L = \mathbb{C} \Leftrightarrow G \simeq \mathbb{E}^\times$; or*
d) *L consists of more than one point and $G \cup L \neq \mathbb{C} \Leftrightarrow G \simeq A_{r1}$.*

The nondegenerate case d) causes the only difficulties. H. Kneser ([Kn], pp. 372–375), uses the logarithm to reduce the proof to the case of simply connected domains.

4. On the structure of the multiplicative group $\mathcal{O}(G)^\times$. The group $\mathcal{O}(G)^\times$ of *all functions that are holomorphic and nonvanishing on G* contains the set $\exp \mathcal{O}(G)$ of all the functions $\exp g$, $g \in \mathcal{O}(G)$, as a subgroup. We have (cf. I.9.3.1)

$$(1) \quad \exp \mathcal{O}(G) = \left\{ f \in \mathcal{O}(G)^\times : \int_\gamma (f'/f) d\zeta = 0 \textit{ for all } \gamma \in Z(G) \right\}.$$

In order to describe the *quotient group* $\mathcal{O}(G)^\times / \exp \mathcal{O}(G)$, we assign every function $f \in \mathcal{O}(G)^\times$ the *$\mathbb{Z}$-linear period map*

$$\lambda_f : H(G) \to \mathbb{Z}, \quad \overline{\gamma} \mapsto \frac{1}{2\pi i} \overline{\gamma}(f'/f) = \frac{1}{2\pi i} \int_\gamma \frac{f'}{f} d\zeta.$$

We denote by $H(G)^*$ the *abelian* group of all $\mathbb{Z}$-linear forms on $H(G)$ *(the dual of $H(G)$)* and note immediately:

(2) *The map $\mathcal{O}(G)^\times \to H(G)^*$, $f \mapsto \lambda_f$, is a group homomorphism with the group* $\exp \mathcal{O}(G)$ *as kernel.*

Proof. Since $(fg)'/fg = f'/f + g'/g$, the map $f \mapsto \lambda_f$ is a homomorphism. Its kernel is $\exp \mathcal{O}(G)$ by (1). □

Now let $L_1, \ldots, L_n$ be distinct holes of G, $n \in \mathbb{N}$. As in 1.3, we choose an *orthonormal* family $\gamma_1, \ldots, \gamma_n$ of paths in G. We fix points $c_\nu \in L_\nu$, $1 \le \nu \le n$. Then the forms $\lambda_{z-c_1}, \ldots, \lambda_{z-c_n} \in H(G)^*$ are well defined and

$$(*) \qquad \lambda_{z-c_\nu}(\overline{\gamma}_\mu) = \operatorname{ind}_{\gamma_\mu}(L_\nu) = \delta_{\mu\nu}.$$

It thus follows:

(3) *The forms $\lambda_{z-c_1}, \ldots, \lambda_{z-c_n}$ are linearly independent. If $b(G) = n$, they form a basis of $H(G)^*$; for all $f \in \mathcal{O}(G)^\times$,*

$$\lambda_f = a_1\lambda_{z-c_1} + \cdots + a_n\lambda_{z-c_n}, \quad \textit{where} \quad a_\nu := \lambda_f(\overline{\gamma}_\nu),\ 1 \le \nu \le n < \infty.$$

Proof. In the case $b(G) = n$, $\overline{\gamma}_1, \ldots, \overline{\gamma}_n$ are a basis of G by Theorem 1. By $(*)$, $\lambda_{z-c_1}, \ldots, \lambda_{z-c_n}$ then form the dual basis of $H(G)^*$. □

(2) and (3) contain an existence and uniqueness theorem.

Product theorem for units. *Let G be a domain with exactly n holes $L_1, \ldots, L_n$; let $\gamma_1, \ldots, \gamma_n$ be a corresponding orthonormal family of paths in G, $n \in \mathbb{N}$. Let $c_\nu \in L_\nu$ be chosen in some way. Then every function $f \in \mathcal{O}(G)^\times$ has a representation*

$$f(z) = e^{g(z)}(z - c_1)^{k_1} \cdot \ldots \cdot (z - c_n)^{k_n},$$

$$\textit{where} \quad g \in \mathcal{O}(G) \quad \textit{and} \quad k_\nu := \frac{1}{2\pi i}\int_{\gamma_\nu} \frac{f'}{f} d\zeta,\ 1 \le \nu \le n.$$

If $f(z) = e^{h(z)}(z - c_1)^{m_1} \cdot \ldots \cdot (z - c_n)^{m_n}$ is another representation with $h \in \mathcal{O}(G)$ and $m_1, \ldots, m_n \in \mathbb{Z}$, then $h - g \in 2\pi i\mathbb{Z}$ and $m_\nu = k_\nu$, $1 \le \nu \le n$.

Proof. By (3), $\lambda_f = \sum k_\nu \lambda_{z-c_\nu}$. Corresponding to $v := (z - c_1)^{\ell_1} \cdot \ldots \cdot (z - c_n)^{\ell_n} \in \mathcal{O}(G)^\times$, $\ell_\nu \in \mathbb{Z}$, we have $\lambda_v = \sum \ell_\nu \lambda_{z-c_\nu}$. By (2), $f = e^g v$ with $g \in \mathcal{O}(G)$ if and only if $\lambda_f = \lambda_v$, i.e. if and only if $\ell_1 = k_1, \ldots, \ell_n = k_n$. □

For domains without holes, the product theorem says (as we already know) that every nonvanishing function $f \in \mathcal{O}(G)$ has a holomorphic logarithm in G.

Another result follows from (2) and (3).

Proposition. *The quotient group $\mathcal{O}(G)^\times / \exp \mathcal{O}(G)$ is isomorphic to a subgroup of $H(G)^*$ (through the map induced by λ).*

The Betti number $b(G) < \infty$ if and only if $\mathcal{O}(G)^\times / \exp \mathcal{O}(G)$ is finitely generated, in which case $\mathcal{O}(G)^\times / \exp \mathcal{O}(G)$ is isomorphic to the group $H(G)^* \simeq \mathbb{Z}^{b(G)}$.

Proof. By (2), the group $T := \mathcal{O}(G)^{\times} / \exp \mathcal{O}(G)$ is isomorphic to the subgroup Image λ of $H(G)^*$. If $b(G) < \infty$, then Image $\lambda = H(G)^* \simeq \mathbb{Z}^{b(G)}$ by (3). Conversely, if T is finitely generated, then T and hence also Image λ have finite rank m. By (3), G then has at most m holes; in other words, $b(G) \leq m$.

5. Glimpses. The product theorem for units can be generalized:

(∗) *Every continuous map* $f : G \to \mathbb{C}^{\times}$ *of a domain* $G \subset \mathbb{C}$ *with exactly* n *holes,* $n \in \mathbb{N}$, *is of the form* $f(z) = e^{g(z)} \prod_1^n (z - c_\nu)^{k_\nu}$, *where* $g \in \mathcal{C}(G)$, $k_\nu \in \mathbb{Z}$.

This statement dates back to Eilenberg's 1936 paper [E]; cf. p. 88 ff. The 1945 paper [Ku] of Kuratowski is also relevant to these topics; cf. p. 332 ff. In textbooks, the theorem can be found in [SZ] (3rd edition, p. 211 ff.) and [Bu] (p. 111 ff.); the reader will find further historical information in [Bu].

It follows from (∗), by setting $f(z,t) := e^{(1-t)g(z)} \prod (z - c_\nu)^{k_\nu}$, $0 \leq t \leq 1$, that the continuous map $f(z) = f(z,0) : g \to \mathbb{C}^{\times}$ is *deformed* by the "*continuous family*" $f(z,t)$ of maps $G \times [0,1] \to \mathbb{C}^{\times}$ to the *holomorphic* map $f(z,1) : G \to \mathbb{C}^{\times}$. Every continuous map $G \to \mathbb{C}^{\times}$ is said to be *homotopic* to a holomorphic map. In this form, (∗) can be considerably strengthened:

If X is a Stein manifold and L is a complex Lie group, then every continuous map $X \to L$ is homotopic to a holomorphic map $X \to L$ (special case of the Oka-Grauert principle; cf. [Gra]*).*

The period homomorphism $\lambda : \mathcal{O}(G)^{\times} \to H(G)^*$ of 4(2) was systematically investigated in 1943 by H. Behnke and K. Stein; they showed that it is *always surjective* (cf. [BS], Satz 10, p. 451). *The groups* $\mathcal{O}(G)^{\times} / \exp \mathcal{O}(G)$ *and* $H(G)^*$ *are thus always canonically isomorphic.* Since $H(G)^*$ has an uncountable basis when $b(G) = \infty$, we have the following phenomenon:

For every domain G, $\mathcal{O}(G)^{\times} / \exp \mathcal{O}(G)$ is a free abelian group; its rank is either finite or uncountably infinite.

The surjectivity of λ is a special case of a theorem on the existence of *additive automorphic* functions with arbitrarily prescribed complex periods on arbitrary noncompact Riemannian surfaces. A modern presentation is given by O. Forster ([F], pp. 214–218).

Bibliography

[BS] BEHNKE, H. and K. STEIN: Entwicklung analytischer Funktionen auf Riemannschen Flächen, *Math. Ann.* 120, 430–461 (1948).

[Bu] BURCKEL, R. B.: *An Introduction to Classical Complex Analysis*, vol. 1, Birkhäuser, 1979.

[E] EILENBERG, S.: Transformations continues en circonférence et la topologie du plan, *Fund. Math.* 26, 61–112 (1936).

[F] FORSTER, O.: *Lectures on Riemann Surfaces*, trans. B. GILLIGAN, Springer, 1981.

[Gai] GAIER, D.: *Konstruktive Methoden der konformen Abbildung*, Springer Tracts in Natural Philosophy 3, Springer, 1964.

[Gra] GRAUERT, H.: Holomorphe Funktionen mit Werten in komplexen Lieschen Gruppen, *Math. Ann.* 133, 450–472 (1957).

[Gru] GRUNSKY, H.: *Lectures on Theory of Functions in Multiply Connected Domains*, Studia Math., Skript 4, Vandenhoeck und Ruprecht, Göttingen, 1978.

[Kn] KNESER, H.: *Funktionentheorie*, Vandenhoeck und Ruprecht, Göttingen, 1958.

[Ku] KURATOWSKI, C.: Théorèmes sur l'homotopie des fonctions continues de variable complexe et leurs rapports à la théorie des fonctions analytiques, *Fund. Math.* 33, 316–367 (1945).

[SZ] SAKS, S. and A. ZYGMUND: *Analytic Functions*, 3rd ed., Elsevier Publ. Co., 1971.

Correction to: Classical Topics in Complex Function Theory

Reinhold Remmert

Correction to:
R. Remmert, *Classical Topics in Complex Function Theory*, Graduate Texts in Mathematics 172, https://doi.org/10.1007/978-1-4757-2956-6

This book was inadvertently published with errors in chapters 2, 5, 8, 10, 12, 13, 14, and in the back matter. The correct versions are given below.

Chapter 2
p. 47, l. 13: "$\mathbb{C}^-$" should be "$\mathbb{C}^- = \mathbb{C} \setminus (-\infty, 1]$".

Chapter 5
p. 116, l. -12: "$\mathcal{O}(U)$" should be "$\mathcal{O}(B)$".

Chapter 8
p. 171, l. -6: "Null homologous" should be "null homotopic".

p. 173, l. 14: "$\mathbb{R}$" should be "$\mathbb{C}$".

p. 177, l. -7: "G to E" should be "G in E".

p. 178, l. 12 and 13: "Since square roots are used in the construction of expansions, it is hardly surprising that..." should be "That square roots are used in the construction of expansions is hardly surprising, since...". The sentence should read: "That square roots are used in the construction of expansions is hardly surprising, since $x \mapsto \sqrt{x}$ is a simple expansion of the interval $[0, 1)$."

The online version of this publication is available at
https://doi.org/10.1007/978-1-4757-2956-6

https://doi.org/10.1007/978-1-4757-2956-6_15

p. 180, l. 6: In statement ii), insert "(has an antiderivative)" between "integrable" and "in". The sentence should read "Every function holomorphic in G is integrable (has an antiderivative) in G."

p. 181, l. -7: "Gedächniss" should be "Gedächtniss".

p. 188, l. -2: "Groups" should be "domains".

Appendix to Chapter 8

p. 195, l. -1 and -2: Replace "...that the limit map in 2) is, by 8.2.5, the uniquely determined map $G \xrightarrow{\sim} \mathbb{E}$" by "...that the limit map in 2) is the map $G \xrightarrow{\sim} \mathbb{E}$, which is uniquely determined by 8.2.5."

p. 196, l. -5: "τ_2" should be "$\tau_2(x)$".

Chapter 10

p. 225, l. 1 (chapter epigraph): Insert "égale" between "jamais" and "ni" and omit the comma. The French sentence should read: "Une fonction entière qui ne devient jamais égale ni à a ni à b est nécessairement une constante."

Chapter 12

p. 268, l. 18: "Pole-shifting" should be "pole-pushing".

p. 272, l. 10 and l. 16: "Pole-shifting" should be "pole-pushing".

p. 273, l. 5: "Pole-shifting" should be "pole-pushing".

p. 274, l. 5: "Pole-shifting" should be "pole-pushing".

p. 277, l. -6: "Pole-shifting" should be "pole-pushing".

p. 284, l. 6–7: Insert "of equal length" after "k segments $[p_1, p_2], [p_2, p_3], \ldots, [p_k, p_1]$". The first part of the sentence should read: "A closed polygon $\tau = [p_1 p_2 \ldots p_k p_1]$ composed of k segments $[p_1, p_2], [p_2, p_3], \ldots, [p_k, p_1]$ of equal length is called...".

Chapter 13

p. 294, l. -14: "Pole-shifting" should be "pole-pushing".

p. 305, l. 20: At the end of the line, "$1 \leq n \leq \infty$" should be "$1 \leq n < \infty$".

p. 305, l. 24: At the beginning of the line, "D" should be "$\mathbb{C}$".

Chapter 14

p. 310, l. 2: "$\mathbb{C}$" should be "$\mathbb{Z}$".

p. 312, l. -13: "Example 13.A.1(4)" should be "13.A.1(3)". That is, "(4)" should be "(3)" and the word "Example" should be omitted.

p. 317, l. 11: "Basis of G" should be "basis of $H(G)$".

Biographies

p. 323, l. 17: "Berl." should be "Verl."

Symbol Index

p. 329: Add "$\mathbb{N}_m$, 8" between "$\Pi_{\nu=k}^{\infty}$, 4" and the next entry.

p. 329: Add "g_c, 176" between "$PD(f)$, 126" and "Aut G, 188".

Name Index

p. 332, l. 23: In the entry for Grunsky, "1936" should be "1986".

Subject Index

Additional page numbers should be inserted in the appropriate numerical order.

p. 337, entry for "admissible expansion": Add "199".

p. 342, entry for "hole": "290" should be "291".

p. 342, entry for "homology group of a domain": Add "210".

p. 344, entry for "modulus of an annulus": Add "171".

p. 345, entry for "null homologous cycle": Add "310".

p. 346: "Pole-shifting technique" should be "pole-pushing technique".

p. 346: "Pole-shifting theorem" should be "pole-pushing theorem".

p. 348: Add "Stirling's formulas, 58" between the entries "step polygons, Jordan curve theorem for" and "Stirling's series".

p. 348, entry for "Šura-Bura's theorem": Add "291".

p. 349, top of page: Add "Schönflies's, 187" immediately after "Runge's, on rational approximation, 292".

p. 349, entry for "Šura-Bura's": Add "291".

Short Biographies

Source, among others: *Dictionary of Scientific Biography*

Lipman Bers, Latvian-American mathematician: b. 1914 in Riga; 1938, dissertation at the German university in Prague; from 1940 on, in the United States; d. 1993 in New Rochelle, New York.

Wilhelm Blaschke, Austrian mathematician: b. 1885 in Graz; professor in Prague, Leipzig, and Königsberg; from 1919 to 1953 in Hamburg; d. 1962 in Hamburg. — Blaschke was a differential geometer, founder of the geometry of webs.

André Bloch, French mathematician: b. 1893 in Besançon; 1913, student at the École Polytechnique; 1914–1915, wounded; 1917, after a bloody family drama, committed to a psychiatric clinic, where he remained until his death in 1948; 1948, posthumously awarded the Becquerel Prize. — Cf. H. Cartan and J. Ferrand: The case of André Bloch, *Math. Intelligencer* 10, 23–26 (1988).

Constantin Carathéodory, Greek-German mathematician: b. 1873 in Berlin; 1891, received his secondary-school diploma in Brussels; 1895, officer in the corps of engineers at the Belgian military school; 1898, in Egypt, with the Nile dam project; 1900, studied mathematics in Berlin; 1905, qualified as a university lecturer in Göttingen; 1909, professor in Hannover; 1913, succeeded F. Klein in Göttingen; 1920, founding president of the Greek university in Smyrna; 1922, flight to Athens; 1924, succeeded F. Lindemann (transcendence of π) in Munich; d. 1950 in Munich.

Leopold Fejér, Hungarian mathematician: b. 1880 in Pécs; 1897–1902, studied in Budapest and Berlin; from 1911 on, professor in Budapest; d.

1959 in Budapest. — Fejér is one of the founders of the great Hungarian school of analysis, which has included P. Erdös, F. and M. Riesz, J. v. Neumann, G. Pólya, T. Radó, O. Szász, G. Szegö, and J. Szökefalvi-Nagy. — Obituary by J. Aczél: Leopold Fejér, In memoriam, in *Publ. Math. Debrecen* 8, 1–24 (1961).

Jacques Hadamard, French mathematician: b. 1865 in Versailles; 1884–1888, student at the École Normale Supérieure; 1897–1909, lecturer at the Sorbonne; 1909–1937, professor at the Collège de France; d. 1963 in Paris. — Obituary by S. Mandelbrot and L. Schwartz in *Bull. Amer. Math. Soc.* 71, 107–1219 (1965).

Friedrich Hartogs, German mathematician: b. 1874 in Brussels; 1905, lecturer and, as of 1927, full professor at the University of Munich; 1935, forced retirement; d. 1943 in Munich, by suicide, because of racial persecution.

Otto Hölder, German mathematician: b. 1859 in Stuttgart; professor in Tübingen, Königsberg, and, from 1899 on, in Leipzig; d. 1937 in Leipzig. — Known for the Hölder inequalities and Hölder continuity as well as the Jordan-Hölder-Schreier theorem on composition series of groups.

Adolf Hurwitz, German-Swiss mathematician: b. 1859 in Hildesheim; secondary school instruction from H. C. H. Schubert, father of the "counting calculus" of algebraic geometry; 1877, studied with Klein, Weierstrass, and Kronecker; 1881, received his doctorate at Leipzig; 1882, qualified as a university lecturer at Göttingen, since graduates of secondary schools emphasizing modern languages were not permitted to qualify as university lecturers at Leipzig; in 1884, at the age of 25, associate professor in Königsberg, and friendship there with Hilbert and Minkowski; 1892, declined an offer to succeed Schwarz at Göttingen and accepted an offer to succeed Frobenius at the Federal Polytechnic School in Zurich; d. 1919 in Zurich. — Worked in function theory, theory of modular functions, algebra, and algebraic number theory.

Carl Gustav Jacob Jacobi, German mathematician: b. 1804 in Potsdam; 1824, received his doctorate and qualified as a university lecturer in Berlin, defended the thesis "All sciences must strive to become 'mathematics'"; 1826, lecturer in Königsberg; 1829, full professor there; 1829, friendship with Dirichlet, whose wife, Rebecca Mendelsohn, described the time they spent together by "they did mathematics in silence"; 1842, member of the order "Pour le Mérite für Wissenschaft und Künste"; 1844, moved to Berlin, ordinary member of the Prussian Academy of Sciences; 1849, financial reprisals because of his conduct after the revolution of March 1848; 1849, called to Vienna; d. 1851 in Berlin, of smallpox. — From about 1830 on, Jacobi was considered the greatest German mathematician after Gauss. Bib.: Gedächtnisrede, delivered in 1852 by L. Dirichlet, in Jacobi's *Ges. Werken* 1, 1–28, or *Teubner-Archiv zur Mathematik*, vol. 10, 1988, ed. H. Reichardt, 8–32; also L. Königsberger: Carl Gustav Jacob Jacobi, *J. DMV* 13, 405–433 (1904).

Robert Jentzsch, German mathematician: b. 1890 in Königsberg; 1914, received his doctorate in Berlin; 1917, lecturer at the University of Berlin; killed in action, 1918.

Paul Koebe, German mathematician: b. 1882 in Luckenwalde, near Berlin; student of H. A. Schwarz; 1907, qualified as a university lecturer at Göttingen; 1914, full professor in Jena; 1926, full professor in Leipzig; d. 1945 in Leipzig. — Koebe was a master of conformal mapping and uniformization theory. He attached great importance to being a famous mathematician; one anecdote recounts that he always traveled incognito, so as not to be asked in hotels whether he was related to the famous function theorist. — Obituaries by L. Bieberbach and H. Cremer: Paul Koebe zum Gedächtnis, *Jahresber. DMV* 70, 158–161 (1968); and R. Kühnau: Paul Koebe und die Funktionentheorie, 183–194, in *100 Jahre Mathematisches Seminar der Karl-Marx Universität* Leipzig, ed. H. Beckert and H. Schumann, VEB Deutscher Berl. Wiss. Berlin, 1981.

Edmund Landau, German mathematician: b. 1877 in Berlin; student of Frobenius; 1909, full professor at Göttingen, as Minkowski's successor; 1905, son-in-law of Paul Ehrlich (chemotherapy and Salvarsan); 1933, dismissed for racial reasons; d. 1938 in Berlin. — Obituary by K. Knopp in *J. DMV* 54, 55–62 (1951), cf. also M. Pinl: Kollegen in einer dunklen Zeit, Part II, *J. DMV* 72, 165–189 (1971). N. Wiener says of the so-called Landau style, "His books read like a Sears-Roebuck catalogue."

Magnus Gustaf Mittag-Leffler, Swedish mathematician: b. 1846 in Stockholm; 1872, received his doctorate in Uppsala; 1873, held a fellowship in Paris; 1874–1875, attended lectures by Weierstrass; 1877, professor in Helsinki; 1881, professor in Stockholm; 1882, founded *Acta Mathematica*; 1886, president of the University of Stockholm; d. 1927 in Stockholm. — Obituaries for Mittag-Leffler were written by N. E. Nörlund, *Acta Math.* 50, I–XXIII (1927); G. H. Hardy, *Journ. London Math. Soc.* 3, 156–160 (1928); and, in 1944, by the first director of the Mittag-Leffler Institute, T. Carleman, *Kung. Svenska Vetenskapsakademiens levnadsteckningar* 7, 459–471 (1939–1948).

Mittag-Leffler was a manager of mathematics. With *Acta Mathematica*, he eased the scientific tensions that had existed since 1870–1871 between the mathematical powers of Germany and France; among those whose work he published in *Acta* was G. Cantor, who faced great hostility. In 1886, he succeeded in getting Sonia Kovalevsky appointed professor; at that time, women were not even allowed as students in Berlin. For more on this, see L. Hörmander: *The First Woman Professor and Her Male Colleague* (Springer, 1991). — Mittag-Leffler's relations with A. Nobel were strained; see C.-O. Selenius: Warum gibt es für Mathematik keinen Nobelpreis?, pp.

613–624 in *Mathemata, Festschr. für H. Gericke*, Franz Steiner Verlag, 1985.

In 1916, Mittag-Leffler and his wife willed their entire fortune and their villa in Djursholm to the Royal Swedish Academy of Sciences (the will was published in *Acta Math.* 40, III–X). The Mittag-Leffler Institute is still an international center of mathematical research.

Paul Montel, French mathematician: b. 1876 in Nizza; 1894, studied at the École Normale Supérieure; 1897, received a fellowship from the Thiers Foundation; 1904, professor in Nantes; 1913, professor of statistics and materials testing at the École Nationale Supérieure des Beaux-Arts; 1956, after the death of É. Borel, director of the Institut Henri Poincaré; d. 1975 in Paris.

Eliakim Hastings Moore, American mathematician: b. 1862 in Marietta, Ohio; 1885, received his doctorate at Yale; 1885, held a fellowship in Göttingen and Berlin; 1892, professor at the recently founded University of Chicago; 1896–1931, permanent chairman of the Mathematical Institute at Chicago; 1899, received an honorary doctorate from Göttingen; d. 1932 in Chicago. — Moore is known, among other things, for Moore-Smith sequences and the Moore-Penrose inverse. L. E. Dickson, O. Veblen, and G. D. Birkhoff are numbered among his students. Moore was probably the most influential American mathematician around the turn of the century; in 1894, for instance, he was one of the founders of the American Mathematical Society. Moore was a member of the National Academy of Sciences.

Alexander M. Ostrowski, Russian-Swiss mathematician: b. 1893 in Kiev; 1912–1918, studied in Marburg with Hensel; 1918–1920, at Göttingen; 1920–1923, assistant at Hamburg; 1923–1927, lecturer at Göttingen; 1927–1958, full professor in Basel; d. 1986 in Montagnola/Lugano. — Obituary by R. Jeltsch-Fricker in *Elem. Math.* 43, 33–38 (1988).

Charles Émile Picard, French mathematician: b. 1856 in Paris; 1889, member of the Académie des Sciences, and from 1917 on its secretary; from 1924 on, member of the Académie Française; d. 1941 in Paris. — Major work in the theory of differential equations and function theory, father of value distribution theory. In his opening address to the International Congress of Mathematicians in 1920, in Strasbourg, he quoted Lagrange's bon mot: "Les mathématiques sont comme le porc, tout en est bon." (Mathematics is like the pig, all of it is good.)

Jules Henri Poincaré, French mathematician: b. 1854 in Nancy; 1879, professor at Caen; 1881, professor at the Sorbonne; d. 1912 in Paris. — Poincaré discovered automorphic functions and did pioneering work in celestial mechanics and algebraic topology. Along with Einstein, Lorentz, and Minkowski, he founded the theory of special relativity. Poincaré was a cousin of Raymond Poincaré, who served several times and for many years as prime minister of France.

George Pólya, Hungarian mathematician: b. 1887 in Budapest; studied in Budapest, Vienna, Göttingen, and Paris; 1912, received his doctorate in Budapest; 1914 to 1940, at the ETH Zurich, from 1928 on as full professor; 1942–1953, full professor at Stanford University; d. 1985 in Stanford. — Pólya enriched analysis and function theory through penetrating and excellently written papers. The books of Pólya and Szegö that appeared in 1925 are among the most beautiful books of function theory. In 1928 Pólya, in "On my cooperation with Gabor Szegö," *Coll. Papers of G. Szegö*, vol. 1, p. 11, gave his own judgment: "The book PSz, the result of our cooperation, is my best work and also the best work of Gabor Szegö." — An obituary can be found in *Bull. London Math. Soc.* 19, 559–608 (1987).

Tibor Radó, Hungarian-American mathematician: b. 1895 in Budapest; 1922, received his doctorate at Szeged, under F. Riesz; from 1922 to 1929, lecturer in Budapest; 1929, emigrated to the United States; from 1930 on, full professor in Columbus, Ohio; d. 1965 in Florida.

Frederic Riesz, Hungarian mathematician: b. 1880 in Györ; studied in Zurich, Budapest, and Göttingen; 1908, high school teacher in Budapest; 1912, professor in Klausenburg (Cluj); from 1920 to 1946, professor in Szeged; from 1946 on, in Budapest; d. 1956 in Budapest.

Marcel Riesz, Hungarian-Swedish mathematician (brother of Frederic): b. 1886 in Györ; studied in Budapest, Göttingen, and Paris; 1911, lecturer in Stockholm; 1926, full professor in Lund; d. 1969 in Lund.

Carl David Tolmé Runge, German mathematician: b. 30 August 1856 in Bremen; beginning in 1876, student in Munich and Berlin, friendship with Maz Planck; 1880, received his doctorate under Weierstrass (differential geometry); 1883, qualified as a university lecturer with work, influenced by Kronecker, on a method for the numerical solution of algebraic equations; 1884, after a visit to Mittag-Leffler in Stockholm, publication of his groundbreaking paper in *Acta Mathematica*; 1886, full professor at the technical college in Hannover, concerned with spectroscopy; 1904, full professor in "applied mathematics" at Göttingen; 1909–1910, visiting professor at Columbia University; d. 3 January 1927 in Göttingen. — Runge was the first advocate of approximate mathematics (numerical analysis) in Germany; his many papers (with Kaiser, Paschen, and Voigt) on spectral physics also earned him an outstanding reputation as a physicist.

Friedrich Schottky, German mathematician: b. 1851 in Breslau; 1870–1874, studied in Breslau and Berlin; 1875, received his doctorate under Weierstrass; 1882, professor in Zurich; from 1892 to 1902, full professor in Marburg; from 1902 to 1922, full professor in Berlin; d. 1935 in Berlin.

Thomas Jan Stieltjes, Dutch mathematician: b. 1856 in Zwolle; 1877–1883, at the observatory in Leiden; 1883, professor at Groningen; from 1886 on, professor in Toulouse; d. 1894 in Toulouse. — A great variety of work in analysis, function theory, and number theory. In 1894 he introduced the integral later named after him.

Gabor Szegö, Hungarian mathematician: b. 1895 in Kunhegyes; 1912–1913, friendship with G. Pólya; 1918, received his doctorate in Vienna; 1921, lecturer in Berlin (with S. Bergmann, S. Bochner, E. Hopf, H. Hopf, C. Löwner, and J. von Neumann); 1926–1934, full professor in Königsberg; 1934, emigrated to St. Louis, Missouri; 1938–1960, full professor at Stanford; d. 1985 in Stanford.

Giuseppe Vitali, Italian mathematician: b. 1875 in Bologna; to a large extent self-taught; 1904–1923, middle school teacher; 1923–1932, professor in Modena, Padua, and Bologna; d. 1932. — Vitali worked mainly in the theory of real functions and is regarded as a precursor of Lebesgue.

Joseph Henry Maclagan Wedderburn, Scottish-American mathematician: b. 1882 in Forfar; 1904, studied in Berlin under Frobenius and Schur; 1905–1909, lecturer at the University of Edinburgh; from 1909 on, at Princeton University, where Woodrow Wilson, later president of the United States, had appointed him preceptor; 1911–1932, editor of *Annals of Mathematics*; d. 1948 in Princeton. — Wedderburn was an algebraist; he classified all semisimple finite-dimensional associative algebras over *arbitrary* ground fields; he showed moreover that finite fields are automatically commutative.

C. Caratheodory 1873–1950

C.D.T. Runge 1856–1927

M.G. Mittag-Leffler 1846–1927

P. Koebe 1882–1945

Pen and ink drawings by Martina Koecher

Symbol Index

(Supplement to the symbol index of *Theory of Complex Functions*)

Name Index

Subject Index

Graduate Texts in Mathematics

continued from page ii

119 ROTMAN. An Introduction to Algebraic Topology.
120 ZIEMER. Weakly Differentiable Functions: Sobolev Spaces and Functions of Bounded Variation.
121 LANG. Cyclotomic Fields I and II. Combined 2nd ed.
122 REMMERT. Theory of Complex Functions. *Readings in Mathematics*
123 EBBINGHAUS/HERMES et al. Numbers. *Readings in Mathematics*
124 DUBROVIN/FOMENKO/NOVIKOV. Modern Geometry—Methods and Applications. Part III.
125 BERENSTEIN/GAY. Complex Variables: An Introduction.
126 BOREL. Linear Algebraic Groups.
127 MASSEY. A Basic Course in Algebraic Topology.
128 RAUCH. Partial Differential Equations.
129 FULTON/HARRIS. Representation Theory: A First Course. *Readings in Mathematics*
130 DODSON/POSTON. Tensor Geometry.
131 LAM. A First Course in Noncommutative Rings.
132 BEARDON. Iteration of Rational Functions.
133 HARRIS. Algebraic Geometry: A First Course.
134 ROMAN. Coding and Information Theory.
135 ROMAN. Advanced Linear Algebra.
136 ADKINS/WEINTRAUB. Algebra: An Approach via Module Theory.
137 AXLER/BOURDON/RAMEY. Harmonic Function Theory.
138 COHEN. A Course in Computational Algebraic Number Theory.
139 BREDON. Topology and Geometry.
140 AUBIN. Optima and Equilibria. An Introduction to Nonlinear Analysis.
141 BECKER/WEISPFENNING/KREDEL. Gröbner Bases. A Computational Approach to Commutative Algebra.
142 LANG. Real and Functional Analysis. 3rd ed.
143 DOOB. Measure Theory.
144 DENNIS/FARB. Noncommutative Algebra.
145 VICK. Homology Theory. An Introduction to Algebraic Topology. 2nd ed.
146 BRIDGES. Computability: A Mathematical Sketchbook.
147 ROSENBERG. Algebraic *K*-Theory and Its Applications.
148 ROTMAN. An Introduction to the Theory of Groups. 4th ed.
149 RATCLIFFE. Foundations of Hyperbolic Manifolds.
150 EISENBUD. Commutative Algebra with a View Toward Algebraic Geometry.
151 SILVERMAN. Advanced Topics in the Arithmetic of Elliptic Curves.
152 ZIEGLER. Lectures on Polytopes.
153 FULTON. Algebraic Topology: A First Course.
154 BROWN/PEARCY. An Introduction to Analysis.
155 KASSEL. Quantum Groups.
156 KECHRIS. Classical Descriptive Set Theory.
157 MALLIAVIN. Integration and Probability.
158 ROMAN. Field Theory.
159 CONWAY. Functions of One Complex Variable II.
160 LANG. Differential and Riemannian Manifolds.
161 BORWEIN/ERDÉLYI. Polynomials and Polynomial Inequalities.
162 ALPERIN/BELL. Groups and Representations.
163 DIXON/MORTIMER. Permutation Groups.
164 NATHANSON. Additive Number Theory: The Classical Bases.
165 NATHANSON. Additive Number Theory: Inverse Problems and the Geometry of Sumsets.
166 SHARPE. Differential Geometry: Cartan's Generalization of Klein's Erlangen Program.
167 MORANDI. Field and Galois Theory.
168 EWALD. Combinatorial Convexity and Algebraic Geometry.
169 BHATIA. Matrix Analysis.
170 BREDON. Sheaf Theory. 2nd ed.
171 PETERSEN. Riemannian Geometry.
172 REMMERT. Classical Topics in Complex Function Theory.
173 DIESTEL. Graph Theory.
174 BRIDGES. Foundations of Real and Abstract Analysis.
175 LICKORISH. An Introduction to Knot Theory.
176 LEE. Riemannian Manifolds.

Printed in the United States
By Bookmasters